Springer Praxis Books

Astronautical Engineering

This book series presents the whole spectrum of Earth Sciences, Astronautics and Space Exploration. Practitioners will find exact science and complex engineering solutions explained scientifically correct but easy to understand. Various subseries help to differentiate between the scientific areas of Springer Praxis books and to make selected professional information accessible for you.

The *Springer Praxis Astronautical Engineering* program covers the very latest applications and systems used in rocket and spacecraft propulsion, spacecraft design, engineering and technology, enabling technologies for current and future space missions both manned and unmanned, including planetary rovers.

Key topics include:

- human missions to the Moon and Mars
- the space debris and radiation environments
- the analysis and design of interplanetary missions, orbital motion, deep space probes and the technologies required for space missions beyond the Solar System into interstellar space and the problems of spacecraft communications

The books are well illustrated with line diagrams and photographs throughout, with targeted use of colour for scientific interpretation and understanding. They feature extensive references and bibliographies, glossaries and appendices.

The books are written for a readership of aerospace and astronautical engineers, space scientists and researchers, spacecraft designers, managers and mission planners in space agencies, space policy makers and postgraduate students in university departments and research institutes in related fields.

Thomas Lund

Spacecraft that Explored the Inner Planets Venus and Mercury

Springer

Published in association with
Praxis Publishing
Chichester, UK

PRAXIS

Thomas Lund
San Diego, CA, USA

Springer Praxis Books
ISSN 2365-9599 ISSN 2365-9602 (electronic)
Astronautical Engineering
ISBN 978-3-031-29837-0 ISBN 978-3-031-29838-7 (eBook)
https://doi.org/10.1007/978-3-031-29838-7

This Springer imprint is published by the registered company Springer Nature Switzerland AG
The registered company address is: Gewerbestrasse 11, 6330 Cham, Switzerland

This book is dedicated to my wife Barbara and to my children Ann, Tom, and Colin

Introduction

Humans have been intrigued by the planets since ancient times. Our distant ancestors likely lay on the ground at night enthralled by the canopy of stars and wondered about those points of light that wandered amid the familiar pattern of stars. As civilization developed, inquiring minds sought to understand the apparent motions of the sun, stars, and visible planets. A towering intellect of the day, Aristarchus of Samos (300 BC–230 BC), proposed a sun-centered (heliocentric) model of the known universe. Samos is a small Greek island. The assertion seemed too preposterous and it was not embraced. As civilization progressed, the Polish astronomer Nicolaus Copernicus finally put it all together in about the year 1510 with the Earth and five visible planets in proper order orbiting the sun.

There was keen interest in learning more about Earth's sister planets through probes lofted by launch vehicles newly available in the early space era. Venus and Mars were within reach of early launch vehicles and those planets were first to be explored. The more distant planets were explored as launcher technology advanced. Both the United States and the Soviet Union sent increasingly capable spacecraft to explore Venus and Mars. Over the years, the United States concentrated on exploring Mars with only a few spacecraft sent to Venus. In contrast, the Soviet Union concentrated on exploring Venus with only a few spacecraft sent to Mars.

Counting only those spacecraft destined for the inner planets that successfully left Earth orbit in the time period 1960–2020, the United States sent 5 spacecraft to Venus, 20 to Mars, and 2 to Mercury. The Soviet Union sent 20 spacecraft to Venus and 10 to Mars. The Soviet Union broke up in 1991 and their missions to the planets stopped for several years. Recent exploration of Venus includes two spacecraft that orbited Venus, one by the European Space Agency and the other by Japan.

The intent of this book is to describe operating systems of a consecutive series of spacecraft that explored Venus and Mercury. Concentrating on exploration of those two planets permitted deserved detail on each of the many spacecraft involved. Major scientific instruments carried by each spacecraft are also described along with findings of the instruments. The spacecraft designed for those explorations were ingenious and reflected the best efforts of talented people working with the technology of the day.

Exploration of Venus was challenging because the planet is covered with dense clouds that precludes photography of the surface from above and it is unbearably hot on the surface. The atmosphere of Venus below the clouds is largely carbon dioxide and the resulting "greenhouse" effect results in a surface temperature of about 462 °C. The atmospheric pressure at the surface of Venus is about 92 times that of Earth.

The planet Mercury was also a challenge to explore because its nearness to the sun required a specially designed spacecraft with good thermal control. In addition, a complicated trajectory with multiple encounters with Earth, Venus, and Mercury was required to position the spacecraft and adjust its velocity to be able to orbit the planet. The nearness to the sun and lack of clouds or atmosphere result in surface temperatures on the sunlit side of Mercury of about 430 °C and, because it does not have an atmosphere, the temperature drops to about −180 °C on the night side.

Exploration of the planets advanced from flybys with picture taking and science probing to orbiting and to soft-landing capsules on the surface. The clear atmosphere of Mercury allowed detailed photographic coverage of its surface. Cloud cover prevented photography of the surface of Venus, so synthetic aperture radar mapping of Venus and topography by radar altimeters was carried out by both the United States and the Soviet Union. Impressively, the Soviet Union soft-landed ten instrumented capsules on the hellish surface of Venus, four of which sent back photographs of the surface. From high-tech to the mundane, two of the capsules that landed did not return pictures because lens caps did not come off of the cameras.

Early US spacecraft were dimensioned in English units. The author converted those values to metric units in keeping with current scientific convention. To aid US readers not as familiar with units in the metric system, conversions are provided when appropriate.

Contents

Chapter 1
The Planets of the Sun

The massive and incredibly hot sun holds eight orbiting planets and several dwarf planets in firm gravitational grasp. It provides life sustaining light and warmth to a special planet, third out from the sun, named Earth.

Four of the planets, referred to as the inner planets, are grouped closer to the sun. Those planets, in order out from the sun, are Mercury, Venus, Earth, and Mars. The average distances of individual inner planets from the sun range from 57.9 million km for Mercury to 227.9 million km for Mars. The average distances of the outer planets from the sun range from 779 million km for Jupiter to 4495 million km for Neptune. Pluto is not included in the list of outer planets since it was reclassified as a dwarf planet in 2006.

A summary of parameters associated with the inner and outer planets are shown in Tables 1.1 and 1.2.

The orbits of the inner planets are shown on the following page (Fig. 1.1). The distance marker shown on the drawing represents 1×10^8 (100 million) km. The orbits are elliptical but, except for Mercury and Mars, the eccentricities of the orbits are small. The eccentricity of Earth's orbit is 0.017 the eccentricity of Venus's orbit is 0.007 and to the scale of Fig. 1.1, it would be difficult to see the difference between the elliptic orbit and the circular average orbits drawn. In the case of Mercury with an eccentricity of 0.205 and Mars with eccentricity of 0.094, the eccentricity of their particular orbits are apparent in the drawing.

All of the planets are unique and intriguing. In the interest of manageable size, this book concentrates on spacecraft that explored the inner planets Mercury and Venus. Later books in this series will treat exploration of Mars and the outer planets.

T. Lund, *Spacecraft that Explored the Inner Planets Venus and Mercury*,
Springer Praxis Books, https://doi.org/10.1007/978-3-031-29838-7_1

Table 1.1 Parameters of the inner planets

Parameter	Mercury	Venus	Earth	Mars
Ave. distance from sun, 10^6 km/10^6 miles	57.9/36.0	108.2/67.2	149.6/93.0	227.9/141.6
Eccentricity of orbit	0.205	0.007	0.017	0.094
Orbital period, days	88	224.7	365.2	687.0
Orbital inclination from ecliptic, degrees	7.0	3.4	0	5.1
Mass, 10^{24} kg/10^{21} tons	0.330/0.364	4.87/5.37	5.97/6.58	0.642/0.708
Diameter, km /miles	4879/3032	12,104/7521	12,756/7926	6792/4221
Gravity, m/s^2/ft./s^2	3.7/12.1	8.9 /29.1	9.8 /32.1	3.7/12.1
Rotation period/length of day[a], hours	1407.6/4222.6	5832.5/2802.0	23.9/24.0	24.6/24.7
Mean temperature, °C/°F	167/333	464/867	15/59	−65/−85

[a]Sunrise to sunrise

Table 1.2 Parameters of the outer planets

Parameter	Jupiter	Saturn	Uranus	Neptune
Ave. distance from sun, 10^6km/10^6 miles	778.6/483.8	1433.5/890.8	2872.5/1784.8	4495.1/2793.1
Eccentricity of orbit	0.049	0.057	0.046	0.011
Orbital period, days	4331	10,747	30,589	59,800
Orbital inclination from ecliptic, degrees	1.3	2.5	0.8	1.8
Mass, 10^{24} kg/10^{21} tons	1898/2093	568/627	86.8/95.7	102/113
Diameter, km/miles	142,984/88,846	120,536/74,897	51,118/31,763	49,529/30,775
Gravity, m/s^2/ft./s^2	23.1/75.9	9.0/29.4	8.7/28.5	11.0/36.0
Rotation period, and length of day, hours	9.9	10.7	17.2	16.1
Mean temperature, °C/°F	−110/−166	−140/−220	−195/−320	−200/−330

Characteristics of Venus

Venus is planet similar in size to Earth that orbits as the second planet out from the sun. Like Earth, it is a terrestrial planet with an iron core. The core is surrounded by a hot mantle of rock covered by a thin crust of very hot rock. Venus is distinguished by an extremely hot average surface temperature of about 462 degrees Celsius. The extremely hot temperature (135 degrees Celsius above the melting pint of lead) is thought to be due to the greenhouse effect of its dense carbon dioxide (CO_2) atmosphere.

Venus rotates nearly upright in its orbit with a tilt of 2.64 degrees. The rotation is very slow, taking 243 earth days for one rotation (one Venus sidereal day). The direction of rotation is opposite to the direction of the revolution about the sun and it is opposite to that of Earth. An observer on the planet would see the sun rise in the

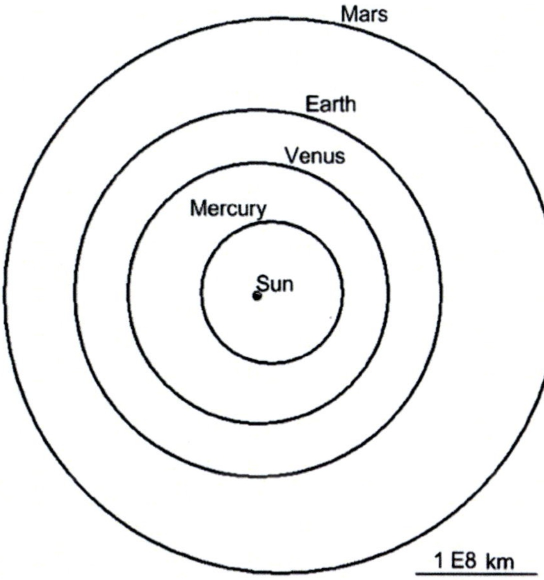

Fig. 1.1 Orbits of inner planets

west. The time for one orbit of Venus around the sun (one Venus year) is 225 Earth days. Quite different from Earth, the Venusian sidereal day is longer than its year. Because of the retrograde rotation, the time for the sun to reappear at the same point in the sky, called a solar day, is about 117 Earth days.

Thick clouds of small droplets of sulfuric acid float above the CO_2 atmosphere. Data from the Pioneer Venus probes indicate that there are three main cloud layers. The upper cloud layer extends between 70 and 57 km altitude, the middle cloud layer extends between 57 and 51 km and the lower cloud layer extends between 51 and 48 km. There is a haze layer between 90 and 70 km above the clouds and another haze layer between 48 and 31 km below the clouds. The region below an altitude of 30 km is clear. The top layers of clouds rush around the planet at about 360 km/h. The clouds circle the planet in a little over four Earth days.

Venus looks nearly white and featureless when reflection from sunlight is viewed at visible wavelengths. A picture taken by the Mariner 10 spacecraft as it was departing Venus is shown on the next page (Fig. 1.2). When viewed at ultraviolet wavelengths, a rich tapestry of cloud patterns is apparent.

Several Soviet Union spacecraft have landed on Venus and some photographed the surface. One such photograph from Venera 13 is shown on the next page (Fig. 1.3). The scene shown is a 90-degree sector of a 180-degree panoramic scan. The geometry of the azimuth scan and camera field of view allowed the horizon to be visible in the upper right corner of the image.

About 98% of the surface of Venus was mapped in detail by the U.S. Magellan spacecraft that orbited Venus and imaged it with cloud-penetrating synthetic

Fig. 1.2 Image of Venus taken in visible light by Mariner 10. (NASA/JPL graphic)

Fig. 1.3 Photograph of surface of Venus taken by Venera 13. (NSSDC Image catalog, Credit USSR)

aperture radar and a radar altimeter. Magellan data revealed that much of the planet is covered with vast planes of lava flow and wide shield volcanos. There are also highland regions with mountains and large ridged plateaus. There are only a few craters in the lava flows suggesting that major volcanism and lava flows occurred

Fig. 1.4 3-D view of Maat Mons developed from Magellan radar data. The vertical scale of the image has been increased ten times. (NASA graphic)

after the late heavy bombardment period that cratered Mercury and the Moon. The largest crater on Venus, named Mead, has diameter of 280 km. The highest mountain, rising about 10.9 km (35,770 feet) above the mean planet radius, is named Maxwell Montes. A 3-D image of another mountain, Maat Mons, developed from Magellan radar data is shown above (Fig. 1.4).

The atmosphere of Venus is made up of 96% carbon dioxide, 3.5% nitrogen and trace amounts of other gasses. Water vapor makes up about 31 parts per million of the atmosphere. The dense CO_2 atmosphere results in very high atmospheric pressure near the surface. A tabulation of the variation of temperature and pressure as a function of altitude is given in Table 1.3. Pressure is given in terms of bars where one bar is the atmospheric pressure on Earth at sea level.

Venus orbits inside of Earth's orbit around the sun and it is closer to Earth than is Mars, which orbits outside of Earth's orbit. Less energy is required of the launch vehicle to reach Venus than to reach Mars so Venus was the first planetary destination by spacecraft from both the United States and the Soviet Union.

Minimum energy launch windows to send spacecraft to Venus occur about every 584 days (19.4 months). These windows correspond to the time when Earth and Venus are near their closest approach as each travel around the sun in elliptical orbits. Each closest approach is different. The average closest approach distance is about 42 million km. Spacecraft launches to Venus by the U.S. and the Soviet Union were grouped a few weeks around minimum energy dates.

Table 1.3 Temperature and pressure vs. altitude over Venus

Altitude, km	Temp., °C	Press., bars
60	−10	0.24
50	75	1
40	143	4
30	222	10
20	306	23
10	385	47
0	462	92

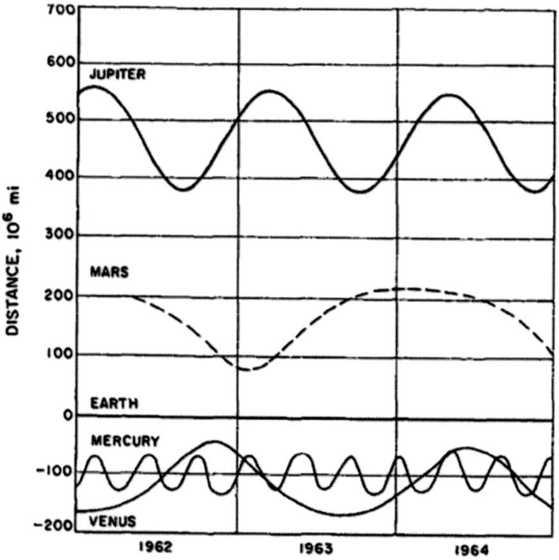

Fig. 1.5 Planetary distances relative to earth. (NASA/JPL graphic)

An informative graphic produced by the Jet Propulsion Laboratory (JPL) showing the distances between Earth and the other planets from 1962 to 1964 as the planets orbited the sun is shown above (Fig. 1.5). The graphic is contained in NASA/JPL technical Memorandum No. 33–212. Mercury and Venus orbit inside of Earth's orbit and they are plotted as negative distances. The ordinate of the drawing is given in miles.

Characteristics of Mercury

Mercury is a small, dense planet that orbits as the first planet out from the sun. It is a terrestrial planet like earth with a metallic core. The core, which is at least partially molten, occupies a substantial portion of the planet. It extends from the center to about 75–85% of the radius of the planet. The core is covered by a mantle or crust several hundred km thick. The average diameter of Mercury is 4878 km, about 38% of the diameter of Earth. The mass of the planet is 3.30114×10^{23} kg and the bulk density is 5429 kg per cubic meter. The bulk density is close to the 5514 kg per cubic meter bulk density of Earth.

Mercury has no atmosphere but instead it has a tenuous exosphere made up of atoms liberated from the surface of the planet by bombardment of the solar wind. The most prevalent atoms in the exosphere are sodium, magnesium, and calcium. Traces of hydrogen, helium, and potassium are also present.

The very thin atmosphere of Mercury allows photographs to be taken of its surface by flyby and orbiting spacecraft. A global mosaic of photographs of Mercury taken by the Messenger spacecraft is shown below (Fig. 1.6). A single photograph of the surface of Mercury taken by Messenger is shown on the following page (Fig. 1.7). The width of that image is 250 km and the resolution is about 245 meters per pixel. Most of the surface of Mercury is heavily cratered with exception of a few large smooth areas. Some craters are very large, likely due to impact of asteroids. The largest crater, named Caloris, is 1500 km in diameter, including ramparts around the crater.

Mercury travels around the sun in an elliptical orbit with periapsis distance of 47 million km and apoapsis distance of 70 million km. The time for one orbit of the sun (a Mercury year) is 88 Earth days. Mercury rotates nearly upright in its orbit with a tilt of about two degrees from perpendicular to the orbital plane. The rotation period is 58.65 Earth days. There is a fixed ratio of orbit period to rotation period of 3:2. The time for the sun to reappear in the sky at the same location on Mercury, a solar day, is 176 Earth days. A solar day on Mercury is twice as long as its year.

Fig. 1.6 Global mosaic of images of Mercury from Messenger spacecraft. Base map on left and color mosaic on right. (NASA/JHUAPL/Carnegie Institution image)

Fig. 1.7 Photograph of surface of Mercury taken by Messenger spacecraft. (NASA/JHUAPL/Carnegie Institution image)

The orbital mechanics of Mercury are such that longitudes of zero degrees and 180 degrees always experience noon at periapsis of the orbit and longitudes of 90 degrees and 270 degrees always experience noon at apoapsis. The solar insolation at Mercury is about 14,464 watts per square meter at periapsis and 6279 watts per square meter at apoapsis. For reference, Earth has insolation of about 1366 watts per square meter. The high solar insolation raises the temperature of the surface in the sunlit region near the equator of Mercury at periapsis to about 430 °C. Since Mercury does not have an atmosphere to distribute heat around the planet, the temperature on the surface on the night side of Mercury falls to about −180 °C.

Bibliography

Couper, Heather, et al. SMITHSONIAN, *Planets*, DK Publishing, New York, 2014

Jet Propulsion Laboratory, NASA, *Planet Compare*, NASA Science Solar System Exploration

Owen, Tobias Chant, *Solar System*, Encyclopedia Britannica, https://www.britannica.com/science/solar-system

Williams, David R., *Planetary Fact Sheet*, NASA, https://gsfc.nasa.gov/planetary/factsheet

Chapter 2
Launch Vehicles for Planetary Spacecraft

The 1960s and 1970s were periods of ambitious space exploration for the Soviet Union and the United States. Launch vehicles, derived from intercontinental ballistic missiles (ICBMs), became increasingly capable and spacecraft technology became evermore sophisticated. Venus and Mars, the two closest planets to earth, were visited by spacecraft several times in that period. The U.S. also sent a spacecraft to fly by Mercury. Exploration of the planets continued by spacecraft from the 1980s to the present time. Japan and the European Space Agency have joined the US and Russia in placing capable spacecraft in orbit around Venus.

Early Soviet Union Space Programs

The Soviet Union conducted a robust space program starting in the mid 1950s. They were first country to orbit a satellite of the earth (Sputnik 1 in October 1957) and they were first to put a man in earth orbit (Yuri Gagarin in April 1961). They also had very ambitious lunar and planetary exploration programs in the 1960s and 1970s. Several probes were sent to the moon in that interval and some were successful. Planetary programs included probes to Venus and to Mars.

Russia was the major entity of the Soviet Union and spacecraft development and space exploration was largely a Russian endeavor. Russia was the country around which the Soviet Union was formed in 1921 and Russia picked up the pieces after the Soviet Union dissolved in 1991.

Management and development of space programs in the Soviet Union were organized in a series of experimental design bureaus known as OKB (Opytno Konstruktorskoye Buro). The lead design bureau was OKB-1 headed by Sergei Korolev. Korolev was both the technical and political leader for space programs in the Soviet Union. Several other design bureaus with expertise in particular areas

T. Lund, *Spacecraft that Explored the Inner Planets Venus and Mercury*,
Springer Praxis Books, https://doi.org/10.1007/978-3-031-29838-7_2

supported OKB-1. The design bureau NPO Lavochkin under Georgy Babakin was at the forefront of design of planetary probes.

Sergei Korolev at OKB-1 was tasked by the Soviet military with developing the first Intercontinental Ballistic Missile (ICBM) in 1953. The missile was known as the R-7. The first artificial satellite of Earth, Sputnik 1, was placed in orbit by a R-7 arranged for Earth orbit. Variants of the R-7 were used to launch several planetary spacecraft.

The Soviet Union launched several spacecraft to Venus and to Mars in the period 1960 to 1991. Spacecraft destined for Venus that were successful or partly successful include: Venera 4 in 1967, Venera 5 and Venera 6 in 1969, Venera 7 in 1970, Venera 8 in 1972, Venera 9 and Venera 10 in 1975, Venera 11 and Venera 12 in 1978, Venera 13 and Venera 14 in 1981, Venera 15 and Venera 16 in 1983, and Vega 1 and Vega 2 in 1984. Those spacecraft and their missions are described in later chapters of this book.

Spacecraft destined for Mars that were successful or partially successful included: Mars 2 and Mars 3 orbiters/landers in 1971, Mars 5 orbiter in 1974, Mars 6 orbiter/lander in 1974, and Phobos 2 in 1989. Following breakup of the Soviet Union in 1991, Russia developed a capable spacecraft destined for Mars called Mars 96. It was launched in 1996 but it was lost due to failure of the launch vehicle. Another spacecraft designed to explore the Martian moon Phobos was launched in 2011 but it was stranded in Earth orbit.

The Soviet Union also conducted an ambitious lunar exploration program. Lunar exploration began with lunar impact missions and continued to lunar flybys. A lunar flyby spacecraft, Luna 3, was the first spacecraft to photograph the far side of the moon in 1959. Exploration of the moon continued with landing capsules on the surface. Two capsules were soft-landed by airbags on the moon in 1966. Both capsules returned scientific data and 360-degree panoramic photographs of the surface. A successful lunar orbiter program followed.

The Soviet Union also undertook a very difficult program to land cosmonauts on the moon. They made a heroic effort to catch up to the United States in the endeavor. The effort was eventually abandoned because of serious problems with their massive launch vehicle, the N1.

Soviet Union Launch Site for Early Planetary Spacecraft

Soviet Union planetary exploration spacecraft were launched from the Baikonur Cosmodrome in the Republic of Kazakhstan. That classic launch site, first used for Soviet ICBMs, has been in constant use for space exploration since the launch of Sputnik-1 in 1957. In modern times, launches to the International Space Station are made from the extensive facilities at the cosmodrome.

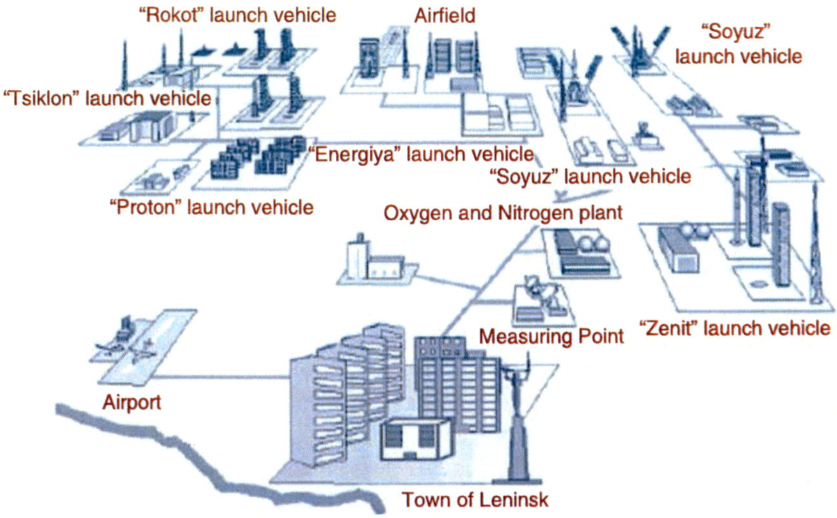

Fig. 2.1 Map of Baikonur Cosmodrome. (NASA graphic collection)

The Cosmodrome is located in an arid, sparsely settled region near the small town of Tyuratam in Kazakhstan. The latitude/longitude coordinates are 43°37′ North/63°19′ East. The latitude is about the same as Montreal, Canada and about 17 degrees higher latitude than Cape Canaveral. As a result, launches from the Cosmodrome do not benefit as much from the earth's rotation as at Cape Canaveral. A map of the Cosmodrome is shown above (Fig. 2.1).

Kazakhstan was a part of the Soviet Union when the cosmodrome was established in the 1950s. The Soviet Union broke up in 1991 and Kazakhstan regained its independence. A leasing arrangement for the land containing the cosmodrome was worked out between Russia and Kazakhstan in 1994.

The Cosmodrome has several launch sites. Perhaps the most famous is Site 1 where Sputnik was launched in 1957 and Yuri Gagarin was launched to orbit the Earth in 1961. It is still an active launch site. The 518th launch from Site 1 took place in July 2019 with launch of Soyuz MS-13 to the International Space Station. Other important launch sites at the cosmodrome include Site 31 for Soyuz-2 launches and Site 81 and Site 200 for Proton launches. Several spacecraft to Venus were launched from Site 81. Those included: Venera 9, Venera 10, Venera 11 and Venera 12. Spacecraft launched to Venus from Site 200 included: Venera 13, Venera 14, Venera 15, Venera 16, Vega 1, and Vega 2.

There were 14 launches from the Baikonur Cosmodrome in 2021. Of those, eight were to the International Space Station. Russian launches were also conducted from the Plesetsk, Kourou, and Vostochny launch sites in 2021.

Soviet Union Launch Vehicles for Early Planetary Spacecraft

Early Russian launch vehicles for lunar and planetary exploration were based on military launchers for ICBMs. The launch vehicles used for planetary exploration are described below.

Soyuz-Molniya 8 K78 Lunch Vehicle

The Soyuz-Molniya 8K78 launch vehicle launched numerous spacecraft to the moon as well as spacecraft to Venus and Mars. A photograph of the Molniya launch vehicle in a horizontal position is shown below (Fig. 2.2). The name Molniya was acquired from a series of communications satellites called Molniya that were successfully placed in orbit by the launch vehicle.

Soyuz-Molniya 8K78 had a central core first stage and four strap-on booster rockets. Two of the booster rockets can be seen on the side of the vehicle facing the viewer in the photograph. The boosters consisted of cylindrical enclosures with propellant tanks and an RD-107 engine at the end. The strap-on booster enclosures were 2.68 meters in diameter and 19 meters long. Propellants for all stages of the launch vehicle were kerosene fuel and liquid oxygen for oxidizer. The thrust of the booster engine was 815 kilonewtons (183,200 pounds) at sea level. The RD-107 engine had four combustion chambers feeding four nozzles and two smaller vernier

Fig. 2.2 Molniya 8K78 launch vehicle. (Wikimedia posting by Matveev Michail)

engines. The vernier engines were steerable. The burn time was about 120 seconds. The boosters were jettisoned after their fuel was exhausted. The four strap-on booster rockets were referred to as blocks B, V, G. and D.

The core first stage, referred to as Block A, had a diameter of 2.99 meters and length of 28 meters. It used a RD-108-8D55 engine that had thrust of 740.7 kilonewtons (166,500 pounds) at sea level. The engine had four combustion chambers and four nozzles. There were also four steerable vernier engines on the stage. The burn time was about 310 seconds. The core stage and the four strap-on boosters fired together at liftoff. Total thrust at liftoff was about 4000 kilonewtons (899,300 pounds).

The next stage in the launch vehicle stack was the Block I stage. It was powered by a RD-0108 engine with a thrust of 294 kilonewtons (66,100 pounds). The stage was 2.66 meters in diameter and 8.1 meters long. The burn time was about 200 seconds.

The launch vehicle up to and including the Block I stage was capable of placing a respectable payload into low Earth orbit. A final stage, the Block L, provided additional thrust to reach escape velocity from the Earth. The Block L stage used an S1.5400 engine that generated 67 kilonewtons (15,000 pounds) of thrust. The stage was 2.56 meters in diameter and 7.1 meters long. The engine was gimballed in pitch and yaw for thrust direction control. Two vernier engines provided roll control. The burn time was about 200 seconds.

The weight of the Molniya 8K78 vehicle at launch was 303,500 kg (669,100 pounds). The length of the vehicle was 40 meters.

Molniya 8K78M Launch Vehicle

An upgrade to the Molniya 8K78 called the Molniya-M or 8K78M was developed in 1963 to improve the reliability of the 8K78 launch vehicle. The Block I stage was redesigned and the Block L stage was improved. The former RD-0108 engine in the Block I stage was replaced by a RD-0110 engine type. The RD-0110 engine generated 298 kilonewtons (67,000 pounds) of thrust. The Molniya 8K78M vehicle weighed 305,460 kg (673,420 pounds) at time of launch. The length was 40 meters.

Proton-K Launch Vehicle

What became known as the Proton launch vehicle was also known as UR-500. It was developed by design bureau OKB-52 under Vladimir Chelomey. The first assignment of the UR-500 was to launch a scientific satellite called Proton into orbit. Three out of four launches of the satellite were successful. The launch vehicle was referred to as Proton after the satellite and that name stuck.

The Proton-K launch vehicle could be configured as either a three-stage or a four-stage vehicle with the fourth stage consisting of the capable Block D stage. The Block D stage served as a space tug for many different lunar and planetary missions. The first launch of a Proton-K with a Block D fourth stage was in March 1967.

Proton-K was used to launch several lunar probes as well as probes to Venus and Mars. It was also used to loft modules to the Russian Mir space station and modules to the International Space Station. Proton-K had made 311 launches before it was supplanted by the Proton-M launch vehicle. A photograph of a Proton-K launch vehicle carrying the Zvezka service module to the International Space Station is shown below (Fig. 2.3).

The first stage of Proton-K contained six booster rockets attached to the end of individual fuel tanks. The fuel tanks with rocket engines attached were clustered about a center oxidizer tank. The diameter of the oxidizer tank was about 4.1 meters and the diameter of each of the fuel tanks was about 1.6 meters. The total length of the first stage was 21.2 meters. The rocket engines were each gimballed to allow steering the first stage.

Fig. 2.3 Russian Proton-K launch vehicle. (Photograph from NASA website KSC-PADIG-028)

The engine for the boosters was the legendary RD-253 designed by Valentin Glushko, head of Design Bureau OKB-456. It was an advanced engine for the time developing 1470 kilonewtons (330,500 pounds) of thrust at sea level. The total thrust of the six booster engines was about 8820 kilonewtons (1.98 million pounds). The burn time was about 120 seconds. The first stage was jettisoned after completion of the burn. Fuel for the engines was unsymmetrical dimethydrazine (UDMH) and the oxidizer was nitrogen tetroxide. That combination of fuel and oxidizer ignited upon contact (hypergolic). The second and third stages of the Proton-K were designed by Glushko and used the same combination of fuel and oxidizer.

The second stage contained four rocket engines mounted to a gimbal structure that allowed steering in flight. The total thrust of the four engines in a vacuum was about 2400 kilonewtons (539,000 pounds). The burn time was 209 seconds. The diameter of the second stage was 4.14 meters and the length was 17.07 meters. The second stage was separated from the third stage after its burn was completed.

The third stage was 4.15 meters in diameter and 4.11 meters long. It contained a single main engine and four gimballed vernier engines. The vernier engines provided steering and fine thrust control. The thrust of the main engine was 574 kilonewtons (129,000 pounds) in vacuum. The burn time was about 236 seconds. The third stage could inject a spacecraft into an earth parking orbit about 200 km high. The third stage also contained a flight control system that controlled the Proton-K vehicle during burn of the first three stages.

The fourth stage, referred to as Block D, was four meters in diameter and 5.5 meters long. Its RD-58 M engine used kerosene fuel and liquid oxygen as oxidizer. The thrust was 85 kilonewtons (19,100 pounds) and the burn time was about 600 seconds. The engine could be started and stopped several times. The Block D served as a space tug for several space missions.

Early United States Space Programs

Several organizations in the U.S. were participating in space related projects in the early 1950s. President Eisenhower favored consolidating all space programs into one civilian controlled organization. This would include non-military space activities of the U.S. Army and the U.S. Air Force. Congress responded and took up the challenge of bringing about the National Aerodynamics and Space Administration (NASA). President Eisenhower signed the bill into law on 29 July 1958. The goal of NASA was to better pursue space activities of national interest and avoid duplication of effort. Forming NASA was also looked on as a way to counter the burgeoning space programs of the Soviet Union.

Much of the original organizations of the entities making up NASA were retained, but several organizations were given new names. The organizations were directed by NASA Headquarters in Washington DC. The new NASA organization incorporated the National Advisory Committee for Aeronautics (NACA), Langley Aeronautical Laboratories, Ames Aeronautical Laboratory, Lewis Flight Propulsion

Laboratory, the Army Ballistic Missile Agency, and the Jet Propulsion Laboratory. The personnel and programs of NACA became the nucleus of the new NASA organization. Names were changed on some of the incorporated organizations and some new organizations were added. The incorporated organizations were referred to as field centers. The field centers and their early directors are listed below.

Field center	First director
Marshal Space Flight Center	Wernher von Braun
Manned Spaceflight Center	Robert Gilruth
Langley Research Center	Henry Reed
Ames Research Center	Smith DeFrance
Goddard Space Flight Center	Harry Goett
Flight Research Center	Paul Bikle
Lewis Research Center	Edward Sharp
Launch Operations Center	Kurt Debus

One organization incorporated into NASA, the Jet Propulsion Laboratory, was called a Federally Funded Research and Development Center (FFRDC) rather than a field center. The Jet Propulsion Laboratory (JPL) was, and still is, managed by the California Institute of Technology (Caltech) under contract from NASA. JPL workers are Caltech employees rather than civil service employees. JPL was led by William Pickering for several years before and after JPL joined NASA.

The United States conducted an ambitious space exploration program in the 1960s and 1970s with several impressive successes. The first U.S. Satellite, Explorer 1, was launched in January 1958. Probes to the planets began with Mariner 2 that reconnoitered Venus by a flyby in August 1962 and Mariner 4 that reconnoitered Mars by a flyby in November 1964. Mariner 5 conducted a flyby of Venus in 1967 and Mariner 6 and Mariner 7 conducted flybys of Mars in 1969. Mariner 9 conducted a Venus orbiting mission in 1971 and Mariner 10 made three flybys of Mercury in 1974.

The United States continued to send sophisticated probes to the planets after 1980. The Messenger spacecraft orbited Mercury from August 1990 to October 1994. Six Mars orbiter missions were conducted and three stationary landers were deposited on Mars: two Viking landers in 1975 and the Insight lander in 2018. Mobile rovers that were landed on Mars include the small rovers Spirit and Opportunity in 2004, the mid-size Curiosity rover in 2012, and the large Perseverance rover in 2021.

Probes to the outer planets included two Jupiter orbiters: Galileo in 1995 and Juno in 2016, the Cassini orbiter of Saturn in 2004, and the New Horizon flyby of Pluto in 2015. The New Horizons spacecraft continued on out to the Kuiper Belt and imaged a double-lobed object named Arrokoth, 6.6 billion km from Earth.

The United States also conducted a methodical exploration of the moon in the 1960s and 1970s. Those explorations included the Ranger, Lunar Orbiter, Surveyor and Apollo programs. Ranger was a photography mission as the spacecraft descended towards impact on the moon. The first impact was in July 1964. The

Lunar Orbiter program orbited five spacecraft around the moon from August 1966 to August 1967 and made photographic maps of 99% of the lunar surface. The surveyor program soft-landed five spacecraft on the moon between May 1966 and January 1968. The impressive Apollo program soft-landed six manned spacecraft on the moon between July 1969 and December 1972. A crew of two astronauts from each mission walked on and explored the moon. The last three missions carried a dune-buggy type vehicle to carry astronauts on explorations farther from the landing site.

Later U.S. missions to the moon included Lunar Prospector in 1988, Lunar Reconnaissance Orbiter in 2009, Grail A and Grail B in 2011, and Lunar Atmosphere and Dust Environment Explorer in 2013. These were all lunar orbiting spacecraft.

United States Launch Sites for Early Planetary Spacecraft

United States probes to the planets were launched from Cape Canaveral in Florida. Construction of the facility began in May 1950. It first served as launch sites for several different military missiles. The first two launches were V-2 rockets.

Cape Canaveral is located on the eastern coast of Florida at latitude of about 28.4 degrees. The Cape's location in a sparsely populated area was an advantage at the time when missile launch accidents were common. The 28.4-degree latitude allowed taking advantage of about 88% of the maximum velocity imparted by earth's rotation when launching to the east. Easterly flying spacecraft would pass over the ocean and stages could be safely dropped is secure areas when they burned out. Another reason for the selection was that there was an Air Force station named the Joint Range Proving Ground a few miles to the south. That facility was renamed Patrick Air Force Base in 1950. Patrick Air Force base supports and administers the Cape Canaveral Air Force Station.

New extensive launch facilities were required for the Apollo program that was planned to land astronauts on the moon. To that end, NASA acquired much of the land of Merritt Island that lies just north of Cape Canaveral. Two launch pads, 39A and 39B were constructed on the island along with the vertical assembly building and other support structures for Apollo. Apollo 11, which landed the first astronauts on the moon, was launched from launch pad 39A in July 1969. NASA's land holding on Merritt Island was named the Kennedy Space Center.

A segment of a map showing Cape Canaveral Air Force Station and the Kennedy Space Center is shown on the next page (Fig. 2.4). A more detailed map of launch complexes at the Cape Canaveral Air Force Station is shown on the following page (Fig. 2.5). Names of active launch complexes are lettered in red on the deailed map. Refurbished launch complexes were given the name Space Launch Complexes (SLC). Launch complexes LC-39A and LC-39B are part of the Kennedy Space Center.

Presently, LC-39A and SLC-40 are leased to SpaceX for launch of their Falcon 9 and Falcon Heavy launch vehicles. SLC-40 is used for launch of SpaceX Falcon

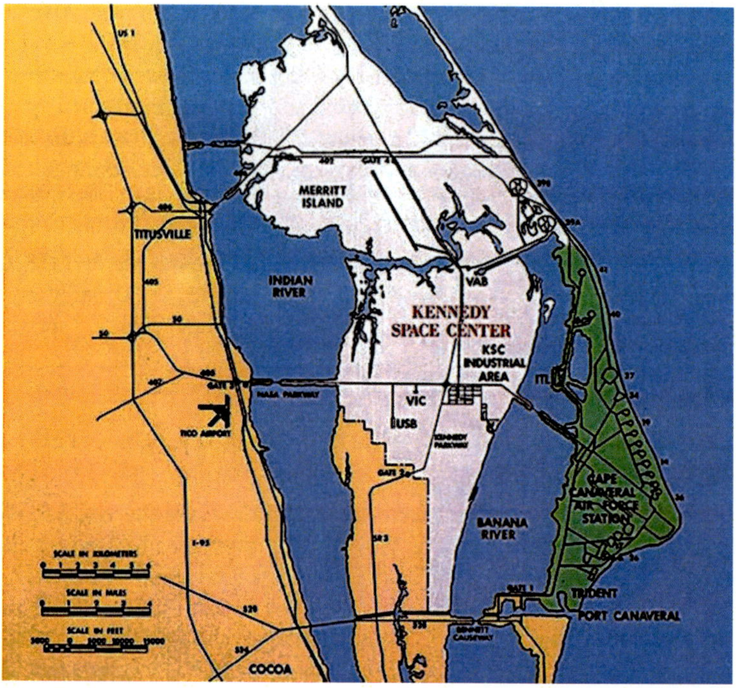

Fig. 2.4 Map showing Cape Canaveral Air Force Station in green and Kennedy Space Center in white. (NASA graphic)

9 launch vehicles. Launch Complex-13 is used to land spent boosters from SpaceX launch vehicles. Delta IV rockets are launched from SLC 37B and Atlas V rockets are launched from SLC-41.

A total of 31 launches from the combined Cape Canaveral Air Force Station and Kennedy Space Center were conducted in 2021. Of those, five were cargo flights to the International Space Station (ISS) and two were crew transport flights to the ISS.

United States Launch Vehicles for Planetary Spacecraft

Launch vehicles for early planetary probes were based on the Atlas, Titan, and Thor intercontinental ballistic missiles with upper stages added. Launch vehicles that originated with the Thor ICBM were given the name Delta. The Mariner series of spacecraft were launched to Venus and to Mars by Atlas-Agena and Atlas-Centaur launch vehicles. The Viking spacecraft that landed on Mars were launched by Titan-3-Centaur launch vehicles. Several spacecraft to Mars, Venus and Mercury were launched by Delta II launch vehicles and a few by Atlas 5 launch vehicles. Some planetary spacecraft including Magellan to Venus and Galileo to Jupiter were launched by the Space Shuttle.

Fig. 2.5 Launch complexes at Cape Canaveral. (Derived from CCAGS.jpg Credit Mark Wade)

Atlas-Agena Launch Vehicles

Atlas was a successful ICBM and it was also well suited as a base vehicle for launch of spacecraft. Early space missions used Agena B and Agena D upper stages to place spacecraft on trajectories to Venus and Mars.

Mariner 1 and Mariner 2 spacecraft to Venus were launched by an Atlas LV-3/Agena B vehicle. Mariner 3 and Mariner 4 destined for Mars were launched by an Atlas LV-3/Agena D vehicle. A photograph of launch of Mariner 1 is shown on the next page (Fig. 2.6). The Agena B stage is the upper portion of the vehicle extending down to the nose fairing of the Atlas first stage. The Agena-B engine nozzle extended down past the black cylindrical area in the photograph to inside the Atlas nose fairing.

The length of the overall vehicle at launch, including the shroud over Mariner, was 31.7 meters. The Atlas was 3.05 meters in diameter and Agena-B was 1.52

Fig. 2.6 Launch of Mariner 1 by Atlas-Agena B. (NASA photo)

meters in diameter. The weight of the overall launch vehicle was about 127,547 kg (281,242 pounds) at liftoff.

The Atlas LV-3 was built by General Dynamics and Agena-B was built by the Lockheed Missiles and Space Company. Atlas was unique in that the fuel and oxidizer tanks that constituted much of the upper structure of the vehicle used thin stainless walls and the shape of the vehicle was maintained by internal pressure of helium gas.

Five engines powered the Atlas at launch; two booster engines, one sustainer engine, and two vernier engines. The engines all burned rocket propellant-1 (RP-1), which was highly refined kerosene. The oxidizer was liquid oxygen. The booster engines were LR89 type built by Rocketdyne and the sustainer engine was a LR105 type also built by Rocketdyne. The vernier engines were Rocketdyne LR101 type.

RP-1 fuel was contained in a cylindrical tank located at the bottom of the vehicle. Liquid oxygen was contained in a cylindrical cryogenic tank above the fuel tank. The engines were ignited by injecting a charge of hypergolic fluid along with RP-1 fuel into the combustion chamber where it spontaneously ignited upon contact with liquid oxygen. Burning continued on RP-1 fuel.

In the cases of launches of Mariner 1 and Mariner 2, the trusts of the engines at sea level were about as follows:

Boosters (two engines)	1,374,555 newtons (309,000 pounds)
Sustainer	253,558 newtons (57,000 pounds)
Verniers (two engines)	8897 newtons (2000 pounds)
Total thrust	1,637,010 newtons (368,000 pounds)

All of these thrust levels were at sea level. The thrust was higher in a vacuum. The total thrust at liftoff of about 368,000 pounds was well above the 281,242-pound weight of the overall launch vehicle and Mariner 1 and 2 spacecraft at launch.

Other versions of the Atlas launch vehicles had slightly different thrust levels. For example, the acceptance test values for thrust of the engines at sea level for the Rocketdyne MA-5 propulsion system totaled 1,721,263 newtons (386,940 pounds)

The booster engines were mounted on gimbals that allowed each engine to pivot five degrees in pitch and five degrees in yaw with respect to the centerline of the Atlas. The pivoted booster engines were used to steer to vehicle to a preprogrammed trajectory after launch. The trajectory followed an arc that gradually tilted from vertical at launch towards the horizontal as the vehicle gained altitude and speed.

In the case of Mariner 2, the booster engines were shut off about 140 seconds after liftoff. The booster engines and their associated fuel pumps were then jettisoned. The sustainer and vernier engines of the Atlas continued to burn until they were cutoff about 240 seconds after launch. The sustainer engine was gimballed and it could be pivoted three degrees in pitch and three degrees in yaw about the centerline. The engine thrusted along the centerline while the boosters were firing. Its pivoting ability was used for steering after the booster engines burned out and were jettisoned. The vernier engines of Atlas could be oriented within a 140-degree arc in pitch and 50-degree arc in yaw. This positioning capability allowed the launch vehicle to be rolled to the desired orientation and to be controlled in pitch and yaw.

The total amount of propellant (fuel and oxidizer) carried by the Atlas was about 104,127 kg (114.8 tons). Of this, about 67,664 kg (74.6 tons) was used during the booster firing and the remainder, 36,463 kg (40.2 tons), was available for use by the sustainer and vernier engines.

The Agena-B upper stage contained a single engine that burned unsymmetrical dimethyl hydrazine as fuel and fuming nitric acid as oxidizer. This combination was hypergolic, igniting upon contact with one another. The engine was developed and built by Bell Aerosystems as their Bell model 8091. The engine generated 16,000 pounds of thrust in a vacuum and it could be shut down and restarted several times in orbit. Agena-B carried about 5533 kg (6.1 tons) of fuel and oxidizer and that gave a total burn time of 240 seconds. The Agena was 1.5 meters in diameter in the propellant and equipment areas and 7.2 meters long. Fully fueled, Agena B weighed about 7.167 kg (15,800 pounds).

Mariner 3 and Mariner 4 spacecraft were launched by Atlas LV-3 /Agena D launch vehicles. Mariner 5 was launched by an Atlas SLV3/Agena D launch vehicle. The SLV-3 model of Atlas was a standardized version of the Atlas D vehicle built for space use. Previous versions of the Atlas D used for space launches had been Intercontinental Ballistic Missiles modified for space vehicle launch. The SLV-3 model also had a new improved autopilot. The Agena D stage was essentially the same as Agena B but it represented a standardized configuration. A slight difference was that it had a burn time of 265 seconds rather than 240 seconds for Agena B.

Atlas-Centaur Launch Vehicles

The Centaur upper stage on Atlas greatly increased the load carrying ability relative to Atlas Agena launch vehicles. The Atlas SLV-3C/Centaur D launch vehicle could place 3500 kg (7718 pounds) in low earth orbit compared with 1825 kg (4024 pounds) for the Atlas SLV-3/Agena D launch vehicle. The Centaur upper stage was developed and built by the Convair Division of General Dynamics.

The Centaur D upper stage was an advanced, high-performance stage that burned liquid hydrogen with liquid oxygen oxidizer. Liquid hydrogen/liquid oxygen propellants give the highest specific impulse of known propellants. The specific impulse (I_{SP}) of Centaur D was 444 seconds. For comparison, the sustainer engine of Atlas that burned RP-1 with liquid oxygen as oxidizer had an I_{SP} of 316 seconds.

The Centaur D stage was cylindrical, 3.05 meters in diameter and about 9.14 meters long. The main propulsion came from two Pratt & Whitney RL10A type engines that provided a total thrust of 133.45 kilonewtons (30,000 pounds). Three different variations of the engine, RL10A-3-1, RL-10A-3CM-1, and RL10A-3-3 could be used depending on the mission. All three engines had essentially the same thrust. The engines were gimballed over ±4 degrees in a square pattern to allow control of pitch, yaw, and roll of the stage during powered flight. A series of small thrusters powered by steam produced by the reaction of hydrogen peroxide with a catalyst allowed attitude control when the main engines were not firing.

Mariner 6, Mariner 7, Mariner 8, and Mariner 9 were launched towards Mars by Atlas SLV-3C/Centaur D launch vehicles. Atlas SLV-3C had a straight upper section to mate with a Centaur upper stage rather than tapered to mate with Agena. The Centaur upper stage had the same diameter as atlas at 3.05 meters. A photograph of the launch of Mariner 8 by an Atlas SLV-3C/Centaur D is shown on the next page (Fig. 2.7).

Fig. 2.7 Launch of Mariner 8 by Atlas/Centaur. (NASA photo)

Titan III/Centaur Launch Vehicles

The Centaur launch vehicle represented a significant step-up in payload weight that could be sent to the planets. The launch vehicle could place 15,400 kg (33,900 pounds) in low earth orbit. A photograph of Titan III/Centaur on the launch pad with a Viking spacecraft destined for Mars is shown on the next page (Fig. 2.8). Titan III was developed and built by Martin Marietta Aerospace.

Fig. 2.8 Titan III/Centaur with Viking spacecraft on launch pad. (NASA photograph)

Titan III was a two-stage core vehicle with two solid-fuel booster rockets attached to the sides. The two solid-fuel booster rockets were built by United Technology Corp. They were 3.05 meters in diameter and 25.9 meters long. Propellants were powdered aluminum fuel with ammonium perchlorate oxidizer. The two boosters generated a total of 10.68 million newtons (2.4 million pounds) of thrust. The burn time was about 118 seconds.

The vehicle was steered during burn of the solid-fuel boosters by a thrust control system that included injectors in four quadrants in the sides of the nozzles of the booster rockets. Nitrogen tetroxide under pressure was applied to selected injectors and that changed the thrust flow angle of the solid fuel engines.

The core portion of the launch vehicle consisted of two stages. The two stages, referred to as stage 1 and stage 2, were 3.05 meters in diameter. Stage 1 was 19.2 meters long and stage 2 was 7.01 meters long. Propellants for the stages consisted of a mixture of equal parts of hydrazine and unsymmetrical dimethyl hydrazine as fuel and nitrogen tetroxide as oxidizer. The fuel mixture was referred to as Aerozine-50.

Stage one was ignited 112 seconds after liftoff, a few seconds before burnout of the booster rockets. It generated about 2.31 million newtons (520,000 pounds) of thrust and burned for 146 seconds. The rocket motor was gimballed to allow steering of the launch vehicle. Stage 2 was ignited after burnout and jettisoning of stage 1. Stage 2 generated about 449,288 newtons (101,000 pounds) of thrust and burned for 210 seconds. The rocket engine was gimballed to allow steering while stage 2 was burning. About 18 seconds after burnout of stage 2, Titan was separated from Centaur. The first burn of Centaur's engine occurred nine seconds after separation.

Centaur D-1 T was an energetic upper stage that burned liquid hydrogen with liquid oxygen as oxidizer. The thrust of Centaur D-1 T was 133,452 newtons (30,000 pounds).

Space Shuttle Launch Vehicle

The Space Shuttle evolved from the desire for a more economical access to space than by the massive, one-time use rockets then in use. The configuration of the new launch system, referred to as the Space Shuttle, was established in 1972. It was a three-element system consisting of an Orbiter, an expendable External Tank carrying liquid hydrogen and liquid oxygen for the Orbiter's engines, and two recoverable Solid Rocket Boosters. The payload was carried in a large bay with hinged covers in the orbiter. A photograph of a launch of Space Shuttle Columbia is shown on the next page (Fig. 2.9). The large red structure in the photograph is the external fuel tank. At the end of a mission, the orbiter reentered into the Earth's atmosphere and glided to a landing at either the Kennedy Space Center in Florida or Edwards Air Force Base in California.

Contracts to develop and build the Space Shuttle were awarded in 1972. Rockwell International was selected to develop the Space Shuttle Orbiter, Martin Marietta developed the External Tank, and Morton Thiokol developed the Solid Rocket Boosters. Rocketdyne, a division of Rockwell, developed the Orbiter's main engines.

There were six orbiters built: Columbia, Challenger, Discovery, Atlantis, Endeavour, and Enterprise. Unfortunately, Challenger and Columbia and their crews were lost due to accidents. Endeavour was built to replace Challenger. The orbiter Enterprise was only used for testing in the atmosphere. The first manned flight of the Space Shuttles occurred in April 1981. The last flight occurred in August 2011. Space Shuttles flew 135 missions during their 30 years of service. The Magellan spacecraft was launched towards Venus by Space Shuttle Atlantis in 1989.

Fig. 2.9 Launch of Space Shuttle Columbia. (NASA photograph)

Orbiter – The Orbiter was about the size of a DC-9 airliner. The major structural sections of the Orbiter were the forward fuselage, which contained the pressurized crew compartment; the mid fuselage, which contained the 4.57 meters wide by 18.29 meters long payload bay and associated doors; the aft fuselage, and the vertical tail. The crew compartment had a flight deck on top of a middeck. The usual crew was seven people. The payload bay was designed to carry cargoes weighing up to 29,000 kg (32 tons) to Earth orbit. Cargoes of up to 14,500 kg (16 tons) could be carried from space back to earth.

The orbiter needed to be protected from the searing heat due to friction when reentering the earth's atmosphere. The thermal protection system kept the temperature of the aluminum skin below 177 °C (350 °F) while temperatures on the leading edge of the wings could rise to 1510 °C (2,750 °F) during reentry. The nose cap and leading edges of the wings were protected with an all-carbon composite consisting of layers of graphite cloth in a carbon matrix. Areas subjected to lower heat were shielded with high-temperature ceramic tiles about six inches square and varying in thickness from one to five inches, depending on the protection needed.

The orbiter contained three main rocket engines that used liquid oxygen and liquid hydrogen for fuel. The thrust of each engine was 1.67 million newtons (375,000 pounds) at sea level. The thrust increased to 2.09 million newtons (470,000 pounds) in a vacuum. The engines could be throttled over a range of 56% to 104% of maximum rated thrust.

External Tank – The External Tank, which was 47 meters long and 8.38 meters in diameter, carried 143,000 gallons of liquid oxygen and 383,000 gallons of liquid hydrogen. As well as carrying fuel, the External Tank formed the structure to which the Orbiter and the two Solid Rocket Boosters were attached. A multi-layered thermal coating was placed on the outside of the External Tank to protect it from extreme temperature variations during pre-launch, launch, and flight. The insulation also reduced boil-off rate of propellants.

Solid-Rocket Boosters – The two solid-rocket boosters each generated 11.79 million newtons (2.65 million pounds) of thrust at sea level. The total thrust of the orbiter's three engines and two booster rockets at lift-off was 28.59 million newtons (6.43 million pounds).

The solid-rocket boosters were 45.5 meters long and 3.71 meters feet in diameter. The solid propellant fuel in each booster weighed 500,000 kg (1.1 million pounds). The burn time was about 127 seconds. After the boosters exhausted their fuel, they were jettisoned from the External Tank. The boosters were slowed by parachutes and fell into the ocean where they were recovered, refueled, and reused.

Delta II Launch Vehicle

The Delta II launch vehicle, built by Boeing, was a very successful vehicle that lofted several space payloads. It sent the Messenger spacecraft on its journey to Mercury. A total of 155 Delta II vehicles were produced. Noteworthy, vehicle number 155 was the 100th consecutive launch success.

Messenger was launched by a three-stage version of the vehicle. A photograph of the launch is shown on the next page (Fig. 2.10). Delta II, with payload fairing, was 38.9 meters long. The vehicle stack consisted of a first stage, second stage, third stage, and a fairing that covered the payload. The first stage included the engine section, liquid oxygen tank, fuel tank, and an interstage that mated to the second stage. The first and second stages and the interstage were 2.4 meters in diameter. The third stage was 1.2 meters in diameter. The first stage was 26.1 meters long. The second stage was 6.0 meters long and the third stage was 2.0 meters long.

Delta II included nine solid propellant booster rockets clustered around the bottom of the first stage to augment the thrust of the first stage engine. Each booster was 14.7 meters long and 1.17 meters in diameter. The thrust of each was 633 kilonewtons (142,300 pounds) and the burn time was about 76 seconds. Six of the solid

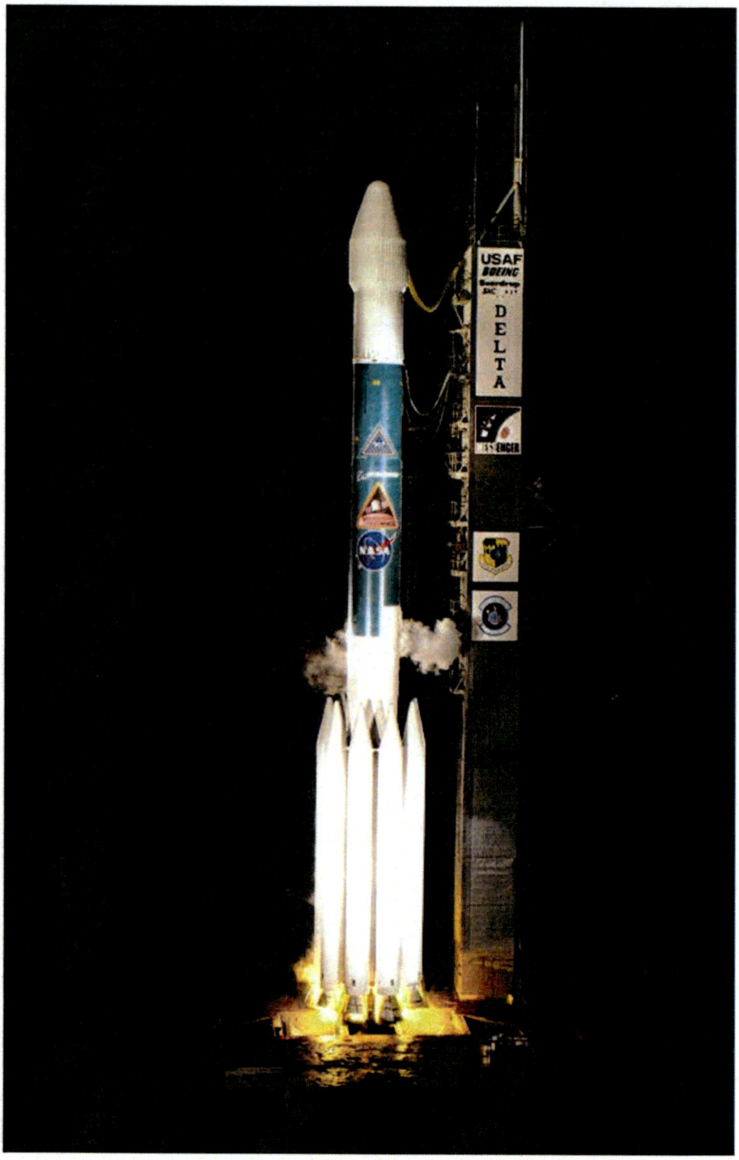

Fig. 2.10 Delta II launch vehicle. (NASA photograph)

rocket boosters were ignited at liftoff and the remaining three were ignited after the first group had burned out.

The main engine in the first stage was a Rocketdyne RS-27A that developed thrust of 890 kilonewtons (200,000 pounds). The burn time was about 260 seconds. The engine was gimballed to provide pitch and yaw control during firing. Fuel for the engine was RP-1 (kerosene) and the oxidizer was liquid oxygen. The stage included two Rocketdyne LR101-NA-11 vernier engines that provided roll control during the main engine burn and attitude control after burnout of the main engine.

The second stage used an Aerojet AJ10-118 K engine that provided thrust of 43.39 kilonewtons (9753 pounds). The burn time was about 430 seconds. The engine was gimballed to provide pitch and yaw control. Fuel for the engine was Aerozine 50, which is an equal mixture of hydrazine and unsymmetrical dimethylhydrazine (UDMH). The oxidizer was nitrogen tetroxide (N_2O_4). The fuel and oxidizer ignited upon contact. The engine could be started and stopped several times by controlling the propellant flow to the engine.

The third stage consisted of a Thiokol Star-48B solid-fuel rocket motor. It provided an average thrust of 68.639 kilonewtons (15,430 pounds) and burn time of about 84 seconds.

Falcon 9 Launch Vehicle

A striking departure from historic government funded launch vehicles, Falcon 9 is a commercial vehicle developed by the Space Exploration Technology Corporation. The corporation, founded by Elon Musk, is universally known as SpaceX. Falcon 9 launch vehicles have been extremely successful with over 130 space launches. While Falcon 9 has not yet been used to launch planetary missions, it will likely be so used in the future and a brief description of it is included in this chapter. SpaceX advertises that a variant of Falcon 9 is capable of delivering a payload weighing 4020 kg to Mars.

Falcon 9 has reached the high distinction of being man-rated by NASA. There have been four launches ferrying crew to the International Space Station (ISS) by Falcon 9 as of November 2021. The launch vehicles to the ISS loft a Crew Dragon spacecraft that can transport seven passengers to the ISS and return them to Earth. Crew Dragon incorporates an ablative aeroshell to survive entry into the atmosphere and slow the spacecraft to allow parachutes to further slow the capsule for a water landing. Crew Dragon was developed and is produced by SpaceX.

Falcon 9 is a two-stage vehicle with a total length of 70 meters and diameter of 3.7 meters. The first stage, 47.7 meters long, is powered by nine Merlin engines. The Merlin engine, developed by SpaceX, generates a thrust of 854 kN. The nine engines firing together develop 7.68 million newtons (1.73 million pounds) of thrust at sea level. Propellants for the engine are RP-1 (refined kerosene) fuel with an oxidizer of liquid oxygen. The available burn time for the first stage is 160 seconds. The second stage is powered by a single Merlin engine that develops thrust of 981 kN in a vacuum. The available burn time for the second stage is 397 seconds.

A significant feature of Falcon 9 is the ability to return the first stage to the surface of Earth for reuse. The stage uses three of the engines to maneuver the spacecraft and slow it for a soft landing on four legs that unfold from the lower body of the stage. The recovered stage is refurbished and reused. SpaceX reported recently that there has been total of 139 launches, 99 landings of the first stage, and 79 recovered stages have been reflown.

A total of seven flights were lofted to the International Space Station by Falcon 9 launch vehicles in 2021. Five of those were cargo supply flights and two transported crew to the station. A photograph of Falcon 9 with a Crew Dragon capsule being prepared for launch from pad 39A at Cape Canaveral is shown on the next page (Fig. 2.11). An arm extending from the gantry provides access for persons to enter the capsule.

Japanese Launch Vehicle for Akatsuki Spacecraft to Venus

The primary Japanese launch vehicle for heavy payloads was the H-IIA that was developed by the National Space Development Agency of Japan (NASDA). The first launch of an H-IIA with satellite payload occurred in August 2001. It has launched several satellites since including the Greenhouse gases Observing Satellite-2 (GOSAT-2) in October 2018, the Selene/Kaguya probe to the moon in September 2007, and the Akatsuki spacecraft sent to orbit Venus in May 2010.

A photograph of an H-IIA launch vehicle carrying the Global Precipitation Measurement spacecraft is shown on the following page (Fig. 2.12). The H-IIA and payload are being transported to the launch pad at the Tanegashima Space Center in the picture. The size of the vehicle can be appreciated by comparing with the size of people below it.

The launch vehicle included a main first stage, which is the yellow cylinder from the base up to the black cylinder section, the second stage, which is the black segment, and two booster rockets attached to the lower portion of the first stage. The first stage was 37.2 meters long and 4 meters in diameter. The second stage was 9.2 meters long and 4 meters in diameter. The solid fuel rocket boosters were 15.1 meters long and 2.5 meters in diameter.

The thrust of the two solid rocket boosters was a total of 5040 kilo newton (kN). The thrust of the first stage was 1098 kN and the thrust of the second stage was 1490 kN. Both the first and second stages burned liquid hydrogen with liquid oxygen oxidizer. The H-IIA was capable of placing a payload of about 10 metric tons in low earth orbit and accelerate about 2.5 metric tons to Earth escape velocity.

Tanegashima Space Center

The Tanegashima Space Center is the main launch complex in Japan. It is located on the southeast side of Tanegashima Island. The island is located about 40 km southeast of Kyushu. A photograph of the launch areas that appears in the Tanegashima Space Center Visitor's Guide is shown on the following page (Fig. 2.13). The Tanegashima Space Center is managed by the Japan Aerospace Exploration Agency (JAXA).

Fig. 2.11 Falcon 9 launch vehicle on Pad 39A. (NASA photograph)

European Space Agency (ESA) Launch Vehicle Soyuz-Fregat

The Venus Express spacecraft was launched from the Russian Baikonur Cosmodrome by a Soyuz-Fregat (Soyuz-FG) launch vehicle. The European Space Agency procured Soyuz-FG vehicles from Starsem, a European-Russian company. Starsem

Fig. 2.12 Japanese H-IIA launch vehicle with GPM satellite being transported to launch pad. (Credit: NASA/Bill Ingalls)

was formed to market Russian made launch vehicles to various users. Shareholders of Starsem are four companies: Aerospatiale, Arianespace, the Russian Space Agency, and TsSKB Samara. Aerospatiale and Arianespace are French companies. The Russian Space Agency (Roskosmos) is responsible for managing Russian Space programs. TsSKB develops, manufactures, and operates launch vehicles and

Fig. 2.13 Launch complex at Tanegashima Space Center. (JAXA photograph)

spacecraft for Russia. TsSKB is located at the Samara Space Center in the city of Samera, Russia. The Starsem Company is also located in Samera.

Soyuz-FG Launch Vehicle

The Soyuz-FG launch vehicle was an upgrade of the Soyuz-Molniya launch vehicle described earlier in this chapter and shown in Fig. 2.2. The Soyuz-FG had a central core first stage and four strap-on booster rockets. The booster rocket enclosures were 19.6 meters long and 2.68 meters diameter. The enclosure held tanks for kerosene fuel and liquid oxygen oxidizer and a RD-117 rocket engine. The thrust was 1021 kN and the burn time was 120 seconds. The boosters were jettisoned after their fuel was exhausted.

The core first stage was 27.8 meters long and 2.95 meters in diameter. The stage held tanks for kerosene fuel and liquid oxygen oxidizer. The engine was a RD-118 type with thrust in vacuum of 1000 kN. The burn time was 286 seconds. The second stage was 6.7 meters long and 2.66 meters in diameter. The rocket engine was a RD-0124 type with thrust of 294 kN. The burn time was 300 seconds. Propellants for the second and third stages were kerosene and liquid oxygen. The third stage was 6.7 meters long and 2.66 meters in diameter. The engine was a RD-0110 type with thrust of 298 kN. The burn time was 230 seconds.

Fregat Upper Stage

Fregat was an upper stage developed by NPO Lavochkin in Russia. It was flight qualified in the year 2000. The version of Fregat used for the launch of Venus Express was 1.5 meters long and 3.35 meters in diameter. It was made up of a flat cluster of six spherical structures grouped about the main engine located in the center of the stage. Two of the spheres held UDMH fuel and two held N_2O_4 oxidizer. The other two spheres held communications equipment and spacecraft electronic systems. UDMH stands for unsymmetrical di-methylhydrazine. The two propellants, UDMH and Nitrogen tetroxide (N_2O_4), ignite upon contact. The large diameter of Fregat required it to be installed within a bulbous fairing at the top of the launch vehicle. The fairing also enclosed the Venus Express spacecraft.

The main engine for Fregat was a S5.92 type that developed thrust of 19.6 kN. The burn time was 1100 seconds. The engine could be stopped and restarted 20 times. The stage also used several small hydrazine fueled thrusters for attitude control.

Fregat was an autonomous upper stage with self-contained guidance, navigation, attitude control, tracking and telemetry systems. It placed the Venus Orbiter spacecraft into an efficient trajectory towards Venus.

Bibliography

Arianespace, *Technical Overview Soyuz*, https://www.arianespacde.com/vehicle/soyuz

Boeing, *Delta II Payload Planners Guide*, Boing document MDC 00H0016, October 2000.

Emery-Riddle Aeronautical University course: Russian Space Operation and Technology.

European Space Agency, *Venus Express Launch Vehicle*, https://sci.esa.int/s/8ZeK5VW

Ezell, Edward Clinton, Exell, Linda Neuman, *On Mars, Exploration of the Red Planet 1958–1978*, NASA SP-4212, 1984 (Appendix E presents details of Atlas-Agena B, Atlas-Agenda D, Atlas-Centaur, and TitaIIIE-Centaur launch vehicles).

Japan Aerospace Exploration Agency, *H-IIA Launch Vehicle*, https://global.jaxa.jp/projects/rockets/h2a/

Kruse, Richard, *R-7 Family of Rockets/Proton Family of Rockets*, https://historic-spacecraft.com/rockets_russian,html

NASA, *The Space Shuttle and its Operations.*

NASA, *Space Shuttle Fact Sheet.*

Russian Space Web, *Fregat Upper Stage*, https://www.russianspaceweb.com/fregat.html

SpaceX, *Falcon User's Guide*, April 2020.

Zak, Anatoly, *Vostok Launch Vehicle*, https://www.russianspaceweb.com

Chapter 3
Soviet Union Spacecraft That Explored Venus 1960–1980

The planet Venus was a tempting target for early space explorers and just within reach of spacecraft lofted by launch vehicles of the day. Venus orbits inside of Earth's orbit around the sun while Mars orbits outside of Earth's orbit, Venus orbits closer to Earth than Mars and the energy required to send a spacecraft to Venus is less than that required to Mars. Consequently, Venus was chosen as the destination for first planetary probes by both the Soviet Union and the United States.

The Soviet Union was first to send spacecraft on a quest to learn more about Venus. The launch vehicles and spacecraft relied on emerging technology at the time and failures of the launcher and of the spacecraft were common. The first nine attempts to reach Venus by the Soviet Union were unsuccessful. Of those, five were caused by failures of launch vehicles. In total, the Soviet Union realized 11 successes or partial successes out of 22 attempts at missions to Venus between 1960 and 1981. It was encouraging that 11 out of the last 13 Russian spacecraft launched to Venus in the period were successes or partial successes. The last eight missions of the time period, Venera 7, 8, 9, 10, 11, 12, 13 and 14, successfully landed instrumented capsules on the surface and Venera 9, 10, 13, and 14 returned photographs of the Venusian terrain.

Minimum energy launch windows to send spacecraft to Venus occur about every 19.2 months. These windows correspond to the time when Earth and Venus are near their closest approach as each travels around the sun in elliptical orbits. Both the United States and the Soviet Union took advantage of minimum energy launch windows. Spacecraft launches to Venus were grouped a few weeks around minimum energy dates.

Communications with Soviet Union planetary spacecraft were carried out using the Pluton Deep Space Communication Facility located on the Crimean Peninsula. The facility used very large receive and transmit antennas that were separated a few kilometers from each other. The transmit and receive antennas each consisted of an array of eight 16-meter diameter parabolic antennas mounted on a large steerable frame.

T. Lund, *Spacecraft that Explored the Inner Planets Venus and Mercury*,
Springer Praxis Books, https://doi.org/10.1007/978-3-031-29838-7_3

A summary of spacecraft missions launched towards Venus by the Soviet Union during early planetary exploration between 1960 and 1981 is given in the table below (Table 3.1).

1VA Spacecraft (Sputnik 7 and Venera 1)

The Sputnik 7 and Venera 1 missions were conducted using a spacecraft known as 1VA. The missions were designed to penetrate the atmosphere of Venus with a probe carrying experiments and return data before the probe impacted the planet. At the time, scientists thought that the atmosphere might be only a few times denser than Earth's atmosphere and they were unaware of the extremely hot temperatures and high pressures to be encountered during descent to the surface. A photograph of a mockup of 1VA in the Memorial Museum of Astronautics in Moscow is shown on the following page (Fig. 3.1). The parabolic antenna for communication with Earth is not fully open on the mockup.

The 1VA spacecraft was developed by the Soviet Union Experimental Design Bureau OKB-1 headed by Sergei Korolev. OKB-1 developed a similar spacecraft called 1M for exploration of Mars. The 1VA spacecraft included a cylindrical structure that necked down to support a dome on top that contained the propulsion system. The main portion of the cylinder was 1.1 meters in diameter. The height of the cylinder with dome was 2.04 meters.

A pressurized metal globe of Earth was contained within the main cylinder. The 70 mm diameter globe was protected by a thermal cover. Water areas on the globe of Earth were painted blue and land masses were gold color. A disk with the Soviet coat of arms on one side and a representation of the inner solar system on the other side was placed inside of the globe. The globe would hopefully survive entry of the cylinder into the atmosphere and impact of the spacecraft on the planet. All being well, it would be examined by impressed inhabitants or future voyagers. The globe did not contain experiments or a transmitter.

Solar panels with total area of about two square meters extended from the sides of the main cylinder of the 1VA spacecraft. The cylinder, which was pressurized by nitrogen gas, contained gyros for attitude control and electronics for various subsystems of the spacecraft. Temperature control was achieved by thermal shutters adjusted as required to radiate heat into space. Internal equipment was cooled by circulating nitrogen gas around the various units by fans.

The spacecraft contained several antennas to communicate with Earth. An omnidirectional antenna supported by a 2.4 meters long boom operated at a wavelength of 1.6 meters. That antenna was used for communications while the spacecraft was still near Earth. Medium gain, semi-directional antennas mounted on the backs of the solar panels were used to transmitted data to Earth and receive commands from Earth during cruise to Venus. Commands were received at a wavelength of 0.39 meters and data was transmitted to Earth at a wavelength of 0.325 meters. The data rate for transmission to Earth was one bit per second.

Table 3.1 Soviet Union spacecraft to Venus from 1960 to 1981

Spacecraft/type	Launch date	Objective	Results
Sputnik 7/1VA	4 Feb 1961	Impact Venus	Launch failure. Spacecraft stranded in earth orbit
Venera 1/1VA	12 Feb 1961	Impact Venus	Contact with spacecraft lost after 5 days of travel
Venera 2MV No. 3/2MV-1	25 Aug 1962	Land a capsule	Launch failure. Spacecraft stranded in earth orbit
Venera 2MV No. 4/2MV-1	1 Sep 1962	Land a capsule	Launch failure. Spacecraft stranded in earth orbit
Sputnik 21/2MV-2	12 Sep 1962	Venus flyby	Launch failure. Spacecraft stranded in earth orbit
Kosmos 27/3MV-1	27 Mar 1964	Land a capsule	Launch failure. Spacecraft stranded in earth orbit
Zond 1/3MV-1	2 April 1964	Land a capsule	Spacecraft failed enroute to Venus
Venera 2/3MV-4	12 Nov 1965	Venus flyby	Spacecraft flew by Venus. No data returned due to communication failure
Venera 3/3MV-3	16 Nov 1965	Land a capsule	Capsule landed on Venus. No data returned due to communication failure
Venera 4/V-67	12 June 1967	Land a capsule	Capsule descended to an altitude of 28 km when high temperature and pressure of atmosphere caused it to fail
Kosmos 167/V-67	17 June 1967	Land a capsule	Launch failure. Spacecraft stranded in earth orbit
Venera 5/V-69	5 Jan 1969	Land a capsule	Capsule descended to an altitude of 18 km when high temperature and pressure of atmosphere caused it to fail
Venera 6/V-69	10 Jan 1969	Land a capsule	Capsule descended to an altitude of 18 km when high temperature and pressure of atmosphere caused it to fail
Venera 7/V-70	17 Aug 1970	Land a capsule	Capsule descended to the surface and transmitted data
Kosmos 359/V-70	22 Aug 1970	Land a capsule	Launch failure. Spacecraft stranded in earth orbit
Venera 8/V-72	27 Mar 1972	Land a capsule	Capsule descended to the surface and transmitted data
Kosmos 482/V-72	31 Mar 1972	Land a capsule	Launch failure. Spacecraft stranded in earth orbit
Venera 9/4V-1	8 June 1975	Orbit Venus and land a capsule	Bus orbited Venus. Capsule descended to surface and returned photographs of surface
Venera 10/4V-1	14 June 1975	Orbit Venus and land a capsule	Bus orbited Venus. Capsule descended to surface and returned photographs of surface
Venera 11/4V-1	9 Sep 1978	Fly by Venus and land a capsule	Bus flew by Venus, Capsule descended to surface and returned data. Photos not returned because lens cover on
Venera 12/4V-1	14 Sep 1978	Fly by Venus and land a capsule	Bus flew by Venus. Capsule descended to surface and returned data. Photos not returned because lens cover on

Fig. 3.1 Mockup of Soviet Union 1VA spacecraft. (Wikimedia Commons photograph posted by Armael)

A parabolic antenna with mesh reflector two meters in diameter was used for communications with Earth when in the vicinity of Venus. The parabolic antenna operated at wavelengths of 8 cm and 32 cm. The wavelength near 32 cm was likely used to transmit data to Earth and 8 cm was likely used to receive commands.

The spacecraft was normally spin stabilized with the solar arrays pointed at the sun during the long cruise to Venus. The spacecraft could also be stabilized in all three axes to communicate with earth with the high-gain parabolic antenna or to observe Venus. An Earth sensor assisted in pointing the antenna towards Earth. The stabilization system included a sun sensor and a star tracker along with a set of gyros. Spacecraft attitude was adjusted by activating nitrogen jets from nozzles around the spacecraft. Nitrogen gas for that purpose was held in a pressurized tank. The spacecraft included a rocket engine for mid-course correction. Propellants for the engine were the hypergolic combination of unsymmetrical dimethyl hydrazine and nitric acid. These propellants ignited upon contact. Thrust of the engine was 2,000 newtons (450 pounds). The weight of the spacecraft, including fuel for the engine, was 644 kg.

Experiments on the 1VA spacecraft included a three-axis magnetometer, ion traps, micrometeorite detector, radiation counter, and an infrared radiometer. The magnetometer, which was mounted near the end of a boom two meters long, was designed to measure magnetic field strength along three orthogonal axes. There were two ion traps on the sun-facing side of the spacecraft. The ion traps set up an electric field and then measured the current developed when ions from the solar wind passed through the field. The magnetometer and ion trap were the only experiments that returned data during the short data gathering time following launch of Venera 1.

Flight of the 1VA Spacecraft

Two 1VA spacecraft were built. The first was launched by a Molniya 8k78 launch vehicle on 4 February 1961. The spacecraft was placed into earth orbit successfully but the Block L fourth stage did not ignite to place the spacecraft into a trans-Venus trajectory. The spacecraft eventually burned up in the Earth's atmosphere.

The second IVA spacecraft was launched on 12 February 1961. The spacecraft was placed in Earth orbit and the Block L fourth stage burned properly to inject the spacecraft into a trans-Venus trajectory. The spacecraft was named Automated Interplanetary Station (English translation) at the time of launch. It was later renamed Venera 1. "Venera" is the Russian name for Venus.

Initial operation of the spacecraft appeared normal and data from the ion traps (solar wind) and magnetometer were transmitted to Earth. Communications were carried our successfully the day after launch and again as scheduled on the fifth day after launch. Communications on the tenth day after launch indicated that uplink commands were received but no data was returned. Communications experienced fading on the fifteenth day after launch. No communication with the spacecraft was realized on 4 March 1961 or thereafter. The unresponsive spacecraft passed by Venus at a distance of about 100,000 km on 19 May 1961. Venera 1, although not successful, demonstrated the ability to send a spacecraft to Venus.

2MV Spacecraft

Engineers and scientists at the Soviet Union Experimental Design Bureau OKB-1 applied what they learned from the 1VA spacecraft to develop a more capable spacecraft that would serve missions to both Venus and Mars. The spacecraft was called 2MV where 2 indicated second generation and MV indicated use for missions to both Mars and Venus. The spacecraft was tailored as required for differences in the Venus and Mars missions.

There were two versions of the spacecraft sent to Venus. Spacecraft 2MV-1 was intended to dispatch a lander to the surface. It consisted of a bus and a landing capsule. Spacecraft 2MV-2 would be used for a Venus flyby mission and it consisted of a bus and an instrument section.

2MV-1 Spacecraft

The 2MV spacecraft consisted of a bus that was similar for all missions and a planetary compartment that was built for a particular mission type. The bus was built around a cylindrical orbital compartment. A rocket motor and propellants were attached to one end of the orbital compartment and the planetary compartment was attached to the other end. In the case of the 2MV-1 spacecraft, the planetary compartment was a spherical landing capsule that would be detached from the orbital compartment and descend to the surface of Venus. The overall length of the spacecraft was about 3.6 meters and it weighed 890 kg.

2MV Bus

The central portion of the bus was the orbital compartment which was cylindrical and about 1.1 meters in diameter around its major portion. Scaling from drawings, the length of the orbital compartment was about 1.4 meters. Electronics for spacecraft control systems and for experiments were mounted within the orbital compartment along with batteries and communications equipment. The compartment was pressurized by nitrogen gas at a pressure of 1.1 Earth atmospheres. The gas was circulated by a fan as part of thermal control of the interior equipment.

A pair of solar arrays with an area of 2.5 square meters was attached to the orbital compartment. The span across the solar arrays was about four meters. Nickel-cadmium batteries with total capacity of 42 ampere-hours at 14 volts provided power until the solar arrays were deployed and supplemented power from the solar arrays during peak demand. The batteries, located within the orbital compartment, were charged by the solar arrays when spacecraft power requirements were modest.

Hemispherical thermal radiators were attached to the end of each solar array. The hemispheres had alternate heat radiator and heat absorbing surfaces. Fluids were circulated around the heating and cooling surfaces and the heating lines and cooling lines were coupled to heat exchangers within the nitrogen atmosphere of the interior of the orbital compartment. Regulators on the heating and cooling lines maintained acceptable temperatures for electronics and other components within the orbital compartment.

The orbital compartment held several antennas for communications. The most prominent was a parabolic antenna 1.7 meters in diameter that was used to send data from the experiments and spacecraft systems to Earth. The spacecraft was maneuvered to point the antenna beam towards Earth for communication sessions. Transmission from the high-gain antennas were conducted at wavelengths of 5 cm, 8 cm, and 32 cm. Whip type antennas attached to the solar panels and operating at a wavelength of 1.6 meters were used for communications when near Earth. There were three conical spiral antennas with semi-directional antenna patterns. One of these was used to receive signals from the landing capsule. The other two received command signals sent from Earth at a wavelength of 39 cm.

The attitude control system included a sun sensor, star tracker, and an Earth sensor along with a set of gyros. Spacecraft attitude was adjusted by activating cold nitrogen gas jets from nozzles around the spacecraft. Nitrogen gas was contained in several small spherical tanks. The sun sensor was used to orient the solar arrays towards the sun during the long cruise to Venus. The star sensor established orientation of the spacecraft in inertial space. The Earth sensor was used to orient the spacecraft to point the high-gain antenna towards Earth during communication intervals.

Midcourse trajectory correction was achieved by first orienting the spacecraft to the attitude that would place the thrust of the rocket motor in proper direction. After ignition, the rocket motor was allowed to burn until integrating accelerometers sensed that the desired velocity change had been achieved. The rocket engine was a KDU-414 type that burned unsymmetrical dimethyl hydrazine (UDMH) with nitric acid as oxidizer. Those propellants ignited upon contact. Thrust of the engine was 1960 newtons (441 pounds). The engine could be stopped and restarted.

Experiments carried on the bus - Experiments carried on the bus included: magnetometer, scintillation detectors, Giger counter, ion traps, micrometeorite sensor, and cosmic ray detector. The magnetometer measured magnetic field strength. The scintillation detectors detected gamma rays by means of sodium iodide scintillation material that emitted photons of visible light when struck by a gamma ray. The Geiger counter responded to charged particles as well as gamma radiation.

There were several ion traps on the exterior of the orbital compartment. The ion traps set up an electric field and then measured the current developed when ions from the solar wind passed through the field. There were two micrometeoroid impact sensors. These were in the form of flat plates attached to the back of the solar panels. The detectors responded with an electrical pulse each time a micrometeoroid with energy above a given threshold struck the plate.

2MV-1 Landing Capsule

The landing capsule was 90 cm in diameter and weighed 350 kg. The capsule was intended to be ejected from the orbital compartment, enter the atmosphere of Venus, and descend to the surface. The capsule contained a heat shield on one side. The center of gravity of the capsule was offset from the geometric center so that the capsule oriented itself with the heatshield in the direction of travel. After being slowed by the head shield, the top cover of the capsule would be jettisoned and parachutes would be deployed to slow the capsule to a survivable landing. Jettisoning of the top cover exposed sensors for experiments and a spiral antenna that transmitted data directly to Earth at a frequency of 922.76 MHz. The capsule also contained electronics for the experiments, a battery, and a transmitter.

Scientific experiments in the landing capsule included: chemical gas analyzer, gamma-ray counter, temperature sensor, density sensor, and pressure sensor. The chemical gas analyzer consisted of five manometer analysis cells. The analyzer was activated by breaking a glass window to let in the atmosphere into the cells and then sealing them. The cells were divided into two volumes by a membrane coated on one side by a chemical that absorbed a particular gas. The pressure difference between the two volumes in one or more cells was measured to determine the amount of a particular gas in those cells. Temperature was determined by measuring the resistance of platinum wire that changed with temperature. The pressure sensor was an aneroid type.

2MV-2 Spacecraft

The 2MV-2 spacecraft was designed for a Venus flyby mission. The bus was essentially the same as that of the 2MV-1 spacecraft but instead of a landing capsule, the planetary compartment consisted of an instrument module. The instrument module was cylindrical in shape about the same diameter as the orbital compartment and about 60 cm long.

The major element of the instrument module was a film-based camera. Film format was chosen because it had better performance than vidicon imaging systems of the day. A further advantage was that film could easily store the large amount of data contained in photographs. The camera was a quite substantial instrument that weighed 32 kg. It looked through a window at the base of the instrument module. Pictures could be taken through a 35 mm (wide angle) lens or through a 750 mm (telephoto) lens. The camera was designed to take and store up to 112 images on 70 mm wide film. The film was developed and dried within the camera after completion of the picture taking session. The developed film was then scanned to convert images to electronic format for transmission to Earth. Scanning could be performed at 1440, 720, or 96 lines per frame.

Image data would be transmitted to Earth using transmitters within the instrument module. There were two transmitters. One transmitter was a short pulse type operating at a wavelength of 5 cm. Data was encoded using pulse position modulation. The other transmitter operated at a wavelength of 8 cm with modulation of a continuous wave carrier. The outputs of the transmitters were connected to the high-gain parabolic antenna mounted on the orbital compartment.

Other optical instruments within the instrument compartment included an infrared spectrometer and an ultraviolet spectrometer. The infrared spectrometer was designed to examine the emission spectrum from Venus in the infrared band of wavelengths. The infrared spectrum would yield information about the temperature and chemical composition of the clouds and atmosphere of Venus. Some of the chemical constituents would generate emission lines in the spectrum and other chemicals would result in deep absorption of the spectrum at particular wavelengths.

Much could be learned about the chemical composition of the region being viewed by analyzing the infrared spectrum.

The ultraviolet spectrometer examined the spectrum resulting from light from the sun passing through the clouds and atmosphere of Venus. Again, emission lines and absorption lines in the ultraviolet spectrum would be used to determine chemical constituents. The ultraviolet spectrometer likely operated in the 3 to 4-micron band as on the later 3MV series of spacecraft.

Flights of the 2MV-1 and 2MV-2 Spacecraft

Two 2MV-1 lander spacecraft were built and one 2MV-2 flyby spacecraft was built. The first 2MV-1 was launched on 25 August 1962. The spacecraft was placed into earth orbit successfully by the Molniya launch vehicle. However, the Block L stage that was to have accelerated the spacecraft into a trans-Venus trajectory malfunctioned and the spacecraft was stranded in earth orbit. The spacecraft was apparently not given an official name but it was referred to as Venera 2MV-1 no. 3.

The second 2MV-1 spacecraft was launched on 1 September 1962. Again, the spacecraft was placed in Earth orbit but the Block L stage failed and the spacecraft was stranded in earth orbit. The spacecraft was referred to as Venera 2MV-1 no 4.

The 2MV-2 spacecraft was launched on 12 September 1962. The third stage of the Molniya launch vehicle exploded after reaching Earth orbit. The spacecraft was stranded in Earth orbit and eventually burned up.

3MV Spacecraft (Venera 2 and Venera 3)

The Soviet Union Experimental Design Bureau OKB-1 continued development of planetary spacecraft to explore both Venus and Mars with a third-generation spacecraft, called the 3MV series. A 3MV-1 spacecraft was intended to land a capsule on Venus and a 3MV-2 spacecraft was intended to gather information during a flyby of Venus. Likewise, the 3MV-3 spacecraft would land a capsule on Mars and the 3MV-4 spacecraft would gather information during a flyby of Mars.

Background of 3MV Spacecraft

Engineers and scientists at OKB-1 sought to understand why the Block L stage, which was the fourth stage of the launch vehicle, had often failed on launches of the 1VA and 2MV series of spacecraft. The failures stranded the spacecraft in Earth orbit. To determine the cause of the problem, they planned to conduct three fully instrumented test flights prior to launches of operational 3MV spacecraft. Those test flights would be given the name Zond, the Russian word for probe.

Two test missions, using spacecraft called 3MV-1A, were planned to travel in an orbit around the sun and return to the vicinity of earth after about six months. One test mission, using a spacecraft called 3MV-4A, was planned to fly a trajectory to Mars. In addition to the test missions, a total of six operational 3MV spacecraft were planned to conduct missions to Venus and to Mars.

As the spacecraft were being built, work by both Soviet Union and U.S. scientists disclosed that the atmospheric pressure on Mars was considerably less than the value of 85 millibars that had been generally accepted up to that time based on scattering of sunlight. NASA began using a value of 10 millibars in mid-1964. For reference, the atmosphere of Earth has an average pressure of 1,013 millibars at sea level. Scientists at OKB-1 realized that the atmosphere of Mars was significantly less dense than they had assumed when the landing capsule was designed. As a result, their Mars landers would descend quickly and would not be slowed enough for a survivable landing. Further, the rapid descent would not allow much time to transmit data during the descent. Launches of operational spacecraft to Mars were put on hold and only the 3MV-4A test spacecraft would be sent to Mars. Three spacecraft, one 3MV-3 and two 3MV-4 versions, that were originally planned for Mars missions were repurposed for Venus missions in 1965.

The responsibility for planetary spacecraft and robotic lunar spacecraft was transferred from experimental design bureau OKB-1 to experimental design bureau OKB Lavochkin in April 1965. OKB-1 had been assigned the daunting task of developing spacecraft for a manned lunar landing and that would take all the efforts of OKB-1's engineers and scientists to fulfil. OKB Lavochkin inherited the design of the 3MV series of spacecraft from OKB-1.

OKB-301 Lavochkin, headed by Semyon Lavochkin, developed fighter airplanes in World War II and jet airplanes and missiles after the war. Semyon Lavochkin died in 1960 and the Lavochkin design bureau was taken over by Korolev's OKB-1. The design bureau was reconstituted under Gorgi Babikin in March 1965. OKB Lavochkin is now known as NPO Lavochkin. OKB Lavochkin had a heritage of airplane and missile design where thorough ground testing before flight was essential. This discipline benefitted the struggling Soviet planetary exploration program.

Unfortunately, none of the 3MV series of spacecraft destined for Venus were successful. Consequently, only a brief description of spacecraft systems and experiments will be given for those spacecraft.

3MV-1 Spacecraft

The 3MV-1 spacecraft was designed to fly by Venus and dispatch a landing capsule to the surface. It consisted of a bus that was similar for all of the 3MV series spacecraft and a landing capsule. The basic bus was also used for several planetary exploration missions that followed. The bus was made up of a cylindrical orbital compartment with solar arrays attached. A rocket engine for midcourse correction was mounted to one end of the orbital compartment and the spherical landing capsule was attached to the other end. A photograph of a 3MV-3 spacecraft that has the same appearance as 3MV-1 appears in NASA NSSDCA/COSPAR ID 1965-092A. The author has not been able to determine the source of the image and it is not reproduced here. Readers are invited to view the excellent image at https://nssdc.gsfc.nasa.gov/nmc/spacecraft/display.action?id=1965-092A.

The spacecraft consisted of a bus and a landing capsule. The general appearance of the spacecraft was similar to that of Venera 4 that is shown in Figure 3.2. A difference was that hemispherical radiators were attached to the outer edges of the solar arrays.

Fig. 3.2 Mockup of Venera 4 spacecraft. (Wikimedia posting by Armael)

3MV Bus

The main structure of the bus was the orbital compartment that was about 1.1 meters in diameter and about 1.8 meters long. The landing capsule was 0.9 meters in diameter. The total length of the spacecraft was 3.6 meters. The overall spacecraft weighed 948 kg and the landing capsule weighed 290 kg.

The interior of the orbital compartment contained electronics for spacecraft control systems and experiments along with communications equipment and batteries. The compartment was pressurized by nitrogen gas. The gas was circulated by a fan as part of thermal control of equipment in the interior of the compartment. A pair of solar arrays was attached to the orbital compartment. When deployed, the solar arrays spanned four meters end-to-end. The solar arrays charged nickel-cadmium batteries with total capacity of 112 ampere-hours at 14 volts.

Thermal radiators in the form of hemispheres were attached to the ends of the solar arrays. The hemispheres had alternate heat radiator and heat absorbing surfaces. Fluids were circulated around the heating and cooling surfaces and the heating lines and cooling lines were coupled to heat exchangers within the nitrogen atmosphere of the interior of the orbital compartment. Regulators on the heating and cooling lines maintained acceptable temperatures for electronics and other components within the orbital compartment.

A collapsible parabolic antenna, two meters in diameter when unfurled, was mounted to the side of the orbital module facing the opposite direction as the solar arrays. The high gain parabolic antenna was used to transmit data to Earth. The communications frequency was 922.76 MHz. The spacecraft was maneuvered to align the antenna with Earth during communication intervals. A broad beamwidth antenna, which could be used to communicate with Earth at 922.76 MHz, was mounted on a boom used to hold the magnetometer. In addition, there was one hemispherical spiral antenna mounted on the side of the orbital compartment and a similar antenna mounted to the end of one solar array to receive commands from Earth at a frequency of 768.96 MHz.

The rocket motor for midcourse correction was a KDU-414 type that burned unsymmetrical dimethyl hydrazine with nitric acid as oxidizer. Those propellants ignited upon contact. Thrust of the engine was 200 kg (441 pounds). The specific impulse was 272 seconds. A total of 35 kg of propellants for the engine were carried.

Experiments carried on the bus included: magnetometer, scintillation detectors, Giger counter, ion traps, micrometeorite sensor, and cosmic wave detector. A bank of ion thruster was carried on the bus as an engineering experiment to see if they would serve for attitude control of the spacecraft. Unfortunately, none of the missions of the 3MV spacecraft were successful and experimental data was not obtained.

Landing Capsule

The landing capsule was spherical in shape, 0.9 meters in diameter, and weighed 290 km. It had an ablative outer shell and a heat shield to protect the systems within the capsule during entry into the Venusian atmosphere. The capsule contained electronics to read out data from the experiments, batteries, transmitter, and an antenna. The capsule was held in place on the bus by straps that allowed it to be released when near Venus.

The upper portion of the capsule was jettisoned after atmospheric breaking had slowed the capsule sufficiently to allow a parachute system to be deployed. Jettisoning of the upper portion of the capsule uncovered a spiral transmit antenna and some experiments. Housekeeping data and data from the scientific instruments was applied to the transmitter and the signal was sent directly to Earth at a frequency of 922.76 MHz.

Experiments contained in the landing capsule included: temperature sensor, pressure sensor, density sensor, micro-organism detector, gas analyzer, gamma ray counter, and an airglow photometer. Although the landing was planned for the night side of the planet, there was interest in measuring the amount of airglow.

Repurposed Spacecraft Originally Planned for Mars

The 3MV-2 spacecraft planned for flyby of Venus did not go forward. A 3MV-4 spacecraft originally intended for a Mars flyby was reconfigured for a flyby of Venus instead. The 3MV-3 spacecraft, designed to place a landing capsule on Mars in 1964 was reconfigured to land a capsule on Venus in 1965. The spacecraft consisted of a bus and a landing capsule similar to the 3MV-1 spacecraft. It was slightly heavier than the 3MV-1 spacecraft weighing 1,042 kg (2,298 pounds).

The 3MV-4 spacecraft consisted of a bus and an instrument compartment. The bus was similar to that used for other 3MV spacecraft. Experiments carried on the bus included a Lyman-α spectrometer that looked for emission from ionized hydrogen at a wavelength of 121.567 nanometers. Other experiments were the same as carried in general on 3MV buses. They included: magnetometer, scintillation detectors, Giger counter, ion traps, micrometeorite sensor, and cosmic wave detector. An ultraviolet spectrometer was added.

The instrument compartment included a new and smaller camera than that used on the 2MV spacecraft. The camera was film based using film 25 mm wide. A 200 mm lens gathered the image which was projected on the film. An ultraviolet spectrometer gathered an ultraviolet spectrum in the 285 to 335 nanometer wavelength range and projected the resulting spectrum on sections of the camera film. The film was developed and dried and then scanned to generate an electronic format to transmit to Earth. A total of 25 camera images and three ultraviolet spectrums could be placed on the film.

The instrument compartment also contained an ultraviolet spectrometer operating in the 190 to 275-nanometer range and two infrared spectrometers operating in the 7 to 39-micron range.

Flights of the 3MV Spacecraft to Venus

The first 3MV-1A test mission was launched on 11 November 1963. The launch vehicle for this mission was the Molniya 8L78 vehicle that had been used for all previous Venus missions. The fourth stage, called Block L, malfunctioned and the spacecraft was stranded in earth orbit. The failed spacecraft was given the name "Kosmos 21." The second 3MV-1A test mission was launched on 19 February 1964. The third stage of the launch vehicle exploded, thus ending the mission. The launch vehicle for this mission was the improved Molniya 8K78M. The Molniya 8K78M would be used for subsequent launches to Venus up to the Venera 9 mission.

The 3MV-4A spacecraft intended as a test mission to flyby Mars was successfully launched on 30 November 1964. Unfortunately, communication with the spacecraft was lost before it reached Mars. The spacecraft was given the name "Zond 2."

The first operational 3MV-1 spacecraft destined for Venus was launched on 27 March 1964. The spacecraft was successfully placed in Earth orbit. However, the attitude control system that would position the spacecraft for proper thrusting direction of the Block L engine did not function and the engine did not fire. The spacecraft was stranded in Earth orbit. It was given the name "Cosmos 27." Instrumentation on the Block L stage for this flight revealed the problem to be a design error involving switching of power to control valves. The design problem with the power switch may have been the cause of several previous launch failures.

The second operational 3MV-1 spacecraft was launched on 2 April 1964. The spacecraft was successfully placed on a trajectory to Venus. Unfortunately, a crack in the pressurized orbital compartment around the sun/star sensor window allowed depressurization of the compartment and the transmitter failed. Soviet Union controllers implemented a work-around that used the transmitter in the descent capsule for communications with the spacecraft. Two separate midcourse corrections were made and the spacecraft was predicted to pass within 100,000 km of Venus. Communication with the spacecraft was lost completely on 24 May 1964 and the unresponsive spacecraft passed by Venus on 19 July 1964. The spacecraft was given the name "Zond 1."

Two of the 3MV spacecraft that had been planned for a flyby of Mars in 1964 were repurposed for a flyby of Venus in 1965. Spacecraft 3MV-4 was launched on 12 November 1965 and it was successfully placed in a trajectory towards Venus. The spacecraft was given the name Venera 2. Injection into the trajectory to Venus was sufficiently accurate that a midcourse correction was not required. Venera 2 was predicted to pass within 24,000 km of Venus.

Venera 2 began overheating as it approached Venus and that seemed to be affecting communications. After arriving in the vicinity of Venus on 27 February 1966, the experiments were turned on. Some reports indicate that the most of the communication system was turned off while the experiments were powered. The plan was to record data from experiments on a tape recorder and then play the recording back for transmission to Earth after the encounter. Unfortunately, contact with the spacecraft could not be regained after the flyby of Venus. The spacecraft was declared lost on 4 March. Investigation indicated that the thermal control system had not worked properly and that the communications system had likely overheated and failed.

The 3MV-3 spacecraft that had been planned to land a capsule on Mars in 1964 was repurposed to land a capsule on Venus in 1965. The spacecraft was launched on 16 November 1965 and it was successfully put on a trajectory to Venus. The spacecraft was given the name "Venera 3." A successful midcourse correction was made on 26 December. Communications with the spacecraft was satisfactory during most of the flight and data from experiments on the bus were relayed to earth. Contact with the spacecraft was lost on 16 February 1966 and could not be regained.

The spacecraft automatically released the landing capsule at about the proper time and the capsule impacted Venus on 1 March 1966. No signals were received from the landing capsule.

Post flight investigation indicated that the probable cause of failure of both Venera 2 and Venera 3 was overheating that resulted in loss of communications. The overheating was thought to be due to degradation of thermal paint exposed to the bright sun during the long trip to Venus.

A second 3MV-4 spacecraft that had been planned for a flyby of Mars in 1964 was also repurposed for a flyby of Venus in 1965. The spacecraft was launched on 23 November 1965. The third stage of the launch vehicle exploded and the spacecraft was stranded in earth orbit. The failed spacecraft was given the name Cosmos 96.

V-67 Spacecraft (Venera 4)

Design bureau NPO Lavochkin headed by Georgi Babskin undertook design of a new spacecraft for the 1967 Venus launch window. The new design series, called V-67, was based on the 3MV-1 type but the spacecraft was improved and it was somewhat larger and heavier than previous models. The V-67 was intended to land a capsule on Venus. Two spacecraft of the type were built.

The final approach to Venus was changed for V-67 spacecraft. Rather than a trajectory that would cause the bus to fly by Venus as was done previously, a direct impact trajectory of the bus was chosen. The bus would eject the landing capsule and then the unprotected spacecraft would enter the atmosphere and burn up. Ejection of the capsule was planned to be initiated by a command from the ground or by a timer if contact with the Earth was lost.

Design of V-67 Spacecraft

A photograph of a mockup of Venera 4, which was the first of the two spacecraft built, was shown as Fig. 3.2 previously. The mockup resides in the Memorial Museum of Astronautics in Moscow.

The spacecraft consisted of a main bus and a landing capsule. The overall length of the spacecraft was 3.5 meters. It weighed 1,106 kg. The main structure of the bus was cylindrical with stepped diameters. A rocket motor and propellants for midcourse correction were attached to one end of the main structure and the landing capsule was attached to the other end. Two deployable solar arrays with a total area of 2.5 square meters were attached to the main structure. A collapsible parabolic antenna was also attached to the main bus structure. Extended, the antenna was 2.3 meters in diameter.

Most of the spacecraft systems and electronics for the experiments on the bus were contained within the cylindrical structure. The compartment was pressurized with nitrogen gas that was circulated for cooling the electronic systems. Thermal control of the new spacecraft received close attention because of overheating problems experienced with previous 3MV spacecraft. The new design used the solid central area of the parabolic antenna as a radiator instead of the hemispherical radiators used in 3MV spacecraft. Pressurized gas within the cylindrical main body of the spacecraft was circulated through a spiral coil that was affixed to the solid rear portion of the antenna. The parabolic antenna was on the opposite side of the spacecraft from the solar arrays and the antenna was largely in the shade of the solar arrays.

The rocket motor for midcourse correction was a KDU-414 type that burned unsymmetrical dimethyl hydrazine with nitric acid as oxidizer. Those propellants ignited upon contact. Thrust of the engine was 200 kg (441 pounds).

Spacecraft attitude was maintained by a series of nitrogen gas jets that were controlled by a stabilization system that included gyros, sun sensor, and an earth sensor. Nitrogen gas was stored under pressure in several small spherical tanks located around the upper cylindrical section in the photograph of the spacecraft in Figure 3.2.

Communications with the bus were carried out using the large parabolic antenna and two cone-shaped spiral antennas. The downlink frequency to earth was 928.4 MHz. The uplink command frequency to the spacecraft was 773.7 MHz. The downlink could be modulated at bit rates of 1, 4, 16, or 64 bits per second.

Experiments carried on the bus included:

- Three-axis flux gate magnetometer to measure magnetic fields
- Four ion traps to detect the solar wind
- Lyman-α ultraviolet spectrometers to detect emission lines of hydrogen and oxygen
- Geiger-Muller gas discharge tube to detect ionizing radiation (Russian STS-5 model)
- Geiger-Muller gas discharge tube with mica end window to detect ionizing radiation (Russian SBT-9 model)
- Two silicon solid state radiation detectors.

Landing Capsule

The landing capsule is located at the bottom of the picture of the Venera 4 spacecraft (Fig. 3.2) with the lettering CCCP on the side. CCCP is the Russian abbreviation for Union of Soviet Socialist Republics in Cyrillic script. The capsule was held by straps to the main body of the spacecraft. The straps were released to deploy the capsule. A photograph of a mockup of the capsule is shown below (Fig. 3.3). The mockup is displayed in the Memorial Museum of Astronautics in Moscow. Both photographs show a smooth painted surface for the landing capsule. For the mission to Venus, the landing capsule was covered with a dark colored ablative heatshield.

The landing capsule, including heat shield, was one meter in diameter. It weighed 383 kg. The photograph of the capsule has the top cover and parachutes removed. During the mission, the cover was released by explosive bolts and a drogue parachute and the main parachutes were deployed in order. The thin tubular appendages on each side of the capsule are transmitting and receiving antennas for the radar altimeter. Those antennas deployed when the main parachute deployed. The spiral at the center of the top of the capsule is the transmitting antenna.

The drogue parachute had an area of about 2.3 square meters when deployed and the main parachute had an area of about 55 square meters. The drogue parachute was deployed after the spacecraft was slowed by aerodynamic braking. The main parachute was deployed when atmospheric pressure increased to 0.6 bars. One bar is approximately equal to the pressure of Earth's atmosphere at the surface.

The interior of the capsule contained a commutation unit, program timing unit, primary transmitter, backup transmitter, temperature control system, two accelerometers, and a radar altimeter. The capsule was powered by a rechargeable battery that was kept charged during travel to Venus by the solar arrays. The battery had sufficient capacity to power the capsule for about 100 minutes.

Fig. 3.3 Mockup of landing capsule for Venera 4. (Wikimedia posting by Rave)

The radar altimeter measured altitude by means of a frequency modulated continuous wave (FM/CW) waveform. This type of radar altimeter was commonly used in airplanes of the day and it was the same type as used on the U.S. Surveyor spacecraft that made soft landings on the moon. The frequency of the transmitted waveform was linearly increased to a peak deviation and then reset to the start frequency and was increased again. The linearly increasing frequency signal was transmitted to the surface of Venus by the transmit antenna and the echo signal was picked up by the receive antenna. The frequency difference between the received signal and the transmitted signal at that time was proportional to altitude. There was an additional frequency component due to Doppler shift. At a transmitted frequency of 770 MHz the Doppler shift amounts to about 5.2 Hz per meter per second of vertical velocity. Particulars on the modulation of the radar altimeter were not in the literature available to the author. It is possible that the Doppler shift was not significant compared to the range frequency.

In the case of Surveyor, the frequency difference was continuously measured giving altitude as a function of time after compensating for Doppler shift. The altitude information was used as part of the guidance system to make a soft landing on the moon. The radar altimeter on Venera 4 was only used to make a single mark when the altitude decreased to 26 km.

The period of the sawtooth FM waveform established the unambiguous range of the radar altimeter. The unambiguous range of the Venera 4 radar altimeter was about 30 km. As a result, the frequency difference at an altitude of 56 km would be about the same as at 26 km. The signal strength at 56 km would be significantly lower than at 26 km and perhaps it was thought that the signal strength would not be sufficient at 56 km to make the 26 km mark. As it turned out, the 26 km mark was made at an altitude of 56 km and that led to confusion until it was sorted out.

Experiments carried by the landing capsule included:

- Two platinum wire type thermometers
- Densitometer using beta ray absorption
- Aneroid barometer
- Gas analyzer (Russian type G-8)
- Gas analyzer (Russian type G-10)

The thermometers operated by measuring the resistance of a length of thin platinum wire. The resistance of the wire increased with temperature. The resistance was accurately measured by an electrical balanced bridge arrangement. One of the thermometers was useful over a range of −63 °C to 457 °C. The other had a range of −3 °C to 327 °C.

The densitometer used an ionization chamber that was coated on the inside with radioactive Strontium-90. Strontium-90 emits electrons when it decays. The electrons struck a detector inside the chamber and were counted. When a material such as the atmosphere of Venus was introduced into the chamber, electrons were absorbed in an amount proportional to the density of the atmosphere. The change in electron count gave an indication of the density. The instrument operated over a density range of 0.0005 to 0.015 grams per cubic centimeter (g/cc). For comparison, the density of the Earth's atmosphere at sea level at 15 °C is 0.00123 g/cc.

An aneroid barometer was used to measure atmospheric pressure. The operating range of instrument was 0.13 to 6.9 bars where one bar is approximately equal to the pressure of Earth's atmosphere at sea level.

The capsule contained two gas analyzer units. One, a Russian type G-8, included five manometer chambers. Atmospheric gas was let into the chambers by breaking a glass window. The individual cells were then sealed. Absorber materials in the cells absorbed select gasses. For example, calcium chloride (CaCl) in a cell absorbed water vapor and the decrease in pressure indicated the amount of water vapor present. Likewise, potassium hydroxide (KOH) in a cell absorbed water vapor and carbon dioxide (CO_2). By subtracting the pressure difference measured in the CaCl cell from that in the KOH cell, the amount of CO_2 in the atmosphere was determined. The second gas analyzer unit, a Russian type G-10, contained two manometer cells and three other experiments designed to measure trace amounts of gasses.

The main constituents of the Venus atmosphere that were intended to be measured by Venera 4 were water vapor, nitrogen, oxygen, and carbon dioxide. Data from the descent capsule indicated that the atmosphere of Venus was composed of 90 to 95% carbon dioxide and less than 2.5% nitrogen. Only trace amounts of water vapor and oxygen were measured.

Flights of the V-67 Spacecraft

Venera 4

Venera 4, the first of the two V-67 spacecraft, was launched from the Baikonur Cosmodrome on 12 June 1967. The fourth stage of the Molniya M launch vehicle was an upgraded Bock L stage referred to as Block VL. The launch and insertion into a trans-Venus trajectory were successful.

Communications with the spacecraft were good and 115 communication secessions were held during the approximately four-month travel time to Venus. The spacecraft was maneuvered to point the high-gain parabolic antenna towards Earth for each communication session. Normally, the spacecraft was oriented with the solar arrays facing the sun.

A course correction was made on 29 July 1967 that resulted in a satisfactory trajectory to impact Venus. The spacecraft approached Venus on 18 October after a smooth trip of 128 days that covered 338 million km. The landing capsule was deployed on 18 October while about 45,000 km from Venus.

The initial approach of the capsule was made on the night side of Venus at a velocity of about 10.7 km per second and at an angle of about 10 degrees from the local vertical. A deceleration of 300 G was recorded as the capsule entered the atmosphere. The top cover of the capsule was jettisoned and the drogue parachute was deployed after the capsule had been slowed to about 300 meters per second by aerodynamic braking. The main parachute was deployed when the atmospheric pressure reached 0.6 bars. The radar altimeter antennas deployed when the main parachute deployed.

The capsule began transmitting shortly after the main parachute deployed. The capsule transmitted for about 93 minutes and 23 sets of data from the instruments were transmitted during that time. Data from the two thermometers, barometer, and densitometer were sequentially commutated onto the transmitter. A data set from the G-10 gas analyzer unit was transmitted soon after data transmission began and a data set from the G-8 gas analyzer was transmitted about six minutes later.

The radar altimeter began operating soon after its antennas deployed. The altimeter was designed to make a mark when it sensed that the altitude decreased to 26 km. It was not designed to provide continuous readout of altitude. The 26 km mark was made and encoded onto the downlink data soon after data transmission began.

The modulation characteristics of the radar altimeter resulted in range ambiguities at 30 km intervals. Apparently, scientists associated with the radar altimeter thought that the signal strength would not be sufficient to make a mark at the first range ambiguity at 26 +30 = 56 km. The actual altitude of the mark was indeed about 56 km. This led to some confusion at first when it was thought that the mark had been made at 26 km, but it was soon straightened out.

The data downlinked from the experiments indicated that the temperature was 31 °C and pressure was 0.75 bars at the time of initial transmission of data. It was later determined that the altitude at first transmission of data was about 55 km. The temperature and pressure increased as the capsule descended. The pressure exceeded the barometer's upper pressure limit of 7 bars about 50 minutes after data transmission began. The density exceeded the 0.015 g/cc upper limit of the densitometer about 70 minutes after data transmission began. The measured temperature had reached 262 °C by the time transmission ceased 93 minutes after beginning. The capsule had descended to an altitude of about 26 km by the time transmission of data ended.

Venera 4 was the first spacecraft to descend into the atmosphere of Venus. A significant finding was that the atmosphere of Venus was composed of over 90% CO_2 and that it had less than 2.5% of nitrogen. Only a trace amount of oxygen was found. Another important finding was that the pressure and density of the atmosphere was much higher than expected. The barometric pressure and densitometer instruments had gone off scale long before transmission ceased at an altitude of about 25 km.

It was not determined why transmission of data only lasted 93 minutes. Possibly, the batteries ran out of power. They were rated to operate the capsule load for 100 minutes. It is possible also that the very high atmospheric pressure cracked the capsule enclosure letting in the scorching atmosphere.

The mission of Venera 4 was a success. It resulted in the first in situ measurements of the atmosphere of Venus.

Kosmos 167

The second of the V-67 spacecraft intended for Venus was launched on 17 June 1967. The launch was successful and the spacecraft was placed in Earth orbit. However, the Block VL stage failed to fire and the spacecraft was stranded in earth orbit. The failed spacecraft was given the name Kosmos 167.

V-69 Spacecraft (Venera 5 and Venera 6)

Design bureau NPO Lavochkin built two spacecraft for the 1969 Venus launch window. Both spacecraft were intended to land a capsule on the surface. The bus was essentially the same as that used for Venera 4 but the landing capsule was improved, taking advantage of information learned from the descent of the Venera 4 capsule into the Venusian atmosphere. The two V-69 spacecraft would be known as Venera 5 and Venera 6.

The approach to Venus was to be a direct impact trajectory rather than a flyby. The bus would eject the landing capsule when near Venus and then the unprotected bus would enter the atmosphere and burn up. Ejection of the capsule was planned to be initiated by a command from the ground or by a timer if contact with the Earth was lost.

Design of V-69 Spacecraft

The appearance of the spacecraft was nearly the same as Venera 4. A photograph of Venera 4 was shown previously as Figure 3.2. The two spacecraft, to be called Venera 5 and Venera 6 consisted of a main bus and a landing capsule. The overall weight of the spacecraft was 1,138 kg. The weight of the improved landing capsule was 405 kg.

The bus portion of the spacecraft was essentially as described for the V-67 spacecraft. The magnetometer experiment on the bus was deleted to save weight.

The landing capsule was redesigned to withstand a deceleration of 450 Gs when entering the atmosphere of Venus and to withstand an external pressure of 25 bars. Much of the 22 kg weight increase of the capsule relative to the capsule used on Venera 4 was due to strengthening of the capsule. As the launch window to Venus approached in 1969, it was suspected that the atmospheric pressure on Venus was much higher than 25 bars. There was not enough time to redesign the capsule, and so the 25-bar pressure design was used to meet the launch window.

The design of the radar altimeter in the capsule was modified to make marks at altitudes of 35 km, 25 km, and 15 km rather than a single mark at 26 km as was done in Venera 4. The area of the main parachute was reduced from 55 square meters used on Venera 4 to 15 square meters on Venera 5 and Venera 6 so that the landing capsule would descend faster and continue operating to a lower altitude.

The experiments within the landing capsule were refined to take advantage of information gained from the Venera 4 mission. The G-10 and G-8 gas analyzers were modified to reduce the dynamic range and thereby improve accuracy of the readings. Separate aneroid barometers were installed with operating ranges of 0.13 to 6.6 bars, 0.66 to 26 bars, and 1 to 39 bars. There were two barometers for each range giving a total of six. The densitometer was changed to a tuning fork type with an operating range of 0.0005 to 40 g/cc. The two platinum wire type thermometers were retained. A photometric sensor was added to measure light levels.

Flight of the V-69 Spacecraft

Venera 5

Venera 5, the first of the two V-69 spacecraft, was launched from the Baikonur Cosmodrome on 5 January 1969. The fourth stage of the Molniya M launch vehicle was the Block VL. The launch and insertion into a trans-Venus trajectory were successful. A midcourse correction was made on 14 March 1967. The spacecraft approached Venus on 16 May 1969 and the landing capsule was released from the bus when about 37,000 km from the planet. The descent was made on the night side of Venus.

The landing capsule descended into the atmosphere and it was slowed by aerodynamic braking. The top cover of the capsule was jettisoned and the drogue parachute was deployed after the capsule had been slowed sufficiently by aerodynamic braking. The main parachute was deployed when the speed had reduced to about 210 meters per second. The transmitter was activated shortly after the parachute deployed and the capsule transmitted data to earth for about 53 minutes. Atmospheric gas chemical analysis was performed at atmospheric pressures of 0.6 and 5.0 bars. Other data was transmitted at 45 second intervals.

About 70 temperature readings and 50 pressure readings were transmitted. Don Mitchell, in *Plumbing the Atmosphere of Venus*, reports the following data from Venera 5:

Altitude	Temperature	Pressure
36 km	177 °C	6.6 atm (bars)
25 km	266 °C	14.8 atm (bars)
18 km	327 °C	27.5 atm (bars)

Another source states that the last transmission indicated an external pressure of 26.1 bars and a temperature of 320 °C. Possibly, the capsule enclosure was cracked open by the high atmospheric pressure and that caused internal systems to fail and ended the data transmission.

Venera 6

Venera 6 was the second of the two V-69 spacecraft. It was launched from the Baikonur Cosmodrome on 10 January 1969, just four days after the launch of Venera 5. The launch and insertion into a trans-Venus trajectory by the Block VL stage were successful. A midcourse correction was made on 16 March 1967. The spacecraft approached Venus on 17 May 1969 and the landing capsule was released from the bus when about 25,000 km from the planet. The descent was made on the night side of Venus.

The landing capsule descended into the atmosphere following the same landing sequence as Venera 5. The main parachute was deployed when the speed had reduced to about 210 meters per second. The transmitter was activated shortly after the parachute deployed and the capsule transmitted data to earth for about 51 minutes. Atmospheric gas chemical analysis was performed at atmospheric pressures of 2.0 and 10 bars. Other data was transmitted at 45 second intervals.

Don Mitchell, in *Plumbing the Atmosphere of Venus*, reports the following data from Venera 6:

Altitude	Temperature	Pressure
34 km	188 °C	6.8 atm (bars)
22 km	294 °C	19.8 atm (bars)

Transmission of data stopped after 51 minutes of transmission at an altitude just below 22 km. Again, the capsule may have been cracked open by the high atmospheric pressure.

Information gathered from the G-10 and G-8 gas analyzers of Venera 5 and Venera 6 yielded the following atmospheric gas constituents: $CO_2 = 95 \pm 2\%$, nitrogen and other inert gases $= 3.5 \pm 1.5\%$, $O_2 < 0.4\%$. Water vapor was 4 to 11 mg per liter at an atmospheric pressure of 0.6 bars.

V-70 Spacecraft (Venera 7)

Design bureau NPO Lavochkin built two spacecraft for the 1970 Venus launch window that were identified as V-70. The spacecraft consisted of a bus and a landing capsule. The bus was essentially the same as used for Venera 4. One addition was cooling tubes that that passed cool nitrogen gas from the bus through a heat exchanger in the capsule. The landing capsule was substantially redesigned from that used on previous missions to survive the extreme pressure of the atmosphere and high temperature of Venus at its surface. The weight of the spacecraft with the redesigned landing capsule was 1,180 kg.

The new landing capsule was designed to withstand an external pressure of 150 bars and temperature of 540 °C. The design for pressure to be survived was conservative, but the actual atmospheric pressure at the surface of Venus was not known at the time. Estimates from measurements at higher altitudes on previous missions put the pressure at about 100 bars.

To achieve the high crush pressure, the capsule used a spherical structure made from titanium that was placed inside of an outer shell and heat shield. The weight of the capsule was 490 kg, an increase of 85 kg over the capsule used for Venera 6. A photograph of a reproduction of the landing capsule for Venera 7 is shown below (Fig. 3.4).

The upper portion of the heat shield has been jettisoned in the photograph and the drogue and main parachutes have been deployed. The shroud lines for the parachutes are draped over the capsule. The small hemisphere with spiral metallic pattern near the center of the top of the structure is the data transmitting antenna. The Yagi type antenna in the foreground is for the radar altimeter. The radar altimeter was a short pulse type that used a single antenna for transmit and receive.

Fig. 3.4 Reproduction of landing capsule for Venera 7. (Wikipedia posting by Stanislav Kozlovsky)

The area of the main parachute was reduced to 2.5 square meters from 15 square meters of Venera 6 to shorten the descent time to the surface. A reefing cord restricted the amount of opening of the main parachute. The reefing cord melted at a temperature of about 200 °C to let the parachute fully open.

Experiments carried in the landing capsule were a thermometer and an aneroid barometer. Data from those instruments and from the radar altimeter were applied to a commutation unit that presented one data reading at a time to the modulator for the transmitter. Unfortunately, a problem developed in the commutator of the Venera 7 landing capsule and only temperature data was transmitted to Earth. The transmitter operated at a frequency of 922 MHz. Data was encoded as frequency shift keying at a rate of one bit per second.

The first of the two V-70 spacecraft was known as Venera 7. The second spacecraft was stranded in Earth orbit because of a malfunction of the upper stage. That spacecraft was given the name Kosmos 359.

Flight of the V-70 Spacecraft Venera 7

Venera 7 was launched from the Baikonur Cosmodrome on 17 August 1970. The fourth stage of the Molniya M launch vehicle was a further improved block L stage referred to as block NVL. The launch and insertion into a trans-Venus trajectory were successful. Two midcourse corrections were made; one on 2 October and the other on 17 November. The spacecraft approached Venus on 15 December 1970. The descent was made on the night side of Venus.

The landing capsule was left attached to the bus during the initial entry into the atmosphere to keep the capsule cool until it was deployed. Cooling of the capsule was accomplished by means of a tube that passed cool nitrogen gas from the bus through a heat exchanger located between the heat shield shell of the capsule and the spherical equipment compartment.

The landing capsule was released from the bus soon after entering the atmosphere and the capsule was slowed by aerodynamic braking. The top cover of the heat shield was jettisoned and the parachutes deployed when the atmospheric pressure reached 0.7 bars at an altitude of about 60 km.

The transmitter was activated after the top cover was jettisoned. A problem developed with the commutator that was intended to sequentially sample temperature, pressure and radar altimeter data and present that data to the transmitter. The commutator hung up and only temperature data was passed on to the transmitter.

The Doppler shift of the transmitted signal was measured on Earth and that gave a velocity history of the landing capsule. The velocity data showed a slowing of the capsule after the main parachute was deployed and a further slowing after the reefing cord on the parachute released. About 19 minutes after transmission began, the velocity abruptly increased. This was thought to have been caused by ripping of the main parachute.

The first temperature reading transmitted was 300 K (27 °C). The descent velocity was about 120 meters per second (m/s) at the time. The atmospheric temperature had increased to about 500 K (227 °C) when the parachute reefing lines released and the velocity slowed rapidly from 27 m/s to 19 m/s.

The capsule landed about 35 minutes after transmission began. The last few minutes of flight appeared to be freefall and the capsule hit the surface at about 17 meters per second. The signal strength decreased by about 15 dB at impact, returned to full strength for about a minute and then decreased by 15 dB from full strength for the rest of the transmission that lasted 23 minutes. It was thought that the capsule bounced after impact and during the bounce time the signal returned to full strength and then the capsule came to rest on its side with the antenna pointed away from Earth resulting on the low received signal strength to the end of the transmission.

Data from the capsule on the surface of Venus indicated a temperature between the adjacent quantized values of 457 °C and 474 °C. Unfortunately, because of the problem with the commutator, temperature measurements were the only data transmitted to Earth.

Kosmos 359

The second of the V-70 spacecraft intended for Venus was launched on 22 August 1970. The launch was successful and the spacecraft was placed in Earth orbit. However, the Block NVL fourth stage shut down early due to a failure in a power supply and the spacecraft was stranded in earth orbit. The failed spacecraft was given the name Kosmos 359.

V-72 Spacecraft (Venera 8)

Design bureau NPO Lavochkin built two spacecraft for the 1972 Venus launch window that were identified as V-72. The first of these was called Venera 8. The second was stranded in Earth orbit and given the name Kosmos 482. The V-72 series would be the last of the 3MV type spacecraft produced. Design was already progressing on the larger and more capable 4V-1 spacecraft to follow.

Venera 8 was the culmination of information learned during previous operation of the 3MV series of spacecraft. The mission of Venera 8 was successful. Profiles of temperature, pressure, and illumination as a function of altitude were transmitted to Earth along with analysis of atmospheric gasses. It also analyzed the chemical composition of the soil on the surface.

The mission was designed to learn more about the clouds of Venus and light levels at various altitudes as the capsule descended. To do this, the descent was planned to be in the sunlit portion of the planet. A restriction was that the spacecraft be in line of sight to the earth to transmit back information from the sensors. This required that the landing be made in the vicinity of the crescent region of Venus as viewed from Earth.

Bus

The bus was essentially the same as used for previous spacecraft of the 3MV family except that it had provisions to cool the interior of the landing capsule with cold air via tubes running between the bus and the capsule. The capsule was cooled to about −15 °C before it was detached from the bus. Experiments on the bus included a cosmic ray detector, solar wind detector and an ultraviolet spectrometer.

Landing Capsule

The V-72 landing capsule was redesigned by taking advantage of the information gathered from previous missions that observed the environment of Venus. The conservative design of the spherical capsule on Venera 7 was able to withstand an external pressure of 150 bars. The capsule was redesigned to withstand a lower pressure of 105 bars. The sphere, which protected electronic systems and was made from titanium, was located within an outer heat shield. The slimmed down spherical vessel reduced the overall weight of the landing capsule of Venera 8 to 495 kg.

The transmit antenna was a cone shaped spiral rather than hemispherical as in Venera 7. Its radiation pattern was a broad peak around the circumference of the capsule. This allowed transmission of data to Earth, which was located at a low elevation angle relative to the capsule during the capsule's descent and its rest on the surface.

A second transmit antenna, with a long cable attached, was ejected over the side of the capsule after landing. It was used for transmission should the capsule tip on its side, as apparently happened with Venera 7. The deployable antenna had a spiral pattern printed on both sides. A gravity switch was used to select the spiral antenna that was pointed upwards. The deployable antenna is located to the right of the conical antenna in the photograph. The radar altimeter antenna was essentially the same as in Venera 7.

Along with high pressure, a formidable environment for the capsule to withstand was very hot temperature. Data from Venera 7 had indicated that the temperature of the atmosphere at the surface of Venus was about 460 °C. The heat was invasive and internal heat from operating electronics in the sealed container contributed to the temperature problem. To increase the operating time on the surface before the heat became excessive for the electronics, the spherical pressure capsule was cooled before the capsule was deployed by porting cooled nitrogen gas around it. The gas at a r temperature of −15 °C was delivered by tubes from the bus for a few days before the capsule was detached. The cooling tubes were located in the upper portion of the capsule that was jettisoned when the parachutes were deployed.

Another method employed in the Venera 8 capsule to slow the rise of temperature of the electronics was the use of a phase change material. The material chosen was lithium nitrate because its melting point of 30 °C was well suited for this application and it had high heat of fusion and high density. Heat of fusion is defined as the heat absorbed by a unit mass of a given solid at its melting point that completely converts the solid to a liquid at the same temperature. The heat of fusion for lithium nitrate is 296 kJ/kg. A Joule is equal to one watt second. Thus, if we had one kilogram of lithium nitrate in the capsule it would absorb the amount of heat generated by 296,000-watt seconds of electrical dissipation while undergoing phase change from solid to liquid. This heat is equivalent to 82 watts of electrical power dissipated for one hour.

A fan within the spherical pressure vessel circulated pressurized nitrogen gas around the electronic components. Thermal insulation around the inside of the spherical enclosure slowed passage of heat from outside of the capsule.

The transmitter produced power of 20 Watts at 922 MHz. Data was encoded by frequency shift keying of the transmitter at a bit rate of one bit per second.

A more extensive set of experiments was carried in the capsule than in Venera 7. The experiment suite included:

- Four resistance type thermometers with ranges of 320–860K, 489–710K, 670–810K, and 290–880K
- Three aneroid barometers with ranges of 0–220 kg/cm^2, 0–150 kg/cm^2, and 0–100 kg/cm^2
- One capacitance barometer with range of 0–80 kg/cm^2
- Two photometers
- Ammonia analyzer
- Gama ray spectrometer

The two photometers were installed in the parachute compartment of the landing capsule. They were installed vertically to view the upper hemisphere relative to the capsule. The two identical photometers fed separate electronic channels located within the capsule. Both channels contained logarithmic amplifiers. One channel covered light flux from 0.002 to 2 W/m^2 and the other channel covered flux from 0.2 to 200 W/m^2. In terms of lux for sunlight, the ranges were 0.253 to 253 lux and 25.3 to 25,316 lux. The ammonia analyzer measured color difference caused by ammonia gas in a bromophenol blue reactant.

The gamma ray spectrometer measured gamma rays naturally emitted by radioactive constituents in the rocks on the surface of Venus. The percentages of naturally radioactive uranium, thorium, and potassium were determined. The spectrometer was mounted within the spherical pressure enclosure but gamma rays were able to penetrate the walls of the enclosure. The spectrometer was mounted near the bottom of the enclosure to be close to the surface of Venus. Background gamma ray impingement from the enclosure was measured and reported back to Earth during the travel to Venus.

The gamma ray spectrometer used a cesium iodide (CsI) scintillator followed by a photomultiplier tube. The CsI crystal emitted a flash of light after being struck by a gamma ray of sufficient energy to raise its energy level after which it fell back to its original state. The flash of light occurred when the excited atom fell back to its original state. The amplitude of the flash was proportional to the energy (frequency) of the incoming gamma ray. Electronics in the spectrometer established a series of sequential bins where each bin represented a particular flash amplitude and hence a particular energy. The number of occurrences in each bin was counted and the amplitude and number of counts were used to identify a particular radioactive material and its percentage of the total. The instrument was calibrated before launch by exposing the capsule to various types of rocks on Earth.

Flight of V-72 Spacecraft Venera 8

Venera 8 was launched from the Baikonur Cosmodrome on 27 March 1972. The fourth stage of the Molniya M launch vehicle was an improved block L stage referred to as block NVL. The launch and the insertion into a trans-Venus trajectory by the block NVL were successful. An accurate midcourse correction was made on 6 April 1972. The spacecraft approached the sunlit crescent of Venus (as seen from Earth) on 22 July 1972 after a journey of 117 days. Communication with the spacecraft was good during the voyage and 86 separate communications sessions were held. Data transmitted back to earth included measurements of the solar wind and spacecraft health. The landing capsule was detached from the bus about one hour before the capsule entered the atmosphere. The pressurized spherical enclosure for the electronics in the capsule had been cooled by ducting cold nitrogen gas (at −15 °C) from the bus around it up until the time of separation.

The landing capsule entered the atmosphere at an angle of about 13 degrees from the local vertical at a speed of about 11.6 km per second. The capsule was slowed to about 250 meters per second by aerodynamic braking by the time the altitude had decreased to 60 km. The parachute and the radar altimeter antenna were then deployed and the instruments and the radar altimeter were turned on. Transmission of sensor data to earth commenced at about 50 km altitude.

The first radar altimeter measurement was at 45 km altitude. A total of 35 altitude measurements were made during the descent with the last measurement at 900 meters. Since the spacecraft was approaching the limb of the planet, tangential velocity resulted in Doppler shift of the transmitted signal that could be accurately measured on Earth. The wind around Venus at high altitude caused the spacecraft drift at about 50 meters per second at an altitude of 40 km. The spacecraft was blown about 60 km sidewise by the wind during its descent.

Measurements of temperature and pressure during the descent were reported in a paper by M. Ya. Marov, et al., of the Academy of Science, Moscow in 1973. The title of the paper was *Venera 8: Measurement of Temperature, Pressure and Wind Velocity on the Illuminated Side of Venus*. The paper plotted the data with temperature and pressure on the ordinate and altitude on the abscissa. The author translated the data to have altitude on the ordinate of the plot and temperature and pressure as abscissas as often used for plots of atmospheric properties in the U.S. In addition, a conversion was made from Kelvin to °C to agree with the convention in this book. The translated plot of temperature as a function of altitude is shown on the next page (Fig. 3.5).

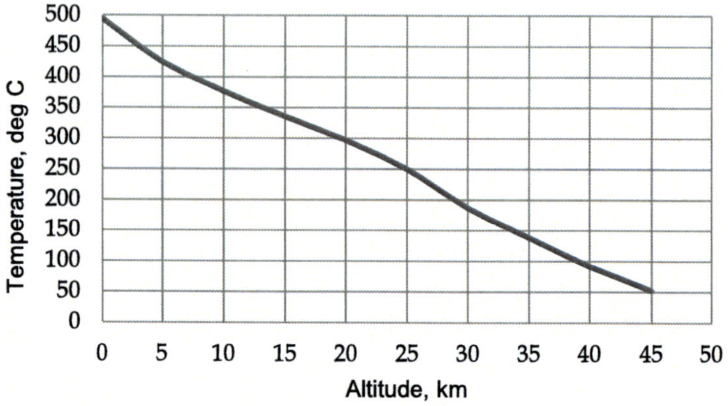

Fig. 3.5 Atmospheric temperature as a function of altitude during descent of Venera 8

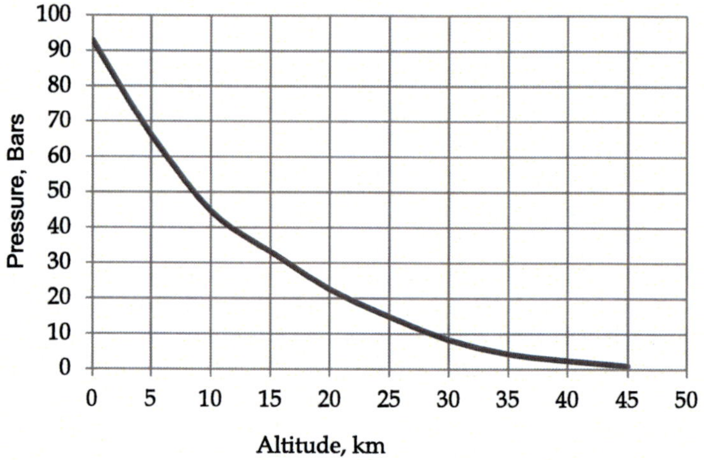

Fig. 3.6 Atmospheric pressure as a function of altitude during descent of Venera 8

A translated plot of pressure as a function of altitude is given above (Fig. 3.6).

The landing capsule contained two photometers to measure light levels as the capsule descended through the clouds surrounding Venus. A smoothed plot of measured light level in terms of Watts per square meter as a function of altitude is given in a paper by V. S. Avduevsky, et al of the Academy of Science, Moscow in 1973. The paper plotted altitude as the abscissa and light level on the ordinate. The author switched these axes to have altitude on the ordinate. The translated plot is shown on the next page (Fig. 3.7).

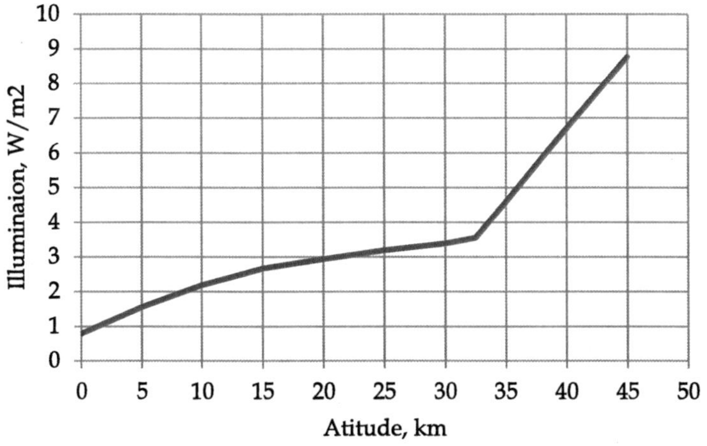

Fig. 3.7 Illumination as a function of altitude during descent of Venera 8

Presence of ammonia was sampled at atmospheric pressures of two bars and eight bars. The ammonia concentration was found to be about 0.1% at an atmospheric pressure of two bars and much less at eight bars. Analysis of atmospheric gases during the descent indicated the following percentages of gasses making up the atmosphere: CO_2 at 94%, nitrogen at 2% and oxygen at 0.1%. The gamma ray spectrometer measured the composition of natural radioactive substances on the surface to be 4% potassium, 2.2 parts per million of uranium, and 6.5 parts per million of thorium. These readings were similar the alkaline basalts on Earth.

The descent time from parachute deployment to landing was about 56 minutes. The capsule operated on the surface for 63 minutes after landing. It likely finally succumbed to the oppressive heat. After landing the deployable transmit antenna with a cable attached was ejected from the spacecraft and landed on the surface to be used in case the capsule toppled over. The capsule did not topple over and good signals were obtained from the transmit antenna on the capsule. The capsule was programmed to transmit data alternately on the main antenna and on the deployable antenna. Good signals were received on Earth from both antennas.

4V-1 Spacecraft (Venera 9 and Venera 10)

The very successful Venera 8 spacecraft was about the limit of performance that could be realized with the 3MV class spacecraft. In particular, the data rate of information transmitted to Earth of one bit per second would not support transmitting images of the surface. A more capable launch vehicle, the Proton-K with a block D upper stage, had become available and that allowed a larger and heavier spacecraft to be sent to explore the planets.

Design bureau NPO Lavochkin developed a new family of larger and more capable spacecraft for the 1975 launch window to Venus. The new spacecraft, identified as 4V-1, was based on the Mars 2 and Mars 3 spacecraft that had been sent to Mars in 1971. The first of the 4V-1 spacecraft, Venera 9, weighed 4,936 kg compared to 1,180 kg for the type 3MV Venera 8 spacecraft. The increased weight was well within the capability of the Proton-K/block D launch vehicle

A major advance in the new series of spacecraft was a camera system in the landing capsule that returned images of the surface of Venus. Early 4V-1 spacecraft generated black and white images and later spacecraft generated color images.

Another difference between the 4V-1 landers and previous landers was use of aerodynamic braking by a disk-like structure 2.1 meters in diameter that slowed the spacecraft to a safe landing speed. Parachutes were used to slow the lander in the upper atmosphere but they were cut loose at about 50 km altitude and the aerodynamic braking slowed the lander in the dense Venus atmosphere the rest of the way down to an acceptably soft landing. This procedure allowed the lander to descend faster through the hot lower atmosphere to lessen the temperature rise within the landing capsule.

The 4V-1 missions were quite ambitious with several new experiments including photography on the landers and photography and experiments on the bus that was placed in orbit around Venus. The bus served as a relay to receive information transmitted by the lander and send the information to Earth by its high-gain antenna. The missions of Venera 9 and Venera 10 returned the most detailed information yet gathered about Venus. The information included photographs of its surface, information about the atmosphere and clouds of Venus and information about the rocks on and under the surface

An informative paper, *Venus Exploration with the Venera 9 and Venera 10 Spacecraft*, by M. V. Keldysh of the Soviet Union Academy of Science was published in 1976. Some of the detail on the spacecraft and much of the information on results of the experiments presented in this section of the book were obtained from that source. Mstislav Keldysh was a famous Russian scientist who contributed to many important scientific endeavors in Russia including aircraft design, atomic processes, and spacecraft science. He had calculated trajectories to Venus and to Mars early in the space exploration era. He was president of the Soviet Union Academy of Science from 1961 to 1975

The 4V-1 spacecraft consisted of a bus and a landing capsule similar to the arrangement used on previous Soviet Union missions to Venus. Both the bus and landing capsule were larger and improved from previous spacecraft. A photograph of a model of the Venera 10 spacecraft in a museum setting is shown on the next page (Fig. 3.8).

The bus for the Venera 9 and 10 series of spacecraft was 2.8 meters long. The center cylindrical section, which contained propellant tanks for the rocket engine, was 1.1 meters in diameter and one meter long. A conical structure at the bottom of the spacecraft in the photograph was 2.35 meters in diameter at its base. The structure necked down below the base of the conical structure to interface with the launch vehicle. The conical structure enclosed the rocket engine as well as containing electronics for spacecraft systems and for experiments. The landing capsule was

Fig. 3.8 Model of Venera 10 spacecraft. (Wikimedia Commons posting by Bekhruzbek Ochilov)

positioned within the spherical entry enclosure mounted at the top of the spacecraft. The entry enclosure was 2.4 meters in diameter. The overall weight of the Venera 9 spacecraft was 4,936 kg at the time of launch. Venera 10, which carried more propellants for the rocket engine, weighed 5033 kg. The spherical entry enclosure with the landing capsule inside weighed 1,560 kg for both Venera 9 and Venera 10.

Bus Obiter

The bus was a capable spacecraft that entered an orbit around Venus after dispatching the landing capsule to the surface. It operated several experiments to gather information about the clouds and upper atmosphere. The extent of the orbit for Venera 9 was 1,500 km by 11,700 km and it was inclined about 34 degrees from the equator of the planet. The bus of Venera 10 was placed into an orbit 1,620 km by 113,000 km around Venus with inclination of 29.5 degrees.

Attitude of the spacecraft while enroute to Venus and during orbit was adjusted by means of nitrogen gas jets mounted at the edges of the solar panels. Nitrogen gas at a pressure of 350 bars was stored in spherical tanks on the outside of the instrument compartment of the bus. The jets were controlled by an attitude control system managed by a digital computer within the instrument compartment. The computer accepted inputs from gyros, a sun sensor, star sensor, and an earth sensor. It also acted on programmed events for the bus as well as responding to commands from Earth.

The rocket engine for course correction and insertion into orbit around Venus was mounted at the center of the spacecraft and protruded through the bottom of the toroidal instrument compartment. The engine was a KTDU-425 type with maximum thrust of 18,890 newtons (4,300 pounds). The engine could be throttled down to about 9,735 newtons (2,216 pounds) thrust. It could be started and stopped seven times. Propellants for the engine were the hypergolic combination of unsymmetrical dimethyl hydrazine (UDMH) and nitrogen tetroxide. The total weight of propellants carried on Venera 9 was 1,093 kg. There was sufficient propellant for an accumulated total firing time of 560 seconds.

Thermal control of the spacecraft was performed in conjunction with hot and cold thermal radiators protruding from each side of the bus. The hot radiator faced the same side as the solar panels (sunlit side) and the cold radiator faced deep space. The thermal radiators extended from the bus to the solar panels. The instrument compartment of the bus was pressurized and pressurized gas from the compartment was circulated through tubes affixed to the back of the thermal panels. The gas was cooled by the cold radiator and it flowed back into the instrument compartment to cool the compartment as required. If the compartment was too cold, gas flowing through the hot radiator was applied to the instrument compartment to warm it. Piping also carried the cooled gas to the landing capsule to cool the capsule before it was detached from the bus.

Solar panels were mounted to structure that was connected to the thermal radiators on the bus. Each panel was 2.1 meters high and 1.25 meters wide. The mounting structure for the solar panels allowed them to be folded in to fit within the launching fairing. When expanded, the solar panels spanned a length of 6.7 meters.

A high gain parabolic antenna with diameter of 1.6 meters was mounted to the bus. It was used to send data from experiments and housekeeping data on the bus to Earth and to relay information from the lander to Earth. The antenna was fixed to the spacecraft and that required the spacecraft to be maneuvered to point the antenna beam towards Earth. At a transmit frequency of 922.7 MHz and parabolic antenna diameter of 1.6 meters, the antenna beamwidth would have been about 13 degrees. An earth sensor aligned with the parabolic antenna was used by the attitude control system to facilitate alignment of the antenna beam with Earth.

The bus included two low-gain, broad beamwidth, conical helical antennas to receive commands from Earth on a frequency of 770 MHz. Broad beamwidth helical antennas were located on the backside of the solar arrays to receive transmitted signals from the lander at frequencies of 122.8 MHz and 138.9 MHz.

The instrument compartment of the bus contained a digital computer that allowed autonomous control of the spacecraft including control of the rocket engine and attitude control jets. The instrument compartment also included electronics for the experiments, the communications system, power supply, and thermal control system. The attitude control system employed redundant sun sensors and redundant star sensors mounted to the outside of the instrument compartment. The star sensors looked for the star Canopus. An Earth sensor was used to align the parabolic antenna beam with Earth during communication events.

The attitude control system maintained orientation of the spacecraft with the solar arrays facing the sun for most of the voyage to Venus. The orientation was adjusted to point the high-gain antenna towards Earth during transmission of data from the spacecraft. The low gain conical spiral antennas, with broad antenna patterns encompassing earth, were used to receive commands from Earth during the voyage.

Experiments on the Bus

Experiments carried on the bus included: two panoramic cameras, two photopolarimeters, infrared radiometer, infrared spectrometer, Lyman-α spectrometer, two ultraviolet photometers, triaxial magnetometer, and ion traps to detect solar wind.

Cameras – Two panoramic cameras were mounted in the orbiter bus to image the clouds around Venus from above. The cameras operated in the ultraviolet region of the spectrum. One camera operated in the band 345 to 380 nanometers and the other camera operated in the band 355 to 445 nanometers.

The panoramic cameras were similar in principle to the camera in the landing capsule that is described in more detail in the next section. The camera used an oscillating mirror that scanned a 30-degree sector under the spacecraft as it skimmed over the planet during the low section of the orbit. Motion of the spacecraft allowed imaging the clouds in a long swath 30 degrees wide as the mirror was scanning. Black and white calibration strips were included in each scan and the calibration data was read out during retrace of the mirror. The 30-degree width of the cross-track scan was mapped into 256 pixels giving each pixel a resolution of about 0.117 degrees.

Photopolarimeters – The photopolarimeters measured the amount of shift of the polarization vector of sunlight reflected from the clouds of Venus. A color wheel allowed measurements to be made at various wavelengths in the 335 to 800 nanometer range. The amount of polarization shift was related to the microstructure of the clouds and the depth through the clouds of the reflection.

Infrared radiometer – The infrared radiometer operated in the 8 to 30-micron infrared band. It gave a measure of the temperature of the upper region of the clouds.

Infrared spectrometers – The infrared spectrometer operated in the 1.5 to 3.0-micron band. The Lyman-α spectrometer operated in the vicinity of 121 nanometers wavelength and it looked for the distinctive hydrogen line.

4V-1 Lander

The lander was contained within a spherical aluminum entry enclosure 2.4 meters in diameter. The entry enclosure, which served as a heat shield, was covered with ablative material to survive high velocity entry into the atmosphere of Venus. A photograph of a model of a 4V-1 lander installed in a cutaway section of the entry enclosure is shown on the next page (Fig. 3.9). The model resides in the Exhibition of Achievements of the National Economy Museum in Moscow.

The inside of the entry enclosure was covered with thermal insulation material. The enclosure had a group of louvers on the side normally facing the sun during the voyage to Venus. The louvers could be opened to allow the sun to shine against an uninsulated portion of the entry shell if the capsule was too cold before being detached from the bus. The heat from the shell radiated to warm the landing capsule.

Fig. 3.9 Cutaway view of model of entry enclosure with 4V-1 lander installed inside. (Wikipedia posting by Don Montgomery)

A photograph of a basic model of the lander in the Cosmos Pavilion in Moscow is shown on the next page (Fig. 3.10). The instrument capsule portion of the lander was a strong spherical enclosure 0.8 meters in diameter made from titanium. The spherical pressure chamber was covered with thermal insulation about 12 cm thick and then by an outer titanium shell. The capsule was supported by a series of legs that were attached to a rim-like landing structure. The rim contained crushable material that absorbed some of the landing shock.

A disk 2.1 meters in diameter, which served as an aerobrake during descent, was mounted just above the spherical capsule. A large helical transmit antenna was mounted above the aerobrake. The hollow inner portion of the cylindrical helix antenna contained the parachutes and some scientific instruments. The narrow cylindrical strip below the aerobrake smoothed the turbulence induced by the lower structure.

The spherical capsule contained electronics and a sequencer to manage capsule operation. It also contained signal processing equipment for the two cameras, electronics for the experiments, two transmitters, and a gamma ray spectrometer.

Fig. 3.10 Model of lander of 4V-1 spacecraft. (Wikimedia posting by Alexxx-Malev)

The spherical capsule was cooled before the lander was deployed by porting cooled nitrogen gas around it. The gas at a temperature of about -10 °C was delivered from the bus for a period of time before the lander was detached. The tubes used to circulate the cold gas are seen rising from the left side of the capsule in the photograph of Figure 3.10. The capsule was well insulated and the precooling extended the operating time of the capsule on the surface before heat on the inside became excessive.

Another method employed in the Venera 9 capsule to slow the rise of temperature of the electronics within the capsule was use of a phase change material. Phase change material had been successfully used in the Venera 8 landing capsule. The material chosen was lithium nitrate trihydrate because its melting point of 30 °C was well suited for this application and it had high heat of fusion and high density.

A fan within the spherical capsule circulated pressurized nitrogen gas around the electronic components. Thermal insulation around the inside and outside of the spherical enclosure slowed passage of heat from outside of the capsule. The thermal control system of the capsule was designed to survive a 75-minute descent through the atmosphere and then a minimum of 30 minutes operating time on the surface.

Lander Experiments

The lander contained a number of scientific instruments that provided data on clouds around Venus and the atmosphere as the lander descended. Those instruments included:

- Two cameras
- Five hot wire thermometers
- Six barometers
- Two anemometers to measure wind speed on the surface
- Spectrometer measuring sunlight levels in visible and infrared wavelengths
- Spectrometer measuring light levels in narrow infrared regions around the absorption bands of CO_2 and water vapor
- Two nephelometers to measure scattering properties of particles in clouds and atmosphere
- Gamma ray densitometer
- Gamma ray spectrometer.

A few of the instruments that have not been described on previous spacecraft are briefly described below.

Camera – The cameras were the most important instrumentation carried by the lander. No other instrument could convey as much information as the first-ever image of the surface of a planet.

The cameras were mounted on opposite sides of the lander and each had a horizontal scanning field of view of about 180 degrees. Taken together, the two cameras provided nearly a full 360-degree panoramic image of the surface of Venus. The camera was a cycloramic type with a mechanically scanning mirror.

The two cameras were mounted on the inside of the spherical pressure vessel. The body of the camera was an irregular rectangular shape about 30 cm long, 14 cm wide and 10 cm high. The optics of the camera included a scanning mirror at the end of a 3.3 cm diameter cylindrical tube about 21 cm long. One end of the tube was attached to the body of the camera. The other end protruded through the spherical pressure vessel and to the inside of a cylindrical quartz window. The quartz window was one cm thick to resist the very high atmospheric pressure of Venus. A lens cap that was jettisoned pyrotechnically protected the viewing port of each camera until the capsule landed.

Don Mitchell, in *Soviet Space Cameras*, states that the angular coverage by the camera was 40 degrees vertical and 180 degrees horizontal. The image was represented by a field of 115 by 512 pixels. The camera was mounted just below the aerodynamic braking disk and about 90 cm above the surface of Venus. The optics of the camera were arranged to direct the center of the vertical coverage at an angle of 45 degrees down from the horizontal.

The image was formed by causing the mirror to nod rapidly in the vertical plane over an included angle of about 40 degrees. A calibrated light source was fed into the optics during retrace of the vertical scan. The mirror assembly was rotated in the horizontal plane so that each vertical scan traced a new region of the surface. The mirror assembly panned 180 degrees in the horizontal plane. The vertical scanning rate was one vertical line each 3.5 seconds. Each line was mapped into 115 pixels. In addition, 13 pixels were required by the calibration signal giving a total of 128 pixels per line.

The external scene was reflected from the mirror down through the objective lens to a pinhole aperture at the image plane. The image from the pinhole, which represented a scanned spot on the Venusian surface, was applied to a photomultiplier tube and subsequently amplified and digitized at six bits per pixel. A seventh bit was added for a parity check. The objective lens was focused at the hyperfocal distance and images at distances from 0.8 meters to infinity were acceptably sharp.

The data rate applied to the transmitter during imaging was 256 bits per second. This comes about from processing 128 pixels per vertical line at 7 bits per pixel in a time of 3.5 seconds. On the basis of a 40-degree vertical scan for 115 pixels, the vertical resolution was about 0.35 degrees. Likewise, on the basis of 180 degrees and 512 pixels, the horizontal resolution was also about 0.35 degrees. At a distance of 10 meters from the camera the resolution would have been about six centimeters.

Gamma ray spectrometer – The gamma ray spectrometer measured gamma rays naturally emitted by radioactive constituents in the rocks on the surface of Venus. The percentages of naturally radioactive uranium, thorium, and potassium were determined. The spectrometer was mounted within the spherical pressure enclosure but gamma rays were able to penetrate the walls of the enclosure. The operation of the instrument was described in the previous section on Venera 8.

Gamma ray densitometer – The gamma ray densitometer was deployed to the surface of Venus after landing. The instrument consisted of a cylinder 4 cm in diameter and 36.2 cm long that contained a cesium-137 radioactive source and three Geiger-Muller tubes that measured the distribution of the reflected gamma rays from the surface. The instrument was attached to an arm that was connected to the lander just above the landing ring. The arm pivoted to stow the cylinder against the landing capsule until after landing. After landing, the arm pivoted down to place the instrument on the surface.

Nephelometers – The nephelometers generated a pulsed source of light and measured scattered light from small particles in the clouds and atmosphere. There were two nephelometers carried by the lander. One measured the backscatter of light from particles in a sample of the atmosphere and the other measured the distribution of scattering at angles of 4, 15, 45, and 180 degrees from the light source. The scattering characteristics were used to estimate the density of particles and their sizes.

Flight of 4V-1 Spacecraft Venera 9 and Venera 10

Two spacecraft of the 4V-1 type were flown in the 1975 launch window to Venus. They were given the names Venera 9 and Venera 10.

Venera 9

Venera 9 was launched from the Baikonur Cosmodrome on 8 June 1975 by a Proton K launch vehicle with a block D upper stage. The launch and insertion into a trans-Venus trajectory were successful. A midcourse correction was made on 16 June 1975 and a second midcourse correction was made on 15 October 1975. The spacecraft approached Venus on 20 October and the lander was released from the bus.

The rocket motor of the bus was ignited for a short period to prevent the bus from impacting Venus. Later, a longer burn was made to place the bus in an orbit around Venus. The final orbit was 1,510 km by 112,200 km at an inclination of 34.2 degrees. The orbital period was 48 hours and 18 minutes. The timing of the orbit was such that during the time period that the lander was descending through the clouds of Venus, the bus had swung around the dark side of the planet and was rapidly rising on the sunlit side. The bus was in good position to pick up signals from the lander and relay them to Earth by its high-gain antenna. The lander, on the sunlit side of Venus, was not in a direct line of sight to Earth.

The Venera 9 lander reached Venus on 22 October and entered the atmosphere on the sunlit side at an angle of 20.5 degrees to the local horizontal. Entry into the atmosphere occurred at a velocity of about 10.7 km per second at an altitude of about 125 km. The shallow entry angle resulted in a relatively low peak deceleration of about 170 gs. The lander was slowed by aerodynamic braking to about 250 meters per second. When the deceleration had decreased to 2 gs at an altitude of about 65 km, explosive bolts separated the top portion of the aeroshell and a small parachute was deployed to pull that portion away. A 2.8-meter diameter braking parachute was then deployed and lander systems were activated and the lander began transmitting data.

A programmer allowed the braking parachute to function for about 15 seconds and then it was detached. An ensemble of three 4.3-meter diameter parachutes was deployed at an altitude of about 62 km. The lower half of the aeroshell was jettisoned a few seconds later. The lander descended under the parachutes for about 20 minutes to an altitude of 50 km. The parachutes were detached at that point and the lander continued its descent while being slowed by the 2.1-meter diameter aerobrake. That arrangement allowed the lander to descend rapidly through the hot atmosphere and lessened heat absorbed by the landing capsule. After a descent of 55 minutes on the aerobrake, the lander reached the surface at a velocity of about seven meters per second.

The landing was at 31.01° North and 291.64° East in the Venusian coordinate system. Landing happened to be in a region about 2.1 km above the average radius of the planet. The average radius of 6,051 km for Venus was established as the zero-elevation datum, akin to sea level on Earth. The lander came to rest on a slope about 20 degrees from horizontal and on an uneven surface that increased the horizontal plane of the camera by another 10 degrees. Shortly after landing, the cameras were activated and the gamma ray densitometer was deployed to the surface.

Venera 10

Venera 10 was launched from the Baikonur Cosmodrome on 14 June 1975 by a Proton K launch vehicle with a block D upper stage. The launch and the insertion into a trans-Venus trajectory were successful. A midcourse correction was made on 21 June 1975 and a second midcourse correction was made on 18 October 1975. The spacecraft approached Venus on 23 October and the lander was released from the bus. The Venera 10 bus was placed into a 1,620 km by 113,000 km orbit around Venus with inclination of 29.5 degrees.

The Venera 10 lander entered the atmosphere of Venus at an angle of 22.5 degrees to the local horizontal. It made a successful landing on the surface on 25 October 1975. The Venera 10 lander performed the same steps as Venera 9 during descent to the surface. It landed at Venus coordinates 15.42 degrees North and 29.51 degrees East. The landing was in an area of Venus that appeared older than the Venera 9 site with chemically weathered rocks on a flatter surface. The elevation at the Venera 10 landing site was estimated to be about 1.5 km above the zero-elevation datum. Venera 10 landed about 2,200 km from the landing point of Venera 9.

Experiments Conducted by the Landers

Photography – The cameras were activated soon after landing. Each lander had two cameras, one on each side of the landing capsule. The lens covers for the cameras were designed to be jettisoned pyrotechnically. The cover of one camera on Venera 9 was so removed but the cover remained on the other camera. Venera 10 suffered the same problem of one of the covers failing to jettison. As a result, images of the surface of Venus were only obtained from one side of each of the landers. The imaging sequence for camera was arranged to make a complete azimuth scan of about 180 degrees and then reverse the direction and rescan in the other direction. The camera scan began about two minutes after landing of Venera 9. The first scan of Venera 9 covered about 174 degrees in 28.8 minutes. The reverse scan had covered 128 degrees when the orbiting bus lost signal from the lander. The lander had transmitted data for 53 minutes after landing. The 174-degree panoramic image is shown on the next page (Fig. 3.11).

The image has been processed to remove vertical bars incurred from periodic transmission of telemetry data. The light covered circular section in the foreground of the image is part of the lander. The geometry was such that with a tilt of 45 degrees from horizontal of the camera, the field of view swept above the horizon near the edges of the scan. The horizontal size of the rocks was estimated to be 50 to 70 cm with thickness of 15 to 20 cm. Receipt of the panoramic image back on Earth represented a major milestone in planetary exploration. It was the first photograph ever taken from the surface of another planet.

A panoramic image taken by the Venera 10 lander from its landing site is also shown on the next page (Fig. 3.12). The Venera 10 site was somewhat smoother

Fig. 3.11 Panoramic image of the surface of Venus taken by the Venera 9 lander. (NASA image catalog)

Fig. 3.12 Panoramic image of the surface of Venus taken by the Venera 10 lander. (NASA image catalog)

than the Venera 9 site with rocks less jagged in appearance. A fine dark material covers some of the surface.

Temperature and pressure measurements – Temperature and pressure of the atmosphere were measured during descent from an altitude of about 60 km to the surface. On the surface of Venus, Venera 9 measured atmospheric pressure of 90 bars and temperature of 460 °C. A model atmosphere of Venus had been postulated in 1972 based on the pressure and temperature measurements as a function of altitude made by Venera 4, 5, 6, 7, and 8. Based on the model atmosphere, it was concluded that Venera 9 landed on a slope in a region about 1.5 to 2 km above the mean surface zero elevation based on a radius of Venus of 6,051 km. Venera 10 landed in a flatter region.

The change of pressure and temperature as a function of altitude measured by Venera 9 and Venera 10 followed closely that measured by Venera 8. The Venera 8 data was plotted in the previous section of this chapter. A few data points from curves plotted by Keldysh in *Venus Exploration with the Venera 9 and Venera 10 Spacecraft*, are given in the table on the next page (Table 3.2). The pressure and temperature measured by Venera 10 on the surface were obtained from another source. In keeping with previous conventions in this book, pressure is given in bars and temperature is given in degrees Centigrade. The data points in the table are not exact since they were obtained by scaling from a chart.

Solar radiation flux – The illumination level at the surface from sunlight filtering through the clouds and atmosphere was about 14,000 lux at the landing site of Venera 9. This is about the same illumination as on a cloudy day at mid-latitudes on

Table 3.2 Pressure and temperature of atmosphere of Venus vs altitude

	Pressure, bars		Temperature, °C	
Altitude, km	Venera 9	Venera 10	Venera 9	Venera 10
57	0.7	0.7	53	35
50	1.5	1.5	99	90
40	4	4	168	154
30	10	10	259	233
20	24	24	327	314
10	50	50	397	393
0	90	92	460	465

Table 3.3 Wind speed around Venus as a function of altitude

Altitude, km	Wind speed, m/s Venera 9	Wind speed, m/s Venera 10
60	58	51
50	58	53
40	58	56
30	39	48
20	21	25
0	0.4 to 0.7	0.8 to 1.3

Earth. The sun was at a solar zenith angle of 30 degrees at the landing site. The total flux at the surface was about 100 watts per square meter.

A plot of solar radiation flux in the upper hemisphere from the Venera 9 lander at five different wavelengths from 0.44 micrometers to 1.16 micrometers is given in the paper by Keldysh. The plots show a nearly linear decrease in radiation flux as the lander descended between 62 km and 48 km and then an abrupt decrease in the rate of change of flux with altitude below 48 km. The rapid decrease in flux between 62 km and 48 km was attributed to scattering from particles in the clouds. From these measurements, the base of the clouds was inferred to be at an altitude of 48 km

Wind speed measurements – The speed of the atmospheric wind was determined by measuring the Doppler shift of the lander's transmitted signal relayed to earth by the orbiter. Scientists that determined the wind speed did an impressive job for they needed to allow for the Doppler shift due to the complex motion of the spacecraft-orbiter-Venus system relative to Earth. A few values from Keldysh's plot of velocity as a function of altitude are given in the table above (Table 3.3). The wind speed was measured by anemometers while the landers were on the surface (zero altitude in the table).

Nephelometer measurements – Two nephelometers were used to investigate the scattering properties of the clouds. The nephelometers contained a light source and they measured scattering at various angles from the light beam. One nephelometer measured scattering at several angles between four degrees and 180 degrees. The other measured scattering at 180 degrees (backscatter). The strongest scattering was found to be at an angle of 45 degrees and the least was at 180 degrees. The maximum scattering occurred at an altitude of about 52 km. The intensity of the scattering decreased rapidly as the lander descended below 52 km and it was essentially zero at 49 km. The measurements indicated that the base of clouds was at an altitude of about 49 km. This result agrees well with the illumination experiment that found the base of clouds at about 48 km.

Data from the nephelometers showed that the clouds had distinct layers with concentrations of 400 to 500 particles per cubic centimeter in some layers and 50 to 100 particles per cubic centimeter in others. Keldysh drew a significant conclusion from the nephelometer data that the range of visibility in the clouds was 1 to 3 km. That implies that the clouds were but a slight haze.

Mass spectrometer measurements – The landers carried a mass spectrometer that measured the ratio of mass to charge of ionized atoms. The device was best suited to measure amounts of inert nitrogen and noble gases in the atmosphere. The spectrometer was not activated until the lander had descended below the base of the clouds to avoid clogging the pores of the device from particles in the clouds. Measurements made by Venera 9 and Venera 10 indicated nitrogen at about 2% of the atmosphere and argon at about 200 parts per million.

Gamma ray densitometer measurements – The gamma ray densitometer was deployed to the surface after landing. It measured scattering from a gamma ray source located in the densitometer. The amount of scattering gave an indication of the density of the underlying rock. Analysis of the measurements made by Venera 10 indicated that the density of rock near the lander was 2.7 to 2.9 grams per cubic centimeter, similar to basalt rocks on Earth.

Gamma ray spectrometer measurements – The gamma ray spectrometer in the landers measured gamma rays naturally emitted by radioactive constituents in rocks. Measurements in percentage by weight made by Venera 9 indicated potassium at 0.9%, thorium at 0.0004%, and uranium at 0.00005%. Measurements by Venera 10 indicated potassium at 0.3%, thorium at 0.0001% and uranium at 0.00007%. Keldysh gives measurements taken of basalt rocks on Earth as follows: Potassium at 0.76%, thorium at 0.00021%, and uranium at 0.000086%. The Earth measurements may have been made by the landing capsule prior to flight.

Operation of the Orbiter

Data relay – The orbiter traveled in a highly elongated orbit 1,510 km by 112,200 km. The orbit positioned the bus on the same side of Venus as the lander while the lander was descending and after it landed. The bus acted as a relay to send lander data to Earth since the lander's view of Earth was blocked by the planet. Conical spiral antennas on the bus picked up transmissions from the lander.

The cameras on the lander were active while the lander was on the surface. Data from one camera was transmitted to the bus on a frequency of 122.8 MHz and the other camera transmitted at a frequency of 138.6 MHz. The data rate from each camera was 256 bits per second. Telemetry from other experiments on the lander were transmitted periodically in short bursts interspersed with the camera data

The orbiter relayed data from the lander to Earth by its high-gain antenna. The transmission frequency to Earth was 992.76 MHz and the data rate was 3,072 bits per second. Lander data was relayed to Earth in real time as well as being recorded on a tape recorder in the bus for later playback. The tape recorder could record two channels of data for 45 minutes.

Photography of clouds -The orbiter contained two cameras that were used to make photographic sweeps of the top of the clouds at ultraviolet wavelengths. One camera operated in the wavelength band 345 to 380 nanometers and the other operated in the band 355 to 445 nanometers. The cameras scanned ±15 degrees in the cross-track direction and used motion of the spacecraft to move the coverage along in the along-track direction. The tops of clouds were imaged in swaths 30 degrees wide and about 1,200 km long while the orbiter was in the low portion of its orbit. At an altitude of about 1,500 km, the 30-degree cross-track scans would encompass about 800 km in width. A total of 17 swaths of camera images were made by the Venera 9 orbiter. The images show considerable structure in the clouds including long dark streaks.

Cloud temperature measurements – The infrared radiometer operated in the 8 to 30 micron infrared band and gave a measure of the temperature of the upper region of the clouds. Temperatures of −42 to −40 °C were measured on the sunlit side and −32 to −31 °C were measured on the dark side of Venus.

Infrared spectrometer measurements – The infrared spectrometer operated in the 1.5 to 3 micron band. It found that the cloud layer had a diffuse upper boundary at an altitude of 68 km. The Lyman-α spectrometer operated in the vicinity of 121 nanometers wavelength and it looked for the distinctive hydrogen spectral line. A hydrogen corona was found around Venus at an altitude of about 275 km. The atomic hydrogen density of the corona was about 10^5 per cubic centimeter.

Magnetic field measurements – No discernable magnetic field was detected in the vicinity of Venus. The upper limit of magnetic field was determined to be less than 0.01% of the magnetic field strength of Earth.

Bistatic radar experiment – The bistatic radar experiment involved transmitting a signal from the high-gain antenna on the bus towards the surface of Venus during the low point of the orbit. Sensitive equipment on earth received the signal reflected from the surface and recorded it for analysis. The result was a one-dimensional line of reflectivity measurements that could be interpreted as degree of surface roughness. Areas of solid rock and of clumps of rocks were identifiable. Mountains with heights of 3 to 5 km were discerned.

Radio occultation measurements – The radio occultation experiment involved processing radio signals received on Earth from the bus as it was passing behind Venus and just emerging from behind Venus in its orbit. Occultation occurred when the bus became hidden from earth by the planet. The atmosphere of Venus caused bending of the radio signal around the planet. Analysis of the signal received on Earth yielded a measure of refractive index and absorption coefficient profiles with depth of the atmosphere. This data in turn was used to determine atmospheric density, pressure, and temperature. Those parameters changed with sun angle.

Radio occultation measurements also allowed probing the ionosphere of Venus and determining the change in height of the ionosphere from day to night on the planet. The electron concentrations in the ionosphere were found to be only about 10% of the electron concentrations in Earth's ionosphere. Higher electron concentrations were found on the day side of Venus. The altitude of the ionosphere at night was found to be about 140 km.

4V-1 Spacecraft, Venera 11 and Venera 12

The Venera 11 and Venera 12 spacecraft of the 4V-1 series were prepared for the 1978 launch window to Venus by Design Bureau NPO Lavochkin. The spacecraft had basically the same configuration as Venera 9 and Venera 10 with a few new instruments. The bus contained two new and significant instruments called SIGNE 2 and KONUS that gathered information about gamma-ray bursts. The lander carried color wheels for the cameras that allowed generating color photographs of the surface at the landing site. Another new experiment was a chemical laboratory to analyze samples of the soil at the landing site. An associated drill gathered samples of soil for the analysis.

The bus did not orbit Venus as in the case of Venera 9 and Venera 10 but rather it deployed the lander during a flyby. The launch occurred towards the end of the 1978 launch window and there was not enough fuel in the bus to sufficiently slow the bus for insertion into orbit around Venus. The flyby trajectory had an advantage in that it allowed a longer period of time for the bus to be within sight of the lander.

Flights of 4V-1 Spacecraft Venera 11 and Venera 12

Venera 11

Venera 11 was launched from the Baikonur Cosmodrome on 9 September 1978 by a Proton K launch vehicle with a block D upper stage. The launch and the insertion into a trans-Venus trajectory were successful. A midcourse correction was made on 16 September 1978 and a second midcourse correction was made on 17 December 1978. The spacecraft approached Venus on 23 December and the lander was released from the bus.

After deploying the lander, the rocket motor in the bus was ignited for short burn to cause the bus to fly by Venus at an altitude of 35,000 km. The trajectory of the bus provided a relatively long time window to relay data from the lander to Earth by the high-gain antenna on the bus.

The Venera 11 lander entered the atmosphere of Venus at a velocity of 11.2 km per second. The lander was slowed by aerodynamic braking by the lander enclosure. A pilot parachute with area of 1.4 square meters was deployed to stabilize the entry enclosure and a parachute with an area of six square meters was deployed to pull off the upper portion of the lander enclosure. The lower half of the landing enclosure was jettisoned shortly afterwards. The main braking parachute with an area of 24 square meters was deployed and the experiments and telemetry system of the sander were powered on. At an altitude of 47 km, which was about the base of the clouds, and at a velocity of 18 meters per second, the braking parachute was jettisoned.

The lander continued its descent being slowed by the aerodynamic braking disk. A successful landing was made at a velocity of about 8 meters per second after a descent lasting 62 minutes. The landing, which occurred on 25 December 1978, was at 14° South and 299° East in the Venus coordinate system. Data transmission from Venera 11 was received for 95 minutes after landing and relayed to Earth. The signal was lost when the bus passed below the lander's horizon.

Venera 12

Venera 12 was launched from the Baikonur Cosmodrome on 14 September 1978 by a Proton K launch vehicle with a block D upper stage. The launch and the insertion into a trans-Venus trajectory were successful. Midcourse corrections were made on 21 September and 14 December 1978. The spacecraft approached the night side of Venus on 21 December and the lander was released from the bus. The rocket motor in the bus was ignited for a short burn to cause the bus to fly by Venus at an altitude of 35,000 km in a trajectory that provided a relatively long time window to relay data from the lander to Earth. Although Venera 12 was launched five days before Venera 11, it made a faster trip and arrived at Venus two days before Venera 11.

The Venera 12 lander entered the atmosphere of Venus on 23 December 1978. After aerodynamic braking, a pilot parachute was deployed followed by a braking parachute and then the descent parachute. The descent parachute was jettisoned at an altitude of about 46,000 km. The lander was slowed for the remainder of the descent by the disk-like aerobrake. It reached the surface with a velocity of about eight meters per second. The successful landing was at location of 7° South and 294° East in the Venus coordinate system. Venera 12 landed about 850 km from the landing site of Venera 11. Data transmission from Venera 12 was received for 110 minutes after landing.

Lander Operations

There were more scientific instruments on the landers for Venera 11 and 12 than on previous landers resulting in a lander weight of about 760 kg.

Experiments on the lander included:

- Two color cameras
- Four temperature sensors

- Three barometers
- Anemometer to measure wind speed on the surface
- Gas chromatograph (Sigma)
- Mass spectrometer (MKh-6411)
- Gamma ray spectrometer
- X-Ray fluorescence spectrometer
- Backscatter nephelometer
- Optical spectrometer
- Penetrometer (PROP-V)
- Low frequency radio sensor (GROZA)
- Soil Analysis unit and drill

Many of the experiments were the same as employed on Venera 9 and Venera 10 that were described in the previous section of this book. The pressure and temperature sensors on Venera 11 yielded measurements of 92.6 bars and 452 °C at the landing site. Venera 12 measured 93.6 bars and 468 °C at its landing site. The optical spectrometer provided data on light intensity as a function of wavelength at altitudes from 65 km to touchdown. The spectrums showed dramatic decrease in intensity of the shorter wavelengths (blue light) at lower altitudes due to Rayleigh scattering in the atmosphere. As a result, a high percentage of shorter wavelengths of light were scattered away and the light reaching the surface of Venus had an orange cast.

New or different experiments carried by the lander are described below.

Color Cameras

Venera 11 and Venera 12 landers each carried two cameras that were improved versions of the cameras carried by Venera 9 and 10. The cameras sequentially photographed through sections of a color wheel having clear, red, green, and blue segments. The resulting images were assembled on Earth into color photographs of the surface of Venus. The basic camera was the same as used in Venera 9 and 10 but with a lower noise photomultiplier tube that resulted in higher signal-to-noise ratio.

The angular coverage by the camera was 37 degrees vertical and 180 degrees horizontal. The panoramic image was represented by a field of 211 by 1024 pixels. The camera was mounted just below the aerodynamic braking disk and about 90 cm above the surface of Venus. The optics of the camera were arranged to direct the center of the vertical coverage at an angle of 45 degrees down from the horizontal.

The image was formed by causing a mirror to nod rapidly in the vertical plane over an included angle of about 37 degrees. A calibrated light source was fed into the optics during retrace of the vertical scan. The mirror assembly was rotated in the horizontal plane so that each vertical scan traced a new region of the surface. The mirror assembly panned 180 degrees in the horizontal plane in a series of 1024 vertical lines. The vertical scanning rate was one vertical line each 0.82 seconds. Each line was mapped into 211 pixels. In addition, 41 pixels were required by the calibration signal during retrace giving a total of 252 pixels of images per line.

The external scene was reflected from the mirror down through the objective lens to a pinhole aperture at the image plane. The image from the pinhole, which represented a scanned spot on the Venusian surface, was applied to a photomultiplier tube and subsequently amplified and digitized at nine bits per pixel. A tenth bit was added for a parity check. The objective lens was focused at the hperfocal distance and images at distances from 0.8 meters to infinity were acceptably sharp.

The data rate applied to the transmitter during imaging was 3,072 bits per second. This comes about from processing 252 pixels per vertical line at 10 bits per pixel in a time of 0.82 seconds. On the basis of a 37-degree vertical scan for 211 pixels, the vertical resolution was about 0.18 degrees. Likewise, on the basis of 180 degrees and 1024 pixels, the horizontal resolution was about 0.18 degrees. At a distance of 10 meters from the camera the resolution would have been about three centimeters.

Unfortunately, the lens caps for the cameras on both Venera 11 and Venera 12 failed to jettison upon command once the lander had come to rest on the surface of Venus and no photographs could be taken. It is one thing for a lens cap to still be on the camera when photographing domestic scenes on Earth, but imagine the angst of Russian scientists when it was discovered that the lens caps were still on the cameras of both Venera 11 and Venera 12 landers with vistas of Venus spread out before them.

Venera 13 and Venera 14 did successfully take panoramic color photographs of the surface of Venus and those photographs are shown in Chapter 5 of this book.

Mass Spectrometer

The landers carried a mass spectrometer that measured the ratio of mass to charge of ionized atoms to identify atoms of constituents in the atmosphere and their concentration. The mass spectrometer was located inside of the spiral transmit antenna.

The mass spectrometer carried on the Venera 11 and 12 landers was a radio frequency (RF) type. Typically, a radio frequency type of mass spectrometer introduces a sample of gas into an ionization chamber where the gas molecules are ionized. The ions are accelerated out of the chamber and focused into a beam by metal lenses with a hole for the beam. A voltage difference between the ionization chamber and the lenses provides the accelerating voltage. The beam at the output of the lens is directed through a gap in the center of a bundle of four rods. Two of the rods have a dc voltage applied and the other two have a radio frequency ac signal applied. A detector is aligned with the gap at the output of the rods. A particular combination of voltages and frequencies causes an ion with a particular mass to charge ratio to spiral through the gap between the rods and strike the detector. Ions with a different mass to charge ratio strike the rods and are not detected.

The mass spectrometer installed in Venera 11 and 12 was contained in a vacuum enclosure with an ion pump to evacuate the enclosure after each sample was analyzed. The instrument was arranged to measure amounts of inert nitrogen and noble gases in atmospheric samples. The spectrometer avoided a problem with clogging fine pores of previous mass spectrometers by opening a good size orifice for only an instant. In addition, the instrument was not activated until the lander had descended to 25 km, which was well below the base of the particulate laden clouds. Measurements were made during the descent with the first measurement made at an altitude of 23 km and the last measurement made about 3 to 1.5 km above the surface. A total of 11 gas samples were acquired and processed by each lander during the descent. After analysis, the old sample was purged from the instrument before the next sample was acquired.

Each sample was scanned eight times, each scan taking about one second. Each scan resulted in a spectrum of amplitude vs mass over charge. The spectrums were encoded and transmitted to the bus during the descent for relay to Earth. Analysis by V. G. Istomin et al., in the paper, *Venera 11 and 12 mass spectrometry of the lower Venus atmosphere*, gave the following results of the mass spectrometer experiment: concentration of nitrogen at 4.5 ± 0.5 %, argon isotopes at ~150 parts per million (ppm), neon at ~ 10 ppm, and krypton at ~0.5 ppm.

Gas Chromatograph (Sigma)

A gas chromatograph was carried on the landers to complement the mass spectrometer in identifying abundances of different gasses in the atmosphere of Venus. The instrument was located inside of the spiral transmit antenna.

A basic gas chromatograph uses thin tubes filled with capillary material that gas molecules adhere to. An inert carrier gas such as neon is used to push the gas sample through the capillary tube. A gas constituent that has weaker adsorption to the walls has a shorter retention time and it is pushed through the capillary tube quickly by the neon carrier gas. A gas constituent that has stronger adsorption to the walls has a longer retention time and appeared later at the output of the tube. A detector at the output of the tube is used to determine retention time and amplitude of the constituents of the gas sample.

The gas chromatograph installed in the Venera 11 and 12 landers used three separate tubes referred to as columns to identify categories of gases. The three columns, each with a detector, were connected in series and the sample of gas passed through all three columns before exiting the instrument. The first column was 2.0 meters long and arranged to identify sulfur compounds, water vapor, and CO_2. The second column was 2.5 meters long and used to identify helium, hydrogen, nitrogen, krypton, methane, and carbon monoxide. The third column, 1.0 meters long was arranged to detect argon. The columns were coiled up into manageable dimensions.

The three columns and detectors were enclosed in a temperature-controlled chamber where the temperature was maintained at 70 °C while the analysis was being performed. In operation, a sample of gas about one cubic centimeter in volume was applied to the inlet of the first column. The outputs of the three detectors were applied to the telemetry system in analog form. A calibration mixture of gas was applied to the instrument and analyzed shortly before the first atmospheric sample was gathered.

A paper by B. G. Gelman, et al., *Gas Chromatograph Analysis of the Chemical Composition of the Venus Atmosphere* gives results of the analysis for the Venera 12 lander. A total of eight samples were taken from an altitude of 42 km down to the surface. The concentrations by volume were found to be: nitrogen (2.5 ± 0.5)%, argon (4 ± 2)10^{-3}%, carbon monoxide (2.8 ± 1.4)10^{-3}%, sulfur dioxide (1.3 ± 0.6)10^{-2}%. The upper limit for oxygen was estimated at 0.002%.

X-Ray Fluorescence Spectrometer

The Venera 11 and 12 landers carried two different X-Ray fluorescence spectrometers. One was used to determine chemical properties of the clouds circling Venus and the other was used to determine chemical properties of a sample of soil on the surface. The soil sample was obtained by a drill.

The X-ray fluorescence spectrometer employed in the clouds used a fan to suck a sample of the atmospheric through a filter with fine pores about four microns in size. The sample was irradiated by gamma rays produced by a radio-isotope Cd-109 source. That cadmium isotope has a half-life of 461 days in its decay to silver-109. It radiates gamma rays with energy of 88 keV during the decay. The gamma radiation induced fluorescence in the sample and emission of x-rays. The energy of the resulting X-rays was measured to give an indication of the chemistry of the sample.

The X-Ray fluorescence spectrometer on Venera 12 operated from the time it was turned on at an altitude of about 64 km down to an altitude of about 49 km when it ceased to operate. Don Mitchell in *Venera: The Soviet Exploration of Venus* gives the results of the measurements as follows:

Element	Density
Sulfur	<0.1 mg/m^3
Chlorine	0.43 ± 0.06 mg/m^3
Iron	0.21 ± 0.06 mg/m^3

The X-Ray fluorescence spectrometer that was designed to determine properties of the soil of Venus was mounted on the inside of the pressure capsule. A series of airlocks was used to transfer soil gathered by a drill to the instrument. Apparently, leaks developed in the tubes associated with the airlocks for both Venera 11 and 12 landers preventing transfer of soil samples. There was conjecture that buffeting and vibration of the lander during the descent because of many appendages caused seals to leak. The apparatus worked well on the Venera 13 lander and a description of the spectrometer and drill is given in Chapter 5 of this book.

Low Frequency Radio Sensor (GROZA)

Previous probes to Venus revealed that there may be lightning flashes in the clouds. A sensor for ultra-low radio frequency (ULF) signals that would be generated by lightning discharges was included on the Venera 11 and 12 landers. A loop antenna and a low frequency receiver to detect these signals was part of an instrument called GROZA, the Russian word for thunderstorm. The loop antenna was 25 cm in diameter and contained in a disk-like enclosure mounted to the supports for the landing capsule. It was set at an angle about 45 degrees to the longitudinal axis of the lander. The receiver covered the band of 8 kHz to 100 kHz. It contained four channels centered at 10 kHz, 18 kHz, 36 kHz, and 80 kHz. The bandwidths of the four channels were 1.6, 2.6, 4.6, and 15 kHz respectively.

A paper by L.V. Ksanfomaliti, et al., *Electrical Discharges in the Venusian Atmosphere*, indicated electrical discharges were quite intense during the landing of Venera 11 and less intense during the landing of Venera 12. The writers concluded that the electrical discharges occur in the clouds and that the energy of the discharges was about the same as lightning discharges on Earth. The measured intensity of the discharges was much stronger at higher altitudes than on the surface.

Data recorded during the descent of Venera 11 showed electrical activity in all four channels. During a particular seven-minute interval, very high amplitudes were recorded in the 10 kHz channel and much lower amplitudes in the 36 kHz and 80 kHz channels. The pulse count rate reached 20 pulses per second during some of the electrical bursts.

The GROZA equipment included a microphone to record sounds of wind and thunder. Brian Harvey writes in *Russian Planetary Exploration*: "After it landed, a massive thunderclap reverberated around the Venera 12 site for 15 minutes and probably affected the whole planet."

Penetrometer (PROP-V)

The Venera 11 and 12 landers carried a penetration device with a rotating conical tip holding a blade that was used to measure the bearing strength of the soil of Venus. The device was referred to as PROP-V, an acronym from the Russian term for an instrument for evaluating the characteristics of the surface of Venus. The rotating conical tip was mounted at the end of a ladder type assembly with two sections that was folded up on the lander prior to landing. After landing it unfolded and the end with the tip fell to the surface of Venus. The device was about 60 cm long when unfolded.

The conical tip, which contained a blade, was rotated by a spring once it was imbedded in the surface. The amount of penetration and amount of rotation of the cone with the blade were observed by one of the cameras on the lander. A dial on the cone indicated the amount of rotation. Electrodes embedded in the cone were used to measure the resistance of the soil.

The penetrometer did not function on either the Venera 11 or Venera 12 landers. There was conjecture that the devices were damaged by hard landings. The PROP-V penetrators deployed successfully on Venera 13 and 14 landers. Photographs of the penetrators extended on the surface are shown in the panoramic photographs of the surface taken by Venera 13 given in Chapter 5 of this book.

Bus

The bus was essentially the same as used for Venera 9 and 10 but with a few different instruments. The instrument group included:

- SIGNE 2 Gamma ray burst detector
- KONUS Gamma ray burst detector
- Ultra violet spectrometer
- Magnetometer
- Solar wind detector

The SIGNE 2 and KONUs gamma ray burst detectors were the first to fly on Venera spacecraft. A description of them is given below.

SIGNE 2 Gamma-ray Burst Detector

The SIGNE 2 gamma-ray detector was developed and built in France. The instrument was used in a joint Russian-French program to investigate the source of gamma ray bursts. The detectors were deployed on Venera 11 and Venera 12 and on a Soviet satellite of Earth called Prognoz 7. Identical detectors were installed on the three spacecraft and triangulation was used to determine the source of the gamma ray bursts. Prognoz 7, launched in October 1978, was a Soviet satellite designed to study solar activity.

Two SIGNE 2 gamma ray burst detectors were installed on each Venera spacecraft to provide full angular coverage. Each detector used a large NaI(T1) crystal with a surface area of 62 square centimeters. The NaI(Tl) crystal consisted of Sodium Iodide (NaI) with a trace of Thallium (Tl). The crystal responded to gamma ray bursts by flashes of light that were detected by a photomultiplier tube. When a gamma ray burst was detected, it triggered a process that opened a window for data gathering. The data stored after a trigger occurred included:

- Count rate with 1/64 second resolution for 16 seconds
- Count rate with 1/512 second resolution for 2 seconds for intense bursts
- Record of five energy spectra each 250 milliseconds for 32 seconds.

The energy of the burst was sorted into 31 different energy channels. The energy range for detection could be selected to be either 50 to 230 kV or 300 to 1,400 kV.

A paper by Barat et al., *A Review of Recent Results from the Franco-Soviet SIGNE Gamma-Ray Burst Experiment*, gives some of the results of the SIGNE 2 experiment involving the Venera 11, Venera 12, and Prognoz 7 spacecraft. The time period covered in the paper was 18 September 1978 to 13 June 1979. A total of 26 confirmed gamma-ray bursts were detected by at least one spacecraft during that period. In five cases the burst was detected in all three spacecraft. The individual times of arrival allowed determining angles to the origin of those five bursts. The accuracy of determining angles to the source of the bursts was a function of the fine structure of the burst and spacecraft separation. The location of bursts detected in November 1978 could be determined within an error box of 1° by 0.1°. Events in April 1979 could be located within an error box of 0.3° by 0.01°. The angles represent latitude and longitude in the galactic coordinate system.

KONUS Gamma-ray Burst Detector

The KONUS system for detecting and measuring characteristics of gamma-ray bursts, designed by Russia, was used effectively on the Venera 11 and 12 busses to gather data on gamma-ray bursts. Versions of the Russian KONUS system were used on later space probes as well, including some U.S. space vehicles.

The KONUS system on Venera 11 and 12 included six scintillation type gamma ray detectors with NaI(Tl) crystals. The crystals were fitted with a lead shield around the sides and leaded glass 16 mm thick on the back such that each crystal had a hemispheric acceptance window in front. The six detectors were arranged with the center of the hemispheric acceptance windows aligned with the plus and minus X, Y, and Z axis of the spacecraft. A photomultiplier tube viewed scintillations through the leaded glass at the back of the crystal. The signals at the output of the photomultiplier tubes were compared to a threshold level. When a gamma-ray event exceeded the threshold, a timing sequence was initiated where the count rate and amplitude were measured. A time analyzer recorded the temporal profile of the burst over three time intervals: two seconds with a resolution of 15.63 milliseconds, 32 seconds with a resolution of 0.25 seconds, and an additional 32 seconds with a resolution of one second. The time analyzer provided a total of 320 channels. The photomultiplier tube outputs were applied to pulse height analyzers that used a 16-channel logarithmic scale.

A comparison of amplitudes in the six detectors allowed computing the angle to the source. The direction of arrival could also be determined by comparing time of arrival of bursts in the Venera 11 and Venera 12 KONUS system as each bus traveled independently in space.

Mazets and Golenetskii in *Recent Results from the Gamma-Ray Burst Studies in the Konus Experiment*, present tables giving of dates of bursts detected in the Venera 11 and 12 busses and their amplitudes and directions to their source. The Goddard Space Flight Center published a catalog, *KONUS: Konus 11 & 12 Gamma-ray Burst Catalog* that gives additional information about each burst.

Mazets and Golenetskii estimated that the energy at the source of stronger bursts to be 10^{40} to 10^{41} ergs. This is a tremendous amount of energy. The total energy output of the sun is about 3.8×10^{33} ergs per second or 1.2×10^{41} ergs per year. Thus, the energy in a gamma ray burst that lasted for a few seconds was about the same as the energy output from the sun in a year. Mazets and Golenetskii concluded that the gamma-ray bursts must have a galactic origin.

Venera 11 and 12 continued gathering gamma ray data until February 1980.

Bibliography

Avduevsky, V. S., et al., *Venera 8: Measurements of Solar Illumination Through the Atmosphere of Venus,* Journal of the Atmospheric Sciences, Volume 30, September 1973

Ball, Andrew, et al., *Planetary Landers and entry Probes,* Cambridge University Press, Cambridge, UK, 2007

Barat, C., et al., *A Review of Recent Results from the Franco-Soviet SIGNE Gamma-Burst Experiments*, Astrophysics and Space Science, Vol 75, 1981

Gelman, V. G. et al., *Venera 12 analysis of Venus atmospheric composition by gas chromatography*, Soviet Astronomy Letters, 1979

Grahn, Seven, *Radio Systems of Soviet Mars and Venus Probes,* http://www.svengrahn.pp.se/radioind/MVradio/MVradio.htm

Harvey, Brian et al., *Russian Space Probes*, Springer-Praxis Books, Praxis Publishing, Chichester, UK. 2011

Harvey, Brian, *Russian Planetary Exploration*, Springer-Praxis Books, Praxis Publishing, Chichester, UK. 2007

Istomin, V. K., et al *Venera 11 and 12 mass spectrometry of the lower Venus atmosphere*, Soviet Astronomy Letters, 1979

Keldysh, M. V., *Venus Exploration with the Venera 9 and Venera 10 Spacecraft,* Icarus Volume 30 Issue 4, 1977

Ksanformaliti, L. V., et al., *Electrical Discharges in the Venusian Atmosphere*, Soviet Astronomy Letters, Vol 5, May-June 1979

LePage, Andrew, J. *You can't fail unless you try: The Soviet Venus Mars missions of 1962,* https://www.drewexmachina.com

LePage, Andrew, J., *If at first you don't succeed—Part 1 and Part 2,* The Space Review, http://www.thespacereview.com

Marov, M. Ys., et al., *Venera 8: Measurements of Temperature, Pressure and Wind Velocity on the Illuminated Side of Venus*, Journal of the Atmospheric Sciences, Volume 30, September 1973

Mazets, E. P., et al., *Recent Results from the Gamma-Ray Burst Studies in the Konus Experiment,* Astrophysics and Space Science 75, 1981

Mitchell, Don, P., *Inventing the interplanetary probe*. https://mentallandscape.com

Mitchell, Don, P., *Venera: The Soviet Exploration of Venus,* http://mentallandscape.com/V_Venus.htm

NASA-NSSDCA, *Cosmos 27*, NSSDCA/COSPAR ID: 1964-014A

NASA–NSSDCA, *Sputnik 19,* NSSDCA/COSPAR ID: 1962-040A

NASA-NSSDCA, *Zond 1*, NSSDCA/COSPAR ID: 1964-016D

NASA-NSSDCA, *Venera 8*, NSSDCA/COSPAR ID: 1972-021A

Siddiqi, Asif, A., *Beyond Earth A Chronicle of deep Space Exploration, 1958-2016*, NASA SP-2018-4041, 2018

Siddiqi, Asif, A., *Deep Space Chronicle*, NASA SP-2002-4524, 2002

Surkov, Yu, A., et al., *Element Composition of Venus Rocks: Preliminary Results from Venera 13 and Venera 14,* Soviet Astronomy Letters vol 8, 1982

Ulivi, Paolo, et al., *Robotic Exploration of the Solar System*, Springer-Praxis Books, Praxis Publishing, Chichester, UK. 2007

Chapter 4
U.S. Spacecraft that Explored Venus 1960–1980

The planet Venus was of great interest to scientists around the world. Thick clouds blanket the planet so the best way to learn about Venus was to explore it remotely by spacecraft. Both the United States and the Soviet Union had capability to send a spacecraft to that intriguing planet in the early 1960s and missions of spacecraft to Venus went forward.

There was much early conjecture on what lay under the clouds. Some scientists believed that Venus was a water world with oceans and a steamy atmosphere with plants. However, examination by Earth based instruments indicated that the atmosphere consisted largely of carbon dioxide with very little oxygen and that seemed to preclude plants. A leading U.S. scientist, Carl Sagan, who was esteemed for his direct talk about space physics to the general public, postulated that the greenhouse effect of the Venusian carbon dioxide atmosphere would make the planet unbearably hot.

Venus was a tempting target for early space explorers and just within reach of spacecraft lofted by launch vehicles of the day. Venus orbits inside of Earth's orbit around the sun and it is closer to Earth than is Mars, which orbits outside of Earth's orbit. Less energy is required of the launch vehicle to reach Venus than to reach Mars so Venus was the first planetary destination by spacecraft from both the United States and the Soviet Union.

Minimum energy launch windows to send spacecraft to Venus occur about every 584 days (19.4 months). These windows correspond to the time when Earth and Venus are near their closest approach as each travels around the sun in elliptical orbits. The closest possible passage of the two planets is about 38 million km (24 million miles). Each closest approach is different. The average closest approach distance is about 42 million km (26 million miles). Both the United States and the Soviet Union took advantage of the minimum energy launch windows. Spacecraft launches to Venus were grouped a few weeks around minimum energy dates.

A summary of spacecraft that were launched towards Venus by the United States during early planetary exploration in the 1960s and 1970s is given in Table 4.1.

T. Lund, *Spacecraft that Explored the Inner Planets Venus and Mercury*, Springer Praxis Books, https://doi.org/10.1007/978-3-031-29838-7_4

Table 4.1 U.S. Spacecraft sent to Venus in the 1960s and 1970s

Spacecraft	Launch date	Objective	Results
Mariner 1	22 July 1962	Flyby Venus	Launch failure. Launch vehicle went off course
Mariner 2	27 Aug 1962	Flyby Venus	Success. Much new information gathered
Mariner 5	14 June 1967	Flyby Venus	Success. New information gathered on atmosphere
Mariner 10	3 Nov 1973	Flyby Venus enroute to Mercury	Success. Both Venus and Mercury explored
Pioneer Venus Orbiter	20 May 1978	Orbit Venus	Outstanding success. 98% of surface mapped by synthetic aperture radar
Pioneer Venus multiprobes	8 Aug 1978	Bus released 4 probes to the surface	Outstanding success. Characteristics of atmosphere measured during descent

The Jet Propulsion Laboratory (JPL) was at the forefront in developing early spacecraft that explored the moon and the planets. Early in the space era they were engaged in developing rocket technology and surface-to-surface ballistic missiles for the U.S. Army. They also built and operated the first series of U.S. satellites called Explorer. Five Explorers were built and three were successfully placed in earth orbit. Explorer 1 was placed in orbit in January 1958. JPL became a part of NASA in December 1958. Development of spacecraft continued at JPL under NASA control.

In December 1959 NASA Headquarters assigned JPL the task of developing and building seven spacecraft to reconnoiter the moon. The program, called Ranger, was eventually successful although failures plagued the first six spacecraft launched. The Ranger mission involved photographing the surface of the moon as the spacecraft descended towards impact.

Planning was also underway at JPL for spacecraft that would travel to Venus and Mars. Those spacecraft were called Mariner A and Mariner B respectively. The launch vehicle planned to loft those spacecraft consisted of an Atlas D first stage and a Centaur second stage. Centaur was an advanced rocket stage that burned liquid hydrogen with liquid oxygen as oxidizer. The Atlas D/Centaur combination would be capable of delivering payloads of 568 kg (1250 pounds) to the vicinity of Venus and Mars. The very capable Mariner A spacecraft planned for Venus was expected to weigh about 500 kg (1100 pounds).

The Centaur development schedule lagged, however, and a compromise launch vehicle and spacecraft were considered to meet the 1962 launch window to Venus. A substitute launch vehicle consisting of an Atlas D first stage and Agena B second stage was pressed into service. A more modest, lightweight spacecraft was proposed by JPL that would use portions of the designs of the Ranger spacecraft and Mariner A. The weight of the new spacecraft, called Mariner R, was limited to about 209 kg (460 pounds) by the capability of the launch vehicle.

Mariner R Spacecraft (Mariner 1 and Mariner 2)

An artist's rendering of the Mariner R spacecraft is shown on the below (Fig. 4.1). The spacecraft had strong resemblance to the Ranger spacecraft that was sent to the moon. The mission for the first two Mariner R spacecraft, Mariner 1 and Mariner 2, was to fly by Venus and probe the planet with scientific instruments.

Management responsibility for the Mariner R program at NASA Headquarters was assigned to the Office of Director of Lunar and Planetary Programs under Oran Nicks. That office reported to the Office of Space Science under Homer Newell. JPL was assigned responsibility for project management of the Mariner R project and for operations during space flight. The Marshal Space Flight Center was responsible for management and conduct of the launch vehicle. The Project Manager at JPL for the Mariner R program was Jack James who had previous engineering and management experience on JPL's ballistic missile programs. He reported to Bob Parks, Planetary Program Director at JPL.

JPL began design of Mariner R in on 8 August 1961. Approval for the program came from NASA Headquarters on 30 August 1961. Impressively, the new Mariner R spacecraft was produced from a blank piece of paper to flight worthy hardware in less than a year. Two spacecraft and a set of spares were built. The set of spares was later assembled into a spare spacecraft. The two spacecraft and the spare were shipped to the launch site at Cape Canaveral the first week in June 1962 in time for

Fig. 4.1 Artist's rendering of Mariner R spacecraft. (NASA image)

the launch window to Venus that opened on 18 July 1962. The flight designations for the two spacecraft were Mariner 1 and Mariner 2.

Mechanical Configuration

A photograph of the Mariner 2 spacecraft in the assembly building at Cape Canaveral is shown below (Fig. 4.2). The solar panels folded up to fit within the shroud of the launch vehicle. A sketch of the Mariner R spacecraft with labels identifying important elements is shown on the next page (Fig. 4.3). Many of the callouts denote locations of the experiments on the spacecraft. The spacecraft was built on a hexagonal magnesium frame 1.04 meters across the diagonals and 36 cm deep. A pyramidal tubular aluminum truss structure was attached to the frame to hold scientific instruments and an omni directional antenna. Structure below the bottom of the hexagonal frame supported two solar arrays and a high-gain parabolic antenna. The span from end-to-end of the solar arrays was 5.0 meters. The height of the spacecraft was 3.66 meters. The total weight of the spacecraft was 202.7 kg. Of that total, 18.6 kg was allocated to scientific experiments.

Each solar array panel was 76 cm wide and 183 cm long. The high-gain parabolic antenna used to communicate with Earth was 1.3 meters in diameter. There were six magnesium chassis attached to the inside structure of the hexagonal frame, one

Fig. 4.2 Mariner 2 spacecraft in assembly building at Cape Canaveral. (NASA photograph)

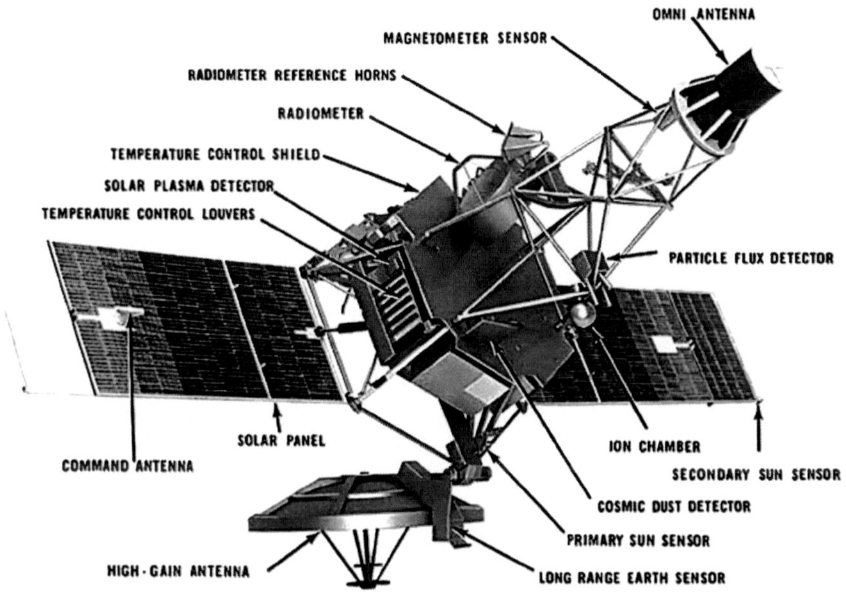

Fig. 4.3 Configuration of Mariner R spacecraft. (NASA graphic)

chassis per side. The six chassis contained electronics for the experiments and for spacecraft systems.

The coordinate system placed the Z-axis along the longitudinal axis of the spacecraft. The X-axis was parallel to the long side of the solar arrays and the Y-axis was perpendicular to the X and Z axes. Spacecraft roll was around the Z-axis, pitch was around the X-axis and yaw was around the Y-axis.

Mariner R Systems

Spacecraft Power

The amount of electrical power required by the spacecraft during the long cruise mode to Venus was about 150 watts. The solar arrays were capable of supplying a minimum of 148 watts and maximum of 220 watts. The power supplied was higher as the spacecraft traveled closer to the sun near the orbit of Venus. The original configuration of each solar array held 4900 cells in an array 76 cm wide and 1.52 meters long. As the spacecraft design progressed, it was found necessary to increase the total number of solar cells by 910. These were added to one of the arrays and the length grew to 1.83 meters. The length of the other solar array panel was also increased to 1.83 meters by a Dacron extension to balance pressure from the solar wind. The nominal output voltage of each solar array was 46 volts.

The solar arrays were oriented with the active side of the panels normal to the longitudinal axis of the spacecraft and facing downward in the photograph of the spacecraft in Fig. 4.2. Sun sensors were used to control gas jets that maintained the longitudinal axis of the spacecraft pointed at the sun during most of the mission.

A rechargeable silver-zinc battery with capacity of 1000 watt hours provided power to the spacecraft during launch and until the spacecraft became oriented with the solar arrays facing the sun. The battery was also used during maneuvering for midcourse correction and as needed when the experiments were powered during encounter with Venus. The battery contained 18 silver-zinc cells and had a nominal output voltage of 28 volts. It could provide a peak power of 510 watts. The substantial battery weighed 15 kg and had a volume of 7702 cubic centimeters.

The outputs from the two solar arrays and from the battery were applied to a power switching and logic unit that accepted power from the solar arrays when available and from the battery when it was not. The battery was maintained in a charged state by a battery charger once solar array power became available.

The power switching and logic unit contained power switching elements that chopped the power at a 2400 Hz rate. The 2400 Hz rate was set by a synchronization signal fed to the unit. The voltage at the output of the power switching and logic unit was applied to a boost regulator that regulated the switched voltage waveform to a set value. The 2400 Hz square wave at the output of the regulator was applied to a power amplifier. The output of the amplifier was a 50-volt square wave that served as power source for spacecraft systems and the experiments.

Each of the six electronic chassis in the spacecraft contained a transformer and rectifier to condition the 2400 Hz square wave power to dc power forms required by its electronic circuits. The electric motors and gyros used in the spacecraft required three-phase sinusoidal power at 400 Hz. This power form was generated by a power synchronizer and power amplifier powered by regulated power from the boost regulator.

Central Computer and Sequencer

Major spacecraft functions were controlled and sequenced by a central computer and sequencer (CC&S). The computer portion of the CC&S was not what we would think of as a computer today but rather it was a timer/sequencer that initiated a series of hard-wired commands. The (CC&S) managed three separate mission sequences: Launch sequence that managed events from launch through the cruise mode, Midcourse propulsion sequence that managed events during midcourse correction, and Encounter sequence that managed events for data collection when in the vicinity of Venus. The CC&S also provided timing for spacecraft systems including the 2400 Hz and 400 Hz reference frequencies for spacecraft power conversion.

Several events were managed by the CC&S during the launch sequence, which extended into cruise flight. Before launch, parameters were inserted into the CC&S that would cause a timer to initiate the encounter sequence 12 hours before approach to Venus. The timer began count when it received a command from launch control three minutes before launch. At 44 minutes after launch, a command to deploy the solar panels was generated by the CC&S. At 60 minutes after launch the CC&S generated a power-up command to the spacecraft attitude control functions. At 116½ hours after launch the CC&S issued a command to begin the Earth acquisition process that trained the high-gain antenna in the direction of Earth.

Attitude Control System

The attitude control system oriented the spacecraft so that the solar arrays were pointed at the sun and the high-gain parabolic antenna was pointed at the Earth during the cruise and encounter modes. The attitude of the spacecraft was adjusted by means of nitrogen cold gas jets. Sun sensors developed angle error signals in pitch and yaw and those signals were processed to control the jets that induced pitch and yaw of the spacecraft to orient the solar arrays at the sun. Likewise, an earth sensor developed error signals in the roll plane that were processed to control gas jets in the roll axis and to control an antenna hinge servo to align the high-gain antenna with the earth. The antenna mount hinge was set at about 72 degrees from the longitudinal axis of the spacecraft during initial Earth acquisition.

During midcourse correction, the attitude of the spacecraft was initially set by the cold gas jets to proper pitch and roll angles for the direction of thrust of the engine. When the rocket motor was firing, spacecraft attitude was maintained by deflecting four small vanes in the exhaust of the rocket motor. The vanes were controlled by the attitude control system. The nitrogen jets did not have sufficient force to counter forces induced by the rocket burn.

The nitrogen gas jets for attitude control were mounted at four locations on the lower portion of the hexagon frame. Two of the locations had a cluster of three jets and the other two had two jets. The two clusters of three jets were mounted on opposite sides of the frame and along the yaw axis of the spacecraft. Two jets of the cluster of three were mounted 180 degrees apart and perpendicular to the longitudinal axis. Those jets were used for roll control. The third jet in the cluster of three was pointed in a direction just outboard of the longitudinal axis and it was used for pitch control. Yaw control was accomplished by two clusters of two jets. The two clusters were mounted on opposite sides of the hexagon frame along the X-axis of the spacecraft.

Nitrogen gas for attitude control was contained in two titanium tanks initially filled to a pressure of 24.13 megapascal (3500 psi). The nitrogen gas was regulated to a pressure of 0.103 megapascal (15 psi) before application to the jets. The thrust of each jet in a vacuum was about 0.0045 kg. The spacecraft carried a total of two kg of nitrogen gas for the attitude control system. About 0.9 kg of nitrogen had been used by the end of the mission of Mariner 2.

The spacecraft had four primary sun sensor and two secondary sun sensors. The secondary sun sensors were used during the initial sun acquisition process. The sun sensors were cadmium-sulfide photodetectors that functioned as light sensitive resistors. The units contained a shade that covered a portion of the detector such that as the spacecraft tilted in the plane of the sensors the amount of light falling on the sensor changed and that changed the resistance.

A positive voltage was applied to one of the sun sensors and a negative voltage was applied to the other. The other terminals of the sensors were connected together and a load resistor was connected from that point to electrical ground. Two of the sun sensors were located along the yaw axis on opposite sides of the frame. When the yaw axis was perpendicular to the sun (pitch angle of zero), both sun sensors had the same illumination and hence the same resistance and the voltage across the load resistor was near zero. If the spacecraft tilted so that one sun sensor had more sunlight than the other, the resistances became unbalanced and a voltage appeared across the load resistor. The sign of the voltage depended on the direction of the tilt away from perpendicular to the sun.

Two other sun sensors were mounted on opposite sides of the frame and they respond to yaw angle of the spacecraft from normal to the sun. The dc voltage outputs of the pairs of sun sensors were applied to conditioning amplifiers that drove valves to open particular nitrogen jets. The accuracy of the sun pointing arrangement was within one degree.

The Earth sensor was used in conjunction with subsystems that controlled roll of the spacecraft and controlled the hinge angle of the high gain antenna to direct the antenna pattern towards Earth. The Earth sensor, which was mounted to the frame of the high-gain antenna, used photodetectors to detect the illuminated Earth against dark space. Earth acquisition was achieved by rotating the spacecraft about its roll (longitudinal) axis with the antenna hinge angle fixed at about 72 degrees until the Earth was detected. The roll was stopped and dc error signals at the roll plane output of the Earth sensor was used to track the Earth in the roll plane. Likewise, error signals in the antenna hinge plane were applied to servo amplifiers that drove a motor to change the hinge angle to center the antenna beam on Earth.

Midcourse Correction Rocket Engine

The trajectory of the spacecraft on its way to Venus was accurately measured by tracking equipment on Earth. These measurements were used to compute the distance of closest approach to Venus. A correction of the trajectory was usually required to refine the trajectory so that the closest approach distance fell within an acceptable range. The correction to the trajectory was made several days after launch by adjusting the velocity vector of the spacecraft by firing the rocket engine. The attitude control system of the spacecraft was used to orient the spacecraft such that the rocket burn would be in the proper direction for the correction. The duration of the burn was controlled to achieve the needed correction.

The rocket engine was mounted in the center of the spacecraft structure and it thrusted along the longitudinal axis of the spacecraft. The nominal thrust of the engine in a vacuum was 226 newtons (59 pounds). The specific impulse of the engine was 231 seconds.

The engine burned anhydrous hydrazine in conjunction with a catalyst. The engine was ignited by injecting 15 cubic centimeters of nitrogen tetroxide oxidizer into the combustion chamber along with the hydrazine. That combination ignited on contact. Once ignited, the rocket engine continued to burn hydrazine in the presence of a catalyst. The catalyst consisted of a bed of 5 mm spherical particles of aluminum oxide impregnated with iron, nickel, and cobalt.

Hydrazine fuel was contained within a bladder in an elliptical shaped tank holding 9.8 kg of fuel. Nitrogen from a spherical tank was applied through a shutoff valve and then ported between the inner walls of the tank and bladder to force the fuel into the combustion chamber. The nitrogen tank was initially pressurized to 22.7 megapascal (3300 psi). The pressure was regulated to 2.1 megapascal (300 psi) before being applied to the fuel tank. The fuel tank was pressurized by nitrogen gas to 1.9 megapascal before launch. The valves were pyrotechnically activated by signals from the central computer and sequencer (CC&S). After the desired velocity increment had been achieved as measured by an accelerometer, the CC&S issued a signal that activated valves to shut off fuel flow to the engine and shut off nitrogen flow to the fuel tank.

In preparation for midcourse correction, the attitude of the spacecraft was set by the nitrogen gas jets to proper pitch and roll angles for the direction of thrust of the engine. When the rocket motor was firing, spacecraft attitude was maintained by deflecting four small vanes in the exhaust of the rocket motor. Any deviation from the set attitude was sensed by the gyros and correcting signals were sent to the vane actuators to tilt the spacecraft to return to the set attitude.

The midcourse correction was performed for the Mariner 2 mission eight days after launch. Based on tracking data from Earth, a necessary velocity change of 31.16 meters per second was computed. This velocity increment would require a burn time of 28.3 seconds. The burn lasted a little longer than commanded and the actual velocity change was determined to be 37.5 meters per second after the burn. The resulting closest approach to Venus was 34,773 km, well within acceptable values.

Telecommunications Subsystem

The telecommunications subsystem accepted commands and data from Earth, provided a coherent return signal that allowed accurate tracking of the spacecraft from Earth, and it transmitted spacecraft performance data and data from scientific instruments back to Earth. Commands and data from Earth were transmitted on a frequency of 890 MHz by one of the 26-meter (85-foot) diameter parabolic antennas of the Deep Space Instrumentation Facility (DSIF). The command information was encoded on a phase shift keyed subcarrier.

There were three DSIF facilities located around the Earth to provide continuous coverage of the spacecraft as the Earth rotated. The locations were at Goldstone, in California, Woomera in Australia, and Harebeesthock in South Africa. The Goldstone facility employed two 26-meter diameter parabolic antennas; one to transmit commands on 890 MHz and one to receive data from the spacecraft on 960 MHz.

Uplink commands – The command signals from Earth were picked up by command antennas on the spacecraft that were located on the solar arrays. There were two command antennas; one was mounted on the front of one of the solar panels and one was mounted on the back. The command antennas had broad angular coverage to accommodate maneuvering of the spacecraft.

The signals at the outputs of the command antennas were combined and applied to a receiver tuned to 890 MHz. The receiver incorporated a phase-lock loop to provide a coherent carrier frequency for retransmission to earth. The coherent arrangement allowed the Doppler shift of the signal received at Earth to be measured and that gave an accurate measurement of the line-of-sight velocity of the spacecraft. The transmitted frequency of 960 MHz was 96/89 times the received frequency of 890 MHz. An output of a phase detector that was part of the phase-lock loop extracted the subcarriers that contained command information from Earth. The subcarriers were demodulated and the resulting signals were applied to a command decoder.

The command decoder decoded the command messages and based on their address sorted them into commands that were to be acted on in real time and those that were to be stored in the CC&S for execution later. There were eleven real time commands plus one spare. The real time commands were prefixed by "RTC" and the stored commands were prefixed by "SC." A tabulation of the commands and their function is given in Table 4.2.

Downlink of data from the spacecraft – The data cycle of information sent back to Earth was made up of frames and subframes containing data words. The engineering data subframe and the scientific experiments data subframe were transmitted sequentially during the cruise mode. Only the experiments data subframe was transmitted during the encounter mode.

A data encoder accepted 51 analog signals from the spacecraft representing engineering data. Temperatures at various portions of the spacecraft were important and several temperature sensors were monitored. The engineering data field was applied to six separate commutator channels referred to as decks. One of the decks had six inputs, two had eight inputs, one had nine inputs, and two had 10 inputs. A master programmer in the data encoder managed the commutation sequence. The outputs of the commutators were sequentially applied to an analog-to-digital converter that generated 7-bit digital words that represented the amplitudes of the input signals.

The data bit rate was 8.33 bits per second and the length of a data word was 0.84 seconds. There were 20 data words in a subframe giving a subframe time of 16.8 seconds. The output of the analog-to-digital converter was applied through a selector switch to a biphase modulator in the transmitted signal path. The selector switch allowed selecting either engineering data or scientific experiments data to be applied to the biphase modulator.

Encoding of data from scientific experiments – Data from the scientific experiments were applied to a Data Conditioning System (DCS) for conditioning and

Table 4.2 Available commands from Earth for Mariner 2

Designation	Command	Action
RTC-1	Roll override	Earth sensor control switched off. Spacecraft will roll until earth sensor detects a bright object
RTC-2	Clockwise hinge override	Changes hinge angle of high-gain antenna. Each command changes hinge angle by 2 degrees.
RTC-3	Counter clockwise hinge override	Same as RC-2 except changes in CCW direction
RTC-4	L-band omnidirectional antenna	Switches transmitter to omnidirectional antenna.
RTC-5	L-band high-gain antenna	Switches transmitter to high-gain antenna.
RTC-6	Initiate mid-course correction	Switches spacecraft attitude to gyro control. Spacecraft turns are initiated according to stored command values. Midcourse motor burn is initiated. After completion of motor burn, spacecraft is reoriented and sun and earth acquisition procedures are initiated.
RTC-7	Command encounter mode	Turns planet science on. Engineering telemetry is turned off. Radiometer scan begins.
RTC-8	Command planet science off	Returns spacecraft to cruise mode. Radiometer is turned off; engineering telemetry is switched into telemetry system. Countermands RTC-10
RTC-9	Command sun acquisition	Unlatches solar panels and removes sun acquisition inhibit
RTC-10	Command cruise science off	Switches science telemetry off; engineering telemetry only transmitted
RTC-11	Spare	
RTC 12	Command Earth acquisition	Removes Earth acquisition inhibit; spacecraft starts roll search. When Earth sensor acquires and nulls an object, telemetry data rate is automatically switched to 8.3 bits per second.
SC-1	Midcourse roll duration	Contains time duration and polarity of mid-course roll maneuver.
SC-2	Midcourse pitch duration	Contains time duration and polarity of mid-course pitch maneuver.
SC-3	Midcourse velocity increment	Contains velocity incremental change required for midcourse motor burn

formatting for transmission to earth by the telemetry system. There were six scientific experiments on the Mariner 2 spacecraft.

The data conditioning system accepted the various inputs from the experiments and conditioned and formatted the data in a form to be telemetered to Earth. The analog voltage data were conditioned, sampled and applied to an analog-to-digital converter. The output of the analog-to–digital converter was a series of 8-bit data words that represented the amplitudes of the input signals. Digital or pulse type signals from experiments applied to the digital conditioning system included magnetometer scale indicator, pulses from the cosmic dust experiment, pulses from the ionization chamber and pulses from the particle detectors.

Data from the experiments was arranged in a master frame divided up into 24 subframes. The duration of the master frame was 483.84 seconds and the duration of the subframes was 20.16 seconds. Each subframe contained 21 8-bit words. The data transmission rate was 8.33 bits per second.

Experiments Carried by Mariner 2

Microwave Radiometer

An important goal of the Mariner 2 spacecraft was to measure the temperature of the surface of Venus during a flyby of the planet. This remote measurement from a substantial flyby distance was accomplished by a two-channel microwave radiometer. A radiometer measures the natural electromagnetic radiation from a surface. Any object or surface above a temperature above absolute zero (zero Kelvin or −273 deg C) radiates with intensity that is a function of the temperature of the body and wavelength of the radiation.

A special state of an object called a black body, which is a perfect radiator and perfect absorber of energy, is the basis for radiometric work. The radiation from an ordinary surface is lower than from a black body by a factor known as the emissivity. The emissivity of bare soil on earth at wavelengths used by the Mariner 2 microwave radiometer is greater than 0.95. Black body characteristics can be used as a first approximation for an unknown body. Black body radiation was investigated by German physicist Max Planck who derived what is now known as Planck's Law in 1900. Planck's law for the spectral radiance per steradian emitted from one square meter of the surface of a black body involves the wavelength of radiation, temperature, velocity of light, Planck's constant and Boltzmann's constant.

Planck's Law can be used to compute radiant energy as a function of temperature for the two wavelengths that were used in the microwave radiometer of Mariner 2. A plot giving radiant energy as a function of temperature for wavelengths of 13.5 mm and 19 mm is shown on the next page (Fig. 4.4). The dimensions of spectral radiance in the plot are watts per square meter per steradian per micron.

Configuration of Microwave Radiometer

The microwave radiometer had the unique ability to probe through the clouds and dense atmosphere of Venus to measure the temperature of the surface of the planet. It was a very important instrument and merited requiring over half of the 18.6 kg total weight allocated to experiments on Mariner 2. The weight of the microwave radiometer with its scanning antenna and signal processing electronics was 10.8 kg. It required 3.5 watts of power when operating and 8.9 watts during calibration.

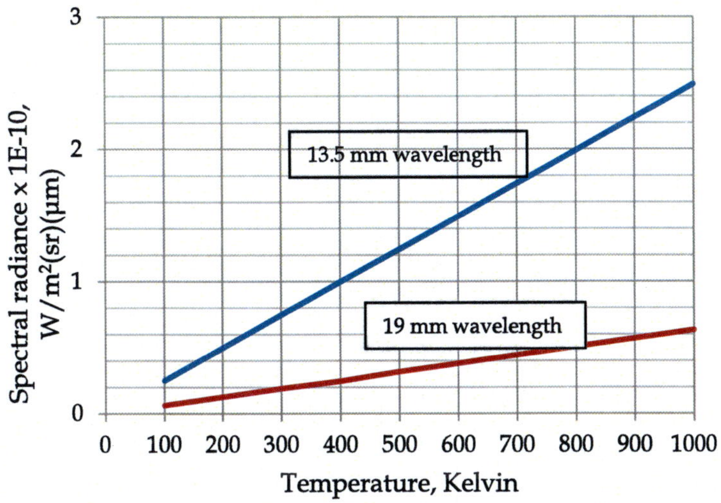

Fig. 4.4 Spectral radiance as a function of temperature for wavelengths of 13.5 cm and 19 cm

The microwave radiometer used a narrow antenna beam to probe through the clouds and dense atmosphere of Venus to the surface. The electromagnetic energy generated by the hot surface was picked up by the antenna and applied to the signal processing portion of the radiometer.

Electromagnetic energy at two wavelengths, 13.5 mm and 19 mm were processed by the radiometer. The corresponding frequencies were 22.22 GHz and 15.79 GHz respectively. These relatively high microwave frequencies allowed narrow antenna beamwidths to be achieved with a modest size antenna. The 13.5 mm wavelength coincides with a water vapor absorption band. Comparing the results with that of the 19 mm radiometer channel, which was not affected by water vapor, would give an indication of the presence of water in the clouds or in the atmosphere.

The antenna was a parabolic dish type 48.5 cm (19.1 inches) in diameter. The beamwidth was 2.2 degrees at 13.5 mm wavelength and 2.5 degrees at 19 mm wavelength. A diplexer and filters separated the 13.5 mm signals from the 19 mm signals and fed each to a separate radiometer channel. Small steps were machined in the parabolic surface to prevent concentration of infrared energy (heat) onto the feed. The small steps did not affect focusing of the much longer microwave wavelength signals onto the feed.

The antenna was arranged to mechanically scan in a plane perpendicular to the longitudinal axis of the spacecraft. The extent of the angular scan was ±60° during search for the planet at the onset of the encounter. The extremes of the scan would place the antenna beam off of the planet to establish baseline data. The scan rate during search was one degree per second. Once the planet was sensed by the radiometer output rising above a threshold, the data mode was entered and the scan rate was reduced to 0.1 degrees per second. The radiometer output was monitored during the data mode and if six consecutive readings were below a set threshold, it was assumed that the antenna beam had passed by the limb of the planet and the antenna scan direction was reversed.

The radiometer was calibrated by inserting signals from a noise tube of known and stable signal level into directional couplers in the two feed lines from the antenna. The calibration signals were at levels corresponding to 1500 K for the 19 mm channel and 800 K for the 13.5 mm channel. Calibration was performed periodically during cruise to Venus and before encounter with the planet.

The radiometer was a Dicke type that used a microwave switch to switch between the signal from the antenna and a reference source. The switching frequency was 950 Hz for the 19 mm channel and 1050 Hz for the 13.5 mm channel. The switches were driven by tuning fork type oscillators. The reference source was a signal from a horn antenna that viewed deep space at essentially zero Kelvin. Separate horns were used for the two channels. The beamwidth of each horn was about 12 degrees. The horns were mounted at the top of the parabolic antenna and they scanned with the antenna. The horns were oriented with the center of their beams about 65 degrees from the center of the beam of the parabolic antenna so the reference horns viewed deep space while the radiometer was operating.

The resulting bi-level level microwave signal at the output of the microwave switch was applied to a detector. The square-wave signal at the output of the detector was amplified and applied to a synchronous detector along with the switching reference signal. The resulting dc voltage at the output of the synchronous detector was integrated with a time constant of 40 seconds. At a scan rate of 0.1 degrees per second in the data mode, the integration time corresponded to about four degrees of scan on the surface. The smoothed dc signal was amplified to become the output signal of the radiometer. The output dc voltage was about 1.0 volts for a source at zero Kelvin and 6.0 volts for a source at 800 Kelvin.

During Mariner 2's encounter with Venus, the scan protocol became disrupted because of a dc level shift of the output of the 19 mm radiometer channel. The baseline had increased to over 1.5 volts which was the voltage level to begin slow antenna scan. The high baseline defeated the sensing level to reverse the direction of scan once the antenna beam passed the limb of the planet. The scan began at the 0.1 degree per second rate and it did not reverse until the end of the mechanical scan range. As a result, one radiometer scan was made on the dark side of the planet, one scan was made near the sunlit terminator, and one scan was made on the sunlit side as the spacecraft passed by the planet.

The altitudes of the spacecraft at mid-scan during the three scans were: 40,200 km for the dark side scan, 37,750 km for the terminator scan, and 35,850 km for the sunlit side scan. The peak temperatures of the surface as determined from analysis of the telemetered data from the radiometer experiment and reported in a paper by Barath et al. (1964), are shown in the table below (Table 4.3). Significant findings were that the surface of Venus is extremely hot and there is not significant difference between the temperature of the sunlit side and dark side.

Table 4.3 Temperatures of surface of Venus derived from microwave radiometer data

Channel	Scan	Location	Peak temperature, K
19 mm	1	Dark side	480
	2	Terminator	590
	3	Sunlit side	460
13.5 mm	1	Dark side	393
	2	Terminator	400
	3	Sunlit side	396

Infrared Radiometer

Mariner 2 carried an infrared radiometer to complement the microwave radiometer discussed in the previous section. It was designed to measure temperature in the range of 200 Kelvin to 600 Kelvin. The infrared radiometer was mounted to the frame of the microwave radiometer antenna such that its optical beam scanned the same target region as the microwave beam. The infrared radiometer was a compact instrument 13 × 14 × 10 cm (5.1 × 5.5 × 3.9 inches) in size and weighed 1.2 kg (2.7 pounds). The infrared instrument had two radiometer channels; one operating at a wavelength of 8.4 μm and the other operating at 10.4 μm. A wavelength of 10.4 μm was selected because it is in an absorption band of CO_2. The other wavelength selected, 8.4 μm, is outside of the high absorption CO_2 band. A comparison of the temperature measured by the two channels would give an indication of carbon dioxide in the paths of the optical beams. Should there be a break in the clouds so that the lower atmosphere or even the surface could be viewed, the temperature readings in the two channels would be substantially different.

The principle of operation of the infrared radiometer was similar to that discussed for the microwave radiometer. The radiometer measured spectral radiance in a target optical beam and a reference optical beam. The reference beam normally viewed deep space at essentially zero Kelvin. The two beams were formed by optical lenses and both had a field of view of 0.9 by 0.9 degrees. The reference beam was directed 45 degrees from the target beam. The mechanism of the radiometer chopped the signal between the target beam and the reference beam using a rotating wheel. The chopping rate was 20 Hertz.

The radiometer contained two thermistor bolometer type detectors which were sensitive to infrared energy. The two detectors were mounted with their sensitive area displaced by 90 degrees. A dichroic mirror was set at a 45-degree angle to the detectors. The mirror allowed 90% of energy with wavelengths longer than 9.5 μm to be passed through to one detector while 90% of the energy with wavelength shorter than 9.5 μm was reflected into the other detector. Filters before the detectors established optical bandwidths of 8.1–8.7 μm for the 8.4 μm channel and 10.2–10.5 μm for the 10.4 μm channel.

The spectral radiance at wavelengths of 8.4 μm and 10.4 μm has a wide dynamic range for temperatures between 200 K and 600 K. For example, the difference in spectral radiance of a black body between 200 K and 600 K at 8.4 μm is a factor of about 300. Amplifiers with a logarithmic response were used following the detectors to accommodate the large dynamic range and still provide acceptable resolution at the output of the radiometer.

The output of the logarithmic amplifier in each channel was applied to a synchronous detector that was excited by a signal at a 20 Hz chopping frequency derived from the rotary optical switch. The resulting dc voltage at the output of the synchronous detector was filtered and amplified to become the output signal of the radiometer.

In-flight calibration of the infrared radiometer was performed during the scanning process. At the extremes of the scan the reference beam passed over a calibration disk whose temperature was measured and telemetered to earth. The target beam viewed deep space during the calibration interval. The radiometer chopped between the target beam at zero Kelvin and the reference beam viewing a plate with known temperature during calibration.

During encounter with Venus, the intended angular scan protocol was to scan a ±60-degree region at a rate of one degree per second until the planet was detected by the microwave radiometer and then reduce the scan rate to 0.1 degrees per second. However, because of the problem with the threshold of the microwave radiometer discussed in the last section, the scan began at a 0.1 degree per second rate and it traveled from mechanical stop to mechanical stop of the scan mechanism.

As a result, one infrared radiometer scan was made on the dark side of the planet, one scan was made near the sunlit terminator, and one scan was made on the sunlit side as the spacecraft passed by the planet. Several measurements of temperature were made during each scan. All of the measurements were near the same value. A summary of results for the 8.4 μm and 10.4 μm channels are shown in the table below (Table 4.4). The readings represent temperature of the clouds.

There was not significant difference between the temperatures measured by the 8.4 μm channel and the 10.4 μm channel. This implied that both channels probed the top of the clouds.

Magnetometer

The purpose of the magnetometer was to measure the strength and direction of the magnetic field in in the vicinity of Venus and measure the magnetic field in interplanetary space during cruise to Venus. The instrument was a three-axis fluxgate type. Measurement of the field strength along the three mutually perpendicular axes allowed determination of the direction of the field. The sensitivity of the

Table 4.4 Temperature measurements by Mariner 2 infrared radiometer

Scan location	Number of measurements	Ave. temp. 8.4 μm	Ave. temp. 10.4 μm
Dark side	5	231 K	224 K
Terminator	7	233 K	237 K
Sunlit side	5	228 K	234 K

magnetometer was about 0.5 gamma where one gamma is equal to one nanotesla. Tesla is the present SI unit for magnetic field. For reference, the strength of Earth's magnetic field is 25,000–65,000 nanotesla.

The magnetometer in Marnier 2 used three magnetic cores aligned along three mutually perpendicular axes. The amplifiers for the dc analog data outputs of each of the three channels automatically switched between a scale of ±64 gamma and ±320 gamma depending on the strength of the field encountered. The output voltage of each of the three channels was in the range of 0–6 volts dc. A voltage of 3.5 volts corresponded to zero external field strength. A voltage less than 3.5 volts indicated that the direction of the magnetic field was negative.

The magnetometer was mounted just below the omni antenna to position it away from the main portion of the spacecraft. Its location can be seen in Fig. 4.3 given previously. The cores and electronics were contained in a metallic cylinder six inches long and three inches in diameter. The magnetometer weighed 4.7 pounds and consumed six watts of power.

The magnetometer responded to several magnetic storms during travel of Marnier 2 to Venus. The largest of such storms are associated with solar coronal mass ejections where very large quantities of plasma with embedded magnetic field are ejected from the sun. Measurements by the magnetometer indicated magnetic field levels ranged from about two gamma during quiet times to five gamma during small magnetic disturbances to over 20 gamma during strong magnetic storms.

Significantly, no magnetic field associated with Venus was detected as the spacecraft passed by the planet. The closest approach to Venus was 34,773 km and the results were qualified to state that there was no measurable magnetic field at that distance from the planet.

High Energy Radiation Experiment

The high energy radiation experiment used an ionization chamber and three Giger-Mueller tubes to measure the number and intensity of energetic particles in interplanetary space and in the vicinity of Venus. The ionization chamber was a unique instrument developed by Dr. Neher of the California Institute of Technology. The particles of interest in the experiment were cosmic rays that included protons, alpha particles, and electrons.

Ionization chamber – The ionization chamber consisted of a five-inch diameter stainless steel sphere filled with argon gas. A slender quartz rod with a conductive coating protruded into the center of the sphere. A quartz fiber, supported at both ends, was suspended very close to the rod. High energy particles would penetrate the walls of the ionization chamber and leave an ionized trail of argon as they passed through the gas. Negative ions accumulated on the quartz rod and imparted a negative static charge to the rod. The suspended quartz fiber was attracted to the charge on the rod and when the static charge built up sufficiently, the fiber contacted the rod. Both the rod and fiber had a conductive coating and upon contact the charge on the rod discharged through the conductive fiber and through a load resistor to form a signal pulse. The fiber returned to its original position and the process repeated. The signal pulse rate was indicative of the flux of particles.

Penetration of the walls of the chamber by cosmic particles required energy greater than 10 million electron volts (MeV) for protons, greater than 40 MeV for alpha particles, and greater than 0.5 MeV for electrons.

During the flight to Venus, the average time between signal pulses was about 500 seconds. The interval between pulses was nearly constant during most of the voyage, changing only about 10%. The interval decreased to 10 seconds during a massive solar flare in October 1962 and then slowly returned to the normal 500-second interval. The instrument did not find any evidence of a trapped radiation belt around Venus.

The ionization chamber was mounted on the pyramidal tubular aluminum truss structure shown in the sketch of the spacecraft given previously in Fig. 4.3.

Geiger-Mueller tubes – Three Geiger-Mueller (GM) tubes provided complementary data to the ion chamber on cosmic particles. Similar GM tubes had been successfully used to explore radiation belts around the Earth by Explorer and Pioneer spacecraft between 1958 and 1960. The collection of three GM tubes was referred to as "particle detectors" in Mariner 2 documentation.

Geiger-Mueller tubes are typically made up of a metallic cylindrical chamber with a wire suspended inside along the axis of the cylinder. The chamber is filled with an inert gas such as neon at less than atmospheric pressure. A small amount of halogen gas is mixed with the inert gas to prevent continuous ionization once started by an energetic particle. The wire is connected to a positive potential of several hundred volts with respect to the walls of the chamber.

When an energetic particle penetrates the walls or end cap of the tube, it creates a string of ion pairs along its path through the neon gas. The electrons of the ion pairs are attracted to the positive potential of the wire and their velocity becomes sufficient to create additional ion pairs due to the collisions and eventually an ionization avalanche occurs. The positive ions are attracted to the case and in their travel collisions with the gas create additional ionization. The ionization process is quenched by the halogen gas. A negative going pulse is generated across a load resistor in series with the positive voltage to the wire at the time of the avalanche. The pulse rate is dependent on the flux of energetic particles.

Two of the GM tubes responded to energetic particle at about the same sensitivity as the ionization chamber. One of the GM tubes was shielded by a sleeve of stainless steel 0.2 mm (0.008 inches) thick. The second GM tube was shielded by a sleeve of beryllium 0.6 mm (0.024 inches) thick. The difference in shields allowed scientists to determine the ratio of electrons to other particles.

The third GM tube was an Anton 213 type with an end-window of mica. The mica window admitted electrons with efficiency of 100 percent for electrons with energy greater than 70 keV and efficiency of 35 percent for electrons with energy of 40 keV. The window admitted protons with energies greater than 500 keV at efficiency of 100 percent. The device had a conical field of view of 90 degrees centered on the axis of the tube at the window end. The field of view was pointed away from

Venus during the encounter. The Anton 213 GM tube was developed by Anton Electronic Laboratories.

The principal investigator for the end-window GM tube experiment was James Van Allen of the State University of Iowa. Van Allen had investigated radiation in space with similar instruments on the Explorer and Pioneer spacecraft. The magnetically trapped radiation belts around the earth, now called Van Allen Radiation Belts, were discovered by Van Allen's investigations.

The end-window GM tube was the most sensitive of the three GM tubes. The number of counts in a 9.6 second interval was read out each 887 seconds during cruise and every 484 seconds during encounter with Venus. The average count rate was 0.6 counts per second during most of the cruise portion of flight except for periods of intense solar flares. The count rate had increased to an average of 1.1 counts per second during the encounter mode on 14 December 1962. Most significantly, there was no discernable change in the count rate during the passage of Mariner 2 past Venus.

The three GM tubes and associated electronics were contained in a box 15.2 × 15.2 × 5.1 cm in size. The overall instrument weighed 0.84 kg and consumed 0.3 watts of power. The instrument was mounted on the pyramidal tubular aluminum truss structure just above the ionization chamber. The location is shown in the sketch of the spacecraft given previously in Fig. 4.3.

Solar Plasma Analyzer

The solar plasma analyzer was designed to gather data on the flux and energy spectrum of positive particles of the plasma emitted by the sun (solar wind). The main components of the solar wind are electrons and protons along with a few alpha particles and other ions. The analyzer operated primarily on protons and spectrums of energies of protons were developed by the analyzer. Since the mass of protons is known, the velocity of protons could be computed from the energy measurements.

The solar plasma analyzer was mounted with its ten-degree field-of-view aligned with the longitudinal axis of the spacecraft. As a result, the field of view was pointed directly at the sun during most of the flight. The instrument used two parallel curved deflection plates as a gauntlet to measure energy of particles. A specific voltage potential between the plates caused a particle with specific energy to curve and follow the curve of the plates. Those particles were captured at the output of the curved plates by a charge collector. Particles with energy less than or greater than the specific amount struck the curved deflection plates and were not counted. The voltage applied to the plates was changed in ten steps to cover a range of particle energies from 240 electron volts to 8400 electron volts.

Positive ions that accumulated in the charge collector were processed by an electrometer amplifier with a logarithmic response. The logarithmic response was needed to accommodate signal currents from the charge collector that could differ by several orders of magnitude. The output of the logarithmic amplifier was a dc voltage that was applied to the data conditioning system that readied data for transmission to Earth. The voltage at the output of the amplifier represented the rate at which positive ions were collected by the instrument for a given differential voltage of the deflection plates.

Most of the particles measured by the analyzer during the flight of Mariner 2 had energies in the range of 750 electron volts to 2500 electron volts. The corresponding velocities were 380 km/sec to 693 km/sec. The data showed peak energies periodically every 27 days corresponding to periods of increased activity of the sun.

Scaling from drawings, the size of the analyzer was about 10.6 × 8.5 × 4.2 inches (27 × 21.6 × 10.7 cm). It weighed 4.8 pounds (2.18 kg) and consumed about one watt of power.

Cosmic Dust Sensor

The purpose of the cosmic dust collector was to measure the flux of interplanetary dust during cruise to Venus and during encounter with the planet. The sensor consisted of a rectangular plate 12.7 × 25.4 cm in size with a crystal microphone mounted at its center. The plate was mounted on top of the hexagonal structure with the short edge in the plane of the longitudinal axis and the wide edge in the plane of the long dimension of the solar arrays. The location of the sensor can be seen in the sketch of the spacecraft given previously in Fig. 4.3.

A particle striking the plate would cause the microphone to respond to the impact and generate a transient voltage pulse. The output of the microphone was fed to two amplifiers, one having higher sensitivity than the other. The amplifier outputs fed separate counters. The counters were read out and reset every 37 seconds during the cruise mode and every 20 seconds during the encounter mode.

Measurements indicated that the flux of particles during cruise was at least a factor of 1000 less than the flux measured by satellites of the Earth. The instrument failed shortly before encounter with Venus and no data on cosmic dust was obtained during the encounter.

Flight of Mariner 1 and Mariner 2 Spacecraft

The flight designations for the two Mariner R spacecraft were Mariner 1 and Mariner 2. The two spacecraft were identical and both were planned to be launched during the 1962 launch window to Venus.

Flight of Mariner 1

Mariner 1 was launched by an Atlas/Agena-B launch vehicle on 22 July 1962. A photograph of the launch is shown on the next page (Fig. 4.5). The launch looked good initially but the spacecraft soon veered off course. Out of caution, the range

Fig. 4.5 Launch of Mariner 1 spacecraft. (NASA photograph)

safety officer transmitted a destruct signal to the spacecraft 293 seconds after launch. The launch vehicle and spacecraft were destroyed. The vehicle was doomed by two faults. One fault was in the Atlas launch vehicle and the other was in the software code in the ground station.

Two interrelated systems were used for guidance of the Atlas launch vehicle. One of the guidance systems was a self-contained autopilot in the vehicle that used a flight programmer, gyro package, and servo based hydraulic actuators to control the spacecraft. This was basically an open loop guidance system that followed pre-programmed events.

The second guidance system used a Mod III Radio Guidance System built by the General Electric Company. A portion of the Mod III system was a rate subsystem that used a beacon in the Atlas to receive two continuous wave signals at different frequencies transmitted from the ground. The two frequencies of the rate subsystem were in L-band. The beacon generated a signal at a frequency that was the average of the two received signals and transmitted that signal to the ground. There were three receiving stations on the ground at different locations. Doppler information and phase relationship of the three received signals allowed determination of vehicle velocity, range, and azimuth.

The Mod III system also contained a track subsystem that used an X-band monopulse antenna on the ground to transmit a coded message to a pulse beacon in the Atlas. The pulse beacon transmitted a return pulse to the ground that was received and used to measure range to the Atlas. The monopulse antenna provided azimuth and elevation data. The coded message sent from the ground included steering and command instructions for the vehicle.

Position and rate information from the track and rate subsystems were applied to data processing and computer facilities on the ground. The rate subsystem provided the most accurate velocity information. The computer on the ground determined if corrections were necessary to the vehicle path. Any such corrections were encoded as steering commands and transmitted to the Atlas where they were decoded to control gimballing of engine thrust.

Operation of the rate beacon on Atlas was intermittent during the launch of Mariner 1. There were four periods of outages ranging from 1.5 seconds to 61 seconds. As a result, there were periods of incorrect data fed to the ground-based computer. The computer code that processed the rate data on the ground intended to smooth the data over "n" readings before acting on it. This operation was to be indicated by an overbar over the function in the code. An overbar is often used in mathematics to indicate arithmetic mean. Unfortunately, in the process of entering the code into the computer, the overbar was omitted. As a result, during the launch of Mariner 1 the computer generated correction commands for the vehicle based on noisy rate data. The vehicle responded to the noisy commands and the trajectory went far enough off course that the vehicle had to be destroyed by the range safety officer.

The New York Times headlined an article on 27 July 1962, "For Want of Hyphen, Venus Rocket Lost." Others picked up the catchy phrase and there are several references to "omission of a hyphen" in writings about Mariner 1. The actual omission in the computer code was the overbar.

Flight of Mariner 2

Mariner 2 was launched by an Atlas/Agena-B launch vehicle on 27 August 1962. The two booster engines and the sustainer engine developed a total of 166,364 kg of thrust. In addition, the two vernier engines were fired to provide an additional 909 kg of thrust for a total of 167,270 kg of thrust. The thrust easily lofted the 127,547 kg total weight of the Mariner 2 launch vehicle and spacecraft off of the ground.

The Atlas performed normally until a few seconds before Booster Engine Cut Off (BECO). At that time one of the vernier engines deflected hard over. Gimballing of the booster engines countered the induced roll. After BECO, the booster engines were jettisoned and booster gimballing no longer countered the induced roll. The other vernier engine was not able to stop the roll and the roll rate reached about one revolution per second. Fortunately, after about one minute, the problem with the errant vernier engine cleared itself and roll stability of the vehicle was regained.

The sustainer engine of Atlas was shut down by a command from the ground (SECO) when a target velocity of the vehicle was reached. Separation of the Atlas from Agena occurred about 300 seconds after launch. The first firing of Agena began 348 seconds after launch and the burn lasted for 147 seconds. The firing of Agena placed the vehicle in a parking orbit 185 km (115 miles) above the Earth. After coasting in the parking orbit for about 16 minutes, Agena was fired again for 97 seconds that inserted the spacecraft on a trajectory to Venus. Insertion occurred 26 minutes, 6 seconds after launch. Mariner 2 was separated from Agena 157 seconds after insertion into a trajectory to Venus.

Several events were managed by the central computer and sequencer (CC&S) in the spacecraft during the flight to Venus. A command to deploy the solar panels was generated by the CC&S 44 minutes after launch. At 60 minutes after launch the CC&S generated a power-up command to the spacecraft attitude control functions. At 116½ hours after launch the CC&S issued a command to begin the Earth acquisition process that trained the high-gain antenna in the direction of Earth. The cruise science experiments were turned on 57.3 hours after launch after normal operation of the spacecraft had been confirmed. This was initiated by sending up the real-time command, RTC-8, to the CC&S.

A midcourse correction had been planned to take up dispersion of the trajectory due to tolerances in the insertion by Agena into the trajectory to Venus. Tracking of Mariner 2 for several days by the Deep Space Instrumentation Facility (DSIF) provided accurate trajectory data from which the flyby distance from Venus was computed.

The DSIF included a series of spacecraft tracking systems located at various points throughout the world so that there would always be at least one tracking system in view of the spacecraft as the Earth rotated. There were three such stations at the time of the flight of Mariner 2. The locations of the three stations were Goldstone Tracking Station in the Mojave Desert in California; Woomera Tracking Station at Island Lagoon, Australia; and Johannesburg tracking Station at Johannesburg, South Africa. In addition, there was a mobile tracking station at Cape Canaveral and a mobile station at Johannesburg.

The main tracking stations were equipped with a 26-meter (85-foot) diameter parabolic tracking antenna and associated transmitter and receiver equipment. The gain of the big antenna was 43.5 dB (factor of 23,387 over an isotropic radiator) at the transmit frequency of Mariner R. The Goldstone and Johannesburg stations had the ability to send commands to the spacecraft. Their transmitter powers could be selected to be either 200 watts or 10,000 watts. The Woomera station had transmitter power of 50 watts. All of the stations had the ability to receive telemetry data from the spacecraft. Each of the stations measured azimuth and elevation data to the spacecraft as well as Doppler shift of the signal received from the spacecraft.

The signal transmitted by the spacecraft was coherent with the signal transmitted from the ground to allow the Doppler shift to be measured. Doppler data provided a very accurate measurement of range rate to the spacecraft and that allowed accurate determination of its trajectory. Tracking data taken during the first seven days of flight indicated that the closest approach distance to the center of the planet without the midcourse correction would be about 386,232 km. The nominal Mariner II aiming point corresponded to a flyby distance above the surface of 13,948 km.

A midcourse correction was initiated about 208 hours (8.9 days) after launch. The spacecraft was about 2.4 million km from Earth at the time. Commands were transmitted from Earth to cause the spacecraft to perform a roll maneuver (SC-1 command) and a pitch maneuver (SC-2 command) to orient the spacecraft so that the rocket engine thrust was in the proper direction to achieve the computed correction. The SC-1 and SC-2 commands were in the form of time. The angular rates were predictable and so a given time of firing the maneuvering engine resulted in a given angular change. A third command, SC-3, defined the velocity increment to be achieved by the midcourse correction. The three commands, SC-1, SC-2, and SC-3 were stored in the central computer and sequencer (CC&S) until time for execution of the midcourse correction. The parameters preloaded were: 51 seconds of minus roll, 795 seconds of minus pitch, and 1013 accelerometer pulses before the rocket engine was shut down. The 1013 accelerometer pulses corresponded to a velocity increment of 31 meters per second.

Two additional commands sent from Earth were the real time commands RTC-4 and RTC-6. RTC-4 commanded switch the transmitter output from the directional antenna to the omnidirectional antenna so that telemetry data could be received during the maneuvers of the spacecraft. The RT-6 command initiated the orientation maneuvers. The sun sensor was turned off before the spacecraft was maneuvered to the orientation needed for midcourse correction. The rocket engine was ignited and burned for about 28 seconds to adjust the trajectory.

Trajectory measurements after the correction predicted a closest approach radius from the center of Venus of 41,000 km. The equivalent predicted flyby distance above the surface was 34,948 km. The actual closest approach to the surface of Venus was 34,854 km during the flyby on 14 December 1962. While the miss distance was larger than the distance of 13,948 km planned, it was well within that required for the mission.

The sun was reacquired and the Earth was reacquired after midcourse correction and the spacecraft resumed its long cruise to Venus with the solar arrays pointed at the sun and the high-gain antenna pointed at Earth.

On 31 October 1962 power was lost from one of the solar arrays and real-time command RTC-10 was transmitted to the spacecraft to turn off the cruise science instruments. The problem with the solar array seemed to clear itself eight days later and the science instruments were turned back on by command RTC-8. The solar array again failed on 15 November but the spacecraft was close enough to the sun by that time that one solar array could carry the electrical load including that of the science instruments.

Mariner 2 initially followed Earth around the sun but then curved inward towards Venus and its speed increased. The spacecraft approached from outside of the orbit of Venus. It passed from behind and above the equator of the planet on the dark side to below and ahead of the planet on the sunlit side. The spacecraft velocity relative to Venus was 6743 km/sec at the time of closest approach on 14 December 1962. The gravitational pull of Venus deflected the path of Mariner 2 by about 40 degrees as it passed by the planet.

A possible problem with the CC&S was detected on 12 December so a real-time command, RTC-7, was transmitted from the goldstone tracking station to the spacecraft early on 14 December to switch from the cruise mode to the encounter mode. Only encounter mode science instruments were telemetered to Earth during the encounter mode. All of the science experiments operated successfully during the encounter. Highlights of the data received were given in the section on "Experiments carried by Mariner 2" earlier in this chapter.

The radiometer scan first impinged on the planet about 65 minutes before the time of closest approach. The distance from the Center of Venus was about 47,000 km at that time. The scan rate of the radiometer antenna in conjunction with the speed of the spacecraft resulted in one radiometer scan on the dark side of the planet, one scan near the sunlit terminator, and one scan on the sunlit side as the spacecraft passed by Venus. The time from the first time that the radiometer scan impinged on the planet until the last surface contact was about 35 minutes.

The encounter mode lasted about seven hours. The switch from encounter mode to the cruise mode was initiated by transmission of real-time command RTC-8 from the Goldstone tracking station. The last communication with the spacecraft occurred on 3 January 1963. At that time Mariner 2 was 86,677 million km from Earth and 8994 million km from Venus.

Mariner 2 was the first spacecraft explore Venus. The mission of Mariner 2 was a success. Significant achievements included measurement of the temperature of the surface of Venus for the first time, measurement of the temperature of the tops of the clouds, and finding that Venus does not have a measurable magnetic field.

Mariner 5 Spacecraft

Background of Mariner 5

Mariner 5 was the third spacecraft launched by the United States to gather information about Venus during a flyby. The first, Mariner 1, suffered a launch failure. The second, Mariner 2, was successful and returned important new information about Venus. Mariner 5 was readied for the 1967 launch window to Venus.

Mariner 5 had its beginning with the Mariner Mars 1964 program. Mariner Mars 1964 was the first ever attempt to reach Mars so the planners were conservative and authorized three spacecraft to be built. The first spacecraft, Mariner 3, failed when the launch shroud failed to jettison. The second spacecraft, Mariner 4, made a successful flyby of Mars in July 1965 and returned the first close-up photographs of the surface of Mars. The third spacecraft, which had been a spare in case Mariner 4 was not successful, was repurposed for a flyby of Venus. The Venus mission was approved by NASA in December 1965.

Funding and time constraints limited modifications to the Mars spacecraft for the Venus mission. A planned launch date of June 1967 gave about 18 months from start of modifications to launch. A significant modification was removal of the digital camera system used to image the surface of Mars. The clouds around Venus would not allow imaging the surface and the camera would have had to be changed to image in the ultraviolet to return useful information about the clouds. The limited amount of information to be learned did not warrant the complexity of the camera subsystem on the spacecraft. Three experiments from the Mars mission were retained. They were: solar plasma probe, magnetometer, and trapped radiation detector. Three new experiments were added: ultraviolet photometer, S-band occultation experiment, and dual frequency occultation experiment.

Mechanical Configuration of Mariner 5

A photograph of a model of the Mariner 5 spacecraft is given on the next page (Fig. 4.6). The spacecraft cruised to Venus with the bottom of the spacecraft in the picture facing the sun.

The spacecraft was built on a hexagonal magnesium frame 127 cm across the diagonals and 46 cm deep. The high-gain parabolic antenna used to communicate with Earth was mounted on top of the frame. A mast 223 cm long that was mounted to the top of the frame held a low gain antenna and a magnetometer. The overall height of the spacecraft was 289 cm. The exit port of the rocket engine used for midcourse correction protrudes from the side of the frame facing the reader in Fig. 4.6.

The octagonal main structure of the spacecraft was constructed from magnesium. Magnesium is a nonmagnetic material and lighter in weight and has a higher tensile strength than aluminum. The main structure was built using upper and lower octagonal rings 1.27 meters across the outer flat diagonals, about 11.7 cm wide, and 1.9 cm thick. Longerons 42 cm long and about 11.7 cm wide were placed at the corners of the octagons between the upper and lower rings. The resulting structure formed eight bays, seven of which held electronics for spacecraft control and for experiments and the eighth held the rocket engine for midcourse correction. The same basic structure was used for many follow-on spacecraft.

Most of the electronic circuits were contained on printed circuit boards that fit within an electronic module 15.2 cm square and 3.8 cm high. The electronic

Fig. 4.6 Photograph of Mariner 5 spacecraft. (NASA/JPL science photo library)

subassemblies were mounted in an electronics chassis that consisted of a flat plate about 51 cm long and 41.9 cm wide with two sets of mounting structure for electronic subassemblies. Each set held 12 electronic subassemblies. The connector end of the electronics subassemblies protruded beyond the sides of the bays in towards the center of the structure. The flat plate of the electronics chassis served as the outer panel of a segment of the octagonal structure. Contents of the bays were as follows:

Bay No.	Electronics assemblies
1	Power and pyrotechnic control
3	Scientific equipment
4	Data encoder and command
5	Communications and tape recorder
6	RF communications
7	Attitude control and CC&S
8	Power and battery

The propulsion rocket engine used during midcourse correction was contained in Bay 2 of the octagonal structure. Fuel for the engine was contained in a spherical tank located in the open center area of the octagon. Fuel was forced out of the tank by nitrogen gas that was contained in a pressurized spherical tank also located in the center area.

A truss type support structure holding the high-gain antenna was made up of aluminum tubing. The truss, which was 43 cm high, was attached to four inner corners of the octagonal frame and terminated in a tubular ring at the upper end. Support structure for the antenna attached to the ring.

The solar arrays were mounted to frames of the same design as for the Mariner 4 spacecraft so that the positions of the nitrogen gas jets used to maneuver the spacecraft could be retained. The jets were located at the outboard edge of the solar arrays. Since Venus is closer to the sun than Mars, the active area of the solar arrays could be less than for the Mars mission. The active portion of the solar arrays occupied the outer portion of the array support frames. The frames holding the solar arrays were mounted on ball joints at the inner corners of the octagonal structure allowing the arrays to be raised up along the longitudinal axis of the spacecraft to fit within the launch shroud. The frames were spring loaded to the deployed position. Pins were pulled by pyrotechnics to release the arrays to their deployed position once the spacecraft was established on a trajectory to Venus. The upper and lower areas of the octagon structure were covered by a thermal shield blanket. The blanket was composed of a 0.0254 mm (0.001 inch) thick sheet of Teflon aluminized on one side and several layers of aluminized Mylar material.

A deployable sun shade octagonal in shape was located on the sun facing side of the octagonal structure to reduce the heat load on the spacecraft from the sun. The shade was made up of 0.0254 mm thick aluminized Teflon with a tubular support ring sewed to the inner perimeter of the shade. Four support tubes were sewed in the outer perimeter of the shade. Lanyards attached to the support rings were connected to the solar array structure such that when the solar arrays deployed the sun shade also deployed.

The outer surfaces of six of the eight octagonal sides of the spacecraft were covered with louvers for temperature control. The louvers were operated by bimetal actuators that automatically opened the louvers as needed to allow heat to radiate into space. The louvers are apparent in of the spacecraft in Fig. 4.6.

The weight of the spacecraft at launch was 244.9 kg. The experiments made up 22.6 kg of the total weight. A photograph of the spacecraft being weighed at Cape Canaveral is shown on the next page (Fig. 4.7). The workers in the photograph give an indication of the size of the spacecraft.

Mariner 5 Systems

Top and bottom views of the spacecraft with elements identified by labels are given on the following page (Figs. 4.8 and 4.9).

Midcourse Correction Rocket Engine

The trajectory of the spacecraft was accurately measured by tracking equipment on Earth. These measurements were used to compute the distance of closest approach to Venus. A correction of the trajectory was usually required to refine the trajectory so that the closest approach distance fell within an acceptable range. The correction

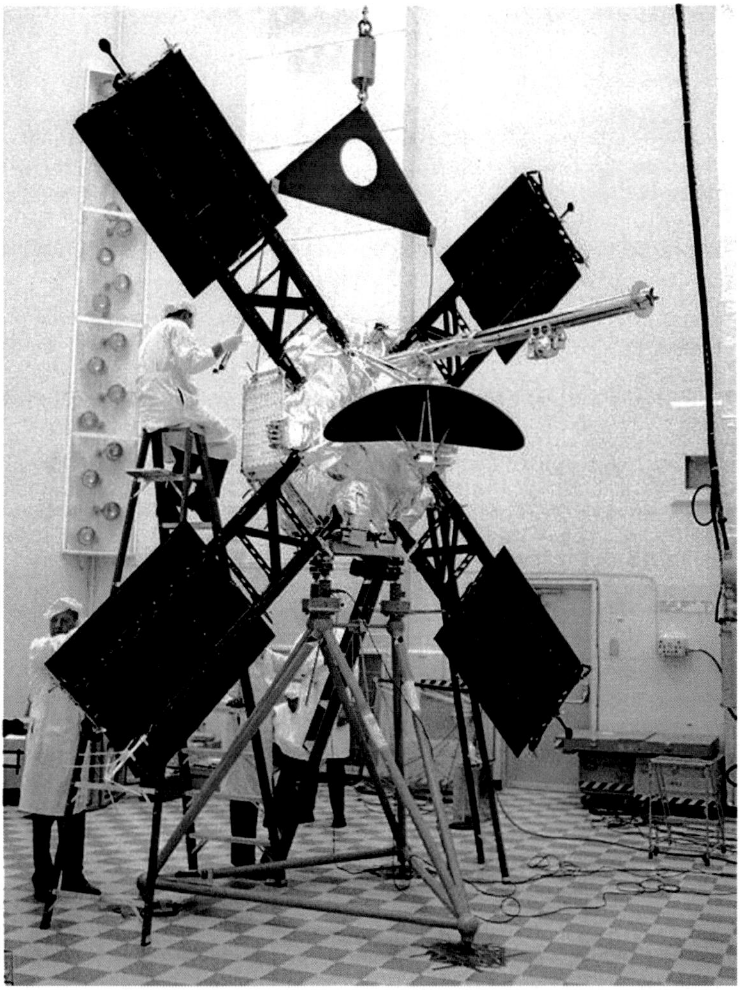

Fig. 4.7 Mariner 5 spacecraft being weighed at Cape Canaveral. (NASA photograph)

to the trajectory of Mariner 5 was made five days after launch by adjusting the velocity vector of the spacecraft by firing the midcourse correction rocket engine. The attitude control system of the spacecraft was used to orient the spacecraft such that the rocket burn would be in the proper direction for the correction. The amount of correction was controlled by the duration of the burn.

The rocket engine was mounted in Bay 2 of the spacecraft structure and it thrusted along a plane approximately 45 degrees from the +X and +Y axes and approximately 90 degrees from the Z axis of the spacecraft. The thrust axis was arranged to pass approximately through the center of gravity of the spacecraft and so the thrust was slightly out of plane of the spacecraft coordinate system.

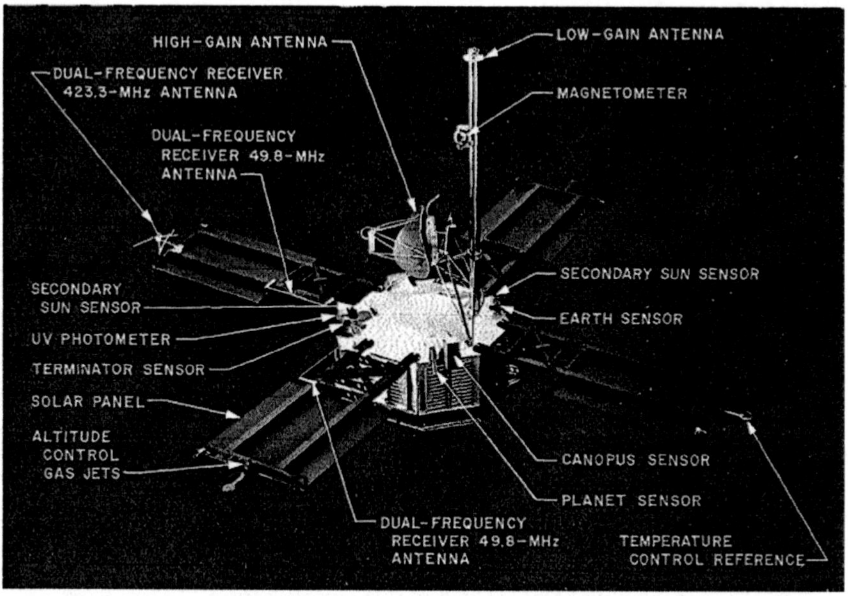

Fig. 4.8 Top view of Mariner 5 with labels. (NASA/JPL graphic)

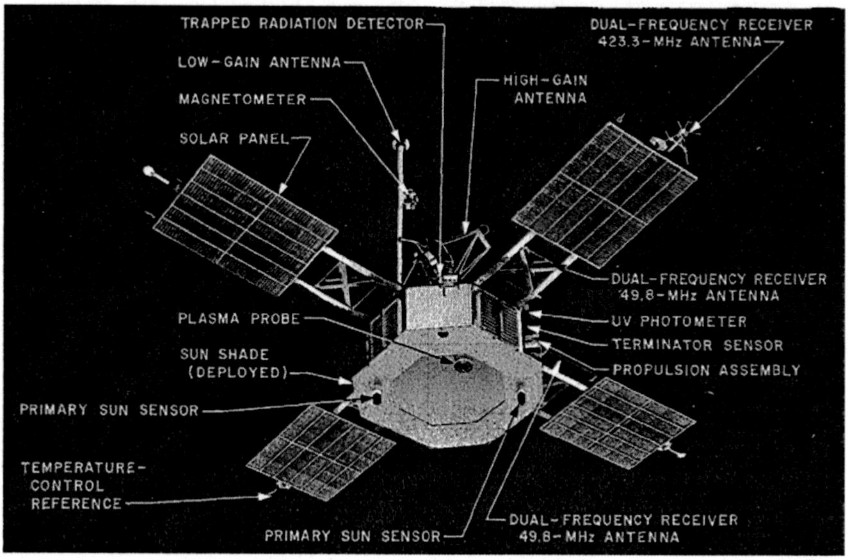

Fig. 4.9 Bottom view of Mariner 5 with labels. (NASA/JPL graphic)

The nominal thrust of the engine in a vacuum was 22.7 kg (50 pounds). The specific impulse of the engine was 233 seconds. The engine burned anhydrous hydrazine in conjunction with a catalyst. The engine was ignited by injecting a small amount of nitrogen tetroxide oxidizer into the combustion chamber along with the hydrazine.

That combination ignited on contact. Once ignited, the rocket engine continued to burn hydrazine in the presence of the catalyst. The catalyst consisted of a bed of small spherical particles of aluminum oxide impregnated with iron, nickel, and cobalt.

Hydrazine fuel was contained within a bladder in a spherical tank that held 9.8 kg of fuel. Nitrogen from a spherical tank was applied through a shutoff valve and then ported between the inner walls of the tank and bladder to force the fuel into the combustion chamber. The nitrogen tank held 0.45 kg of nitrogen under pressure. The pressure was regulated to 2068 kilopascals (300 psi) before being applied to the fuel tank. Valves in the fuel line and in the oxidizer line were opened simultaneously for ignition.

The valves were pyrotechnically activated by signals from the central computer and sequencer (CC&S). After the engine had fired for the required interval of time, the CC&S issued a signal that fired pyrotechnic activated valves to shut off fuel flow to the engine and shut off nitrogen flow to the fuel tank. Spacecraft attitude was maintained when the rocket motor was firing by deflecting four small vanes in the exhaust of the rocket motor. These vanes had the ability to deflect the thrust vector by ±5 degrees. The vanes were deflected by signals from the midcourse autopilot to maintain a constant spacecraft attitude.

Spacecraft Power

Electrical power for the spacecraft was provided by a battery during launch and early portion of flight and by solar arrays during the long cruise to Venus. The solar arrays were deployed and acquisition of the sun was achieved about 30 minutes after launch. The attitude of the spacecraft was oriented by the attitude control system to point the solar arrays directly at the sun after sun acquisition. The solar arrays then provided power for the spacecraft during cruise and encounter with Venus.

The battery was used to supplement power from the solar arrays during the midcourse maneuver while the solar arrays were not pointed directly at the sun. The battery was a silver-zinc type with capacity of 1200 watthours. The battery was used for 46 minutes during the midcourse event and supplied about 83 watts of power during that time. A battery charger, powered by the solar arrays, charged the battery after use. The battery was not used after the midcourse maneuver.

There were four solar array panels, each with an area of one square meter. Each panel contained 1470 solar cells with 14 cells connected in parallel and 105 parallel groups of 14 cells connected in series. The output voltage from the array was about 44 volts at the expected spacecraft electrical load at a temperature of 58 °C. The voltage increased when lightly loaded so a shunt arrangement of six 8.33-volt Zener diodes in series was used to limit the voltage to 50 volts. Zener diodes are specially doped silicon PN junction devices that exhibit an abrupt increase in current flow when a reverse bias voltage is increased to the diode's breakdown value. Further increase in voltage results in a constant voltage across the Zener diode and increasing voltage drop across the impedance of the voltage source.

The solar arrays were capable of providing a total of about 405 watts of power at the vicinity of Earth and about 615 watts of power at the vicinity of Venus. The increase in power comes about from the increased solar intensity at Venus countered by higher array temperature that lowered the output voltage of the array. The solar intensity was 135 W/m^2 at Earth and 268 W/m^2 at Venus during the flight of Mariner 5. However, the temperature of the solar arrays increased from 45 °C at launch to 115 °C near Venus. The voltage output from the array decreased from 45.5 volts at launch to 32.8 volts near Venus due to the higher temperature.

The outputs from the four solar arrays and from the battery were applied to a power regulator assembly which accepted power from the solar arrays when that was available and from the battery when power from the arrays was not available or not sufficient. The spacecraft was able to seamlessly switch from battery power at 27.6 volts to solar array power at 44 volts by use of boost regulators. The boost regulator stepped up voltage at its input to closely regulated 52-volt DC output. There were two boost regulators referred to as the main boost regulator and the maneuver boost regulator. The main regulator provided power to the spacecraft during cruise and encounter with Venus. The maneuver boost regulator provided power to the maneuver loads. The maneuver boost regulator served as a backup to power the spacecraft should the main boost regulator fail.

The 52-volt output of the main boost regulator was applied to the main inverter. That device received a synchronizing signal from the central computer and sequencer that established a 2.4 kHz switching frequency. The inverter developed a 50-volt square wave power form at a frequency of 2.4 kHz. Likewise, the maneuver inverter converted the 52-volt DC power form at the output of the maneuver boost regulator to a 50-volt square wave at a frequency of 2.4 kHz. The maneuver boost regulator also powered a three-phase, 400 Hz inverter used to power the gyros. Each of the six electronic chassis in the spacecraft contained a transformer and rectifier to condition the 2400 Hz square wave power to dc power forms required by its electronic circuits.

Guidance and Control

The guidance and control system included the Central Computer and Sequencer, attitude control subsystem, autopilot and planet sensor, terminator sensor, and Earth sensor.

Central Computer and Sequencer – Major spacecraft functions were controlled and sequenced by a Central Computer and Sequencer (CC&S). The CC&S functioned as a timer/sequencer that initiated a series of hard-wired commands to control the spacecraft during the mission. Only a few commands, including parameters for midcourse correction, were required from the ground. Parameters for midcourse correction were computed on the ground after several days of tracking the spacecraft. Those parameters were transmitted up to the spacecraft and loaded into the CC&S for execution later by the CC&S. The commands from the ground pertaining to midcourse correction included:

Pitch turn duration and polarity (QC-VI-1)
Roll turn duration and polarity (QC-VI-2)
Motor burn duration (QC-VI-3)
Initiate maneuver sequence (DC-V27)

Some of the commands issued by the CC&S are given in Table 4.5. Times given are times from liftoff (T = 0), time from receipt of the initiate maneuver command from the ground (M = 0), and time of encounter with Venus (E = 0). The time of closest

Table 4.5 Commands issued by CC&S

Command	Command Identifier	Time	Destination
Deploy solar arrays	L-1	T + 53 min	Pyrotechnics
Turn on attitude control subsystem	L-2	T + 57 min	Attitude control subsystem
Turn on Canopus sensor and initiate roll search	L-3	T + 997 min	Attitude control subsystem
Maneuver sequence start (Midcourse maneuver)	M-1	M = 0	Attitude control subsystem
Spacecraft to inertia control	M-2	M + 59 min	Attitude control subsystem
Set turn polarity	M-3		Attitude control subsystem
Start pitch turn/Stop turn	M-4		Attitude control subsystem
Start roll turn/Stop turn	M-5		Attitude control subsystem
Start midcourse motor	M-6		Pyrotechnics
Stop midcourse motor	M-7		Pyrotechnics
Reacquisition of attitude references	M-1 and M-2		Attitude control subsystem
Switch bit rate to 8½ bits per second	MT-6	E – 87 days	Data encoder
Set Canopus sensor cone angle No. 1	MT-1	E – 56.4 days	Attitude control subsystem
Set Canopus sensor cone angle No. 2	MT-2	E – 39.7 days	Attitude control subsystem
Set Canopus sensor cone angle No. 3	MT-3	E – 23.0 days	Attitude control subsystem
Set Canopus sensor cone angle No. 4	MT-4	E – 9.2 days	Attitude control subsystem
Transmit on high-gain antenna, receive on low gain antenna	MT-5	E – 18.5 days	Radio subsystem
Begin encounter sequence	MT-7	E – 12.5 hours	Power
Begin data automation subsystem encounter mode	MT-8	E – 6 hours	Data automation subsystem
Switch to data mode 4	MT-9	E + 14 hours	Data encoder

approach was otherwise known as the encounter time (E = 0). Some of the times of midcourse commands were variable so they are blank in the table.

Attitude control system – The attitude control system oriented the spacecraft so that during cruise to Venus the solar arrays were directed towards the sun and the Canopus sensor was pointed towards the star Canopus. Gas jets controlled by the attitude control system were affixed to the outer edges of the solar arrays to provide maneuvering capability. The jets developed thrust by emitting cold jets of nitrogen gas. Pairs of jets were used to induce rotation about a given axis of the spacecraft.

The spacecraft coordinate system was such that the Z-axis was along the longitudinal axis of the spacecraft and the long dimension of the solar arrays was parallel to the X-axis. Steering signals used to point the +Z axis of the spacecraft towards the sun were provided by sun sensors. Steering signals acting in the roll plane to align with Canopus were provided by a Canopus tracker.

Mariner 5 contained two independent gas jet systems for redundancy. Each system included a spherical tank holding nitrogen gas, a pressure regulator, and six jet nozzles with attendant solenoid operated gas valves. Individual gas valves could fail and indeed one whole gas system could fail and the redundancy would allow the mission to be successfully flown.

The gas jets for the attitude control system were mounted at the outer edges of the solar arrays. Each of the two solar arrays lying along the Y-axis of the spacecraft had two jets pointed in the opposite directions in the plane of the Z axis. The jets on one array would fire in the +Z direction and the jets on the other array would fire in the –Z direction to cause the spacecraft to rotate around the X-axis (pitch). Likewise, the solar arrays lying along the X-axis fired pairs of jets to cause rotation about the Y-axes (yaw). The two arrays lying along the X-axis also contained pairs of jets oriented in the plane of the X-Y axis, one pointed in the +Y direction and the other pointed in the –Y direction. These jets on opposite sides of the spacecraft were fired in pairs to induce roll of the spacecraft about the longitudinal or Z axis.

Nitrogen gas for attitude control was contained in spherical titanium tanks 22.8 cm in diameter initially filled to a pressure of 17,236 kilopascals. Each of the redundant gas systems had its own nitrogen tank holding 1.13 kg of nitrogen. The nitrogen gas was regulated to a pressure of 103,421 pascals (15 psi) before application to solenoid operated gas valves that applied gas to the jets. The thrust of each jet in a vacuum was 0.010 to 0.017 newtons (0.0023 to 0.0039 pounds) depending on the size of the opening in the jet. The gas jet maneuvering system was quite efficient as attested to by the fact that only 0.19 kg of nitrogen out of 2.27 kg available had been used by the end of the mission

Signals used to align the Z-axis of the spacecraft with the sun were developed by sun sensors. A pair of sun sensors referred to as the secondary sun sensors were located on the side of the spacecraft facing away from the sun during cruise and a pair of sun sensors referred to as the primary sun sensors were located on the sun-facing side. The two primary sun sensors were located on opposite sides of the frame below electronics bays #2 and #6. Segments of the detectors were aligned in the X and Y planes. The two secondary sun sensors were located above electronics bays #2 and #6.

Each of the primary sun sensors had two narrow field detectors and two wide field acquisition detectors. Each of the secondary sun sensors had four wide field acquisition detectors. The acquisition sun sensors in total provided complete angular coverage about the spacecraft and they were used during the initial sun acquisition process to orient the sun-facing side of the spacecraft towards the sun.

The sun sensors used cadmium-sulfide photodetectors that functioned as light sensitive resistors. The primary sun sensors contained shade structures that covered a portion of each detector such that as the spacecraft tilted in the plane of the sensors the amount of light falling on the photodetector changed and that changed the resistance. The effective half angle of the narrow field sensor was ±3 degrees. The operation of the sun sensors was the same as described previously for the Mariner 2 spacecraft.

The viewing angle of the Canopus sensor involved two angles: a clock angle and a cone angle. The clock angle of the sensor was fixed to the spacecraft in the X-Y plane. The zero clock angle was at 45 degrees from the –Y and + X axes of the spacecraft. The cone angle was the angle between the line to the sun and the line to Canopus from the spacecraft. The initial cone angle to Canopus was set into the Canopus sensor before launch. Periodic updates to the cone angle were made during the mission.

Steering error signals from the Canopus sensor were used by the attitude control system to rotate the spacecraft about the longitudinal or Z axis to maintain track of Canopus in the roll plane. The cone angle of the sensor had a field of view of 11 degrees. The center of the cone angle was adjustable from 75.5 degrees to 111.1 degrees in 5.125-degree steps to keep Canopus within the field of view of the sensor. Initially, the center of the cone angle was set at 79.7 degrees. The CC&S initiated four changes to the center of the cone angle of the sensor as the spacecraft traveled towards Venus. The last update, made nine days before encounter, switched the center of the cone angle to 100.3 degrees.

Autopilot – An important task performed by the autopilot was to maintain attitude control of the spacecraft while the midcourse engine was firing. Attitude control was achieved by deflecting vanes in the exhaust stream of the rocket motor. The nitrogen jets associated with the attitude control system did not have sufficient force to counter rotation forces induced by the rocket burn.

The autopilot included a set of gyros that provided angular position and angular rate data. That data was used by the autopilot to control moveable vanes in the exhaust of the rocket motor. The vanes generated torques in response to their deflection. There were four moveable vanes in the exhaust stream, each positioned by an actuator. The autopilot determined the amount of deflection required by each vane to correct any attitude error of the spacecraft. It then drove the actuators until the attitude of the spacecraft was returned to its set value. Angular rate information from the gyros was also employed for damping the servos during the sun and Canopus acquisition modes.

Special Sensors

Special sensors in Mariner 5 keyed certain events during the mission. These sensors included: planet sensor, terminator sensor and Earth sensor.

Planet sensor – A planet sensor was required to detect the presence of Venus a minimum of 38 minutes before encounter. The actual time of detection of Venus by the planet sensor in Mariner 5 was 67 minutes before encounter. The output of the planet sensor initiated start of the tape recorder and started data acquisition timer B, which in turn started sending data to track 1 of the tape recorder.

The planet sensor used a lens to focus reflected sunlight from Venus onto a cadmium sulfide photocell. Electronic circuitry associated with the photocell triggered a threshold detector when the light intensity reached a given level. A rectangular aperture at the focal point of the lens restricted the light falling on the detector to a field of view of 4.55 degrees by 1.5 degrees. Baffles prevented sunlight and reflected light off of the structure from entering the sensor.

The planet sensor was mounted next to the Canopus sensor on Bay 8 of the octagonal structure. The center of the viewing angle was at a clock angle of 45 degrees and a cone angle of 155 degrees. The 4.55 degree by 1.5-degree field of view of the sensor was rotated by 41 degrees clock angle to match the shape of the limb of Venus as seen by the approaching spacecraft.

Terminator sensor – The terminator sensor detected the passage over the terminator from darkness to sunlight on the planet. Operation of the terminator sensor was similar to that of the planet sensor. The field of view of the terminator sensor was 1.5 degrees in clock angle and 2.5 degrees in cone angle. The center of the field of view was at a clock angle of 110 degrees and cone angle of 110 degrees. The terminator sensor was mounted to Bay 2 of the octagonal structure. The output of the sensor was used to fire squibs that released pins on the high-gain antenna mount to let springs move the antenna pointing angle to the second position used during the encounter. The terminator was detected 12 seconds after the time of closest approach of Mariner 5 to Venus.

Earth sensor – The earth sensor was a backup to the Canopus sensor. In case the Canopus sensor failed, the spacecraft would be rotated about its roll axis until the Earth was acquired in the Earth sensor. The Earth would then serve as a reference for the spacecraft roll angle. The earth sensor provided useable output out to distances traveled up to 20 days after launch. The Earth sensor's function would be very important if the Canopus sensor was not operable before the midcourse correction maneuver. The field of view of the Earth sensor was three degrees clock angle and 50 degrees cone angle. It had an accuracy of about ±1 degree in clock angle. The Earth sensor was not needed on the flight of Mariner 5.

Telecommunications Subsystem

The telecommunications subsystem received and demodulated command signals from Earth, transmitted telemetry data to Earth, provided two-way coherent signal for spacecraft tracking from Earth, and received and retransmitted pseudo-random sequences for accurate ranging from Earth. As part of a science experiment, the spacecraft transmitted a signal used for occulting measurements of the Venusian atmosphere.

Antennas – The telecommunications subsystem operated in S-Band, receiving at a frequency of 2116 MHz and transmitting at a frequency of 2295 MHz. Mariner 5 used a low-gain antenna with nearly omnidirectional antenna pattern to receive transmissions from Earth during the entire mission. The omnidirectional antenna was also used to transmit telemetry data to Earth during the early portion of flight, including midcourse correction maneuvers. A high-gain antenna was provided to transmit data to earth during the later portion of the mission and for data playback from the tape recorder in the post encounter period.

The low-gain antenna was mounted on top of a mast to position it above the solar arrays when they were folded up for launch. That allowed communications after launch and before the solar arrays were deployed. The 2.1 meters high mast consisted of a circular waveguide 10 cm in diameter that fed the antenna. The antenna was a crossed slot type formed by crimping the top of the waveguide in a four-section crimp. The antenna pattern was tailored by a circular ground plane mounted to the mast a quarter of a wavelength under the crossed slots. The radiation pattern of the antenna was a cardioid of revolution. The peak gain was 6.4 dB at 2116 MHz and 5.4 dB at 2.298 MHz.

The high-gain antenna was reflector type with an elliptical parabolic sector. The reflector was formed from thin aluminum honeycomb 1.17 meters wide and 0.53 meters high. The gain of the antenna was 21.8 dB at 2116 MHz and 23.3 dB at 2298 MHz. The antenna had a beamwidth of about 7.6 degrees in the plane of the wide reflector dimension and 15.8 degrees in the plane of the narrow dimension.

The high-gain antenna was supported above the octagon structure by a truss structure of aluminum tubing. The pointing angle of the antenna was fixed at 107.4 degrees clock angle and 110.7 degrees cone angle from the time of launch to the time of passing over the terminator of sunlight on Venus. The clock angle was measured from the line to Canopus and the cone angle was measured from the line to the sun (Z-axis of the spacecraft). After passing the terminator, the antenna was rotated by 17.7 degrees about a line passing through the wide axis of the antenna. The direction of Earth at the time of encounter of Venus was a clock angle of about 102 degrees and a cone angle of about 104 degrees.

Receiver and transmitter – Commands sent up from Earth were intercepted by the low-gain antenna and fed through a circulator to the receiver. A circulator provided isolation between the receiver and transmitter that also used the low-gain antenna. The receiver was a dual conversion superheterodyne type with a phase tracking loop. The phase tracking loop included a voltage-controlled oscillator (VCO) whose output frequency of 19.125 MHz was multiplied by a factor of 180 and fed to the first mixer to close the phase tracking loop. The VCO output frequency was multiplied by a factor of five to drive the second detector. A phase detector that was part of the phase tracking loop detected the phase shift keyed modulation on the signal sent up from Earth and passed the modulation output to the command decoder.

The output frequency of the VCO in the phase tracking loop was applied to two redundant exciter units when the phase tracking loop in the receiver was in lock with the signal from Earth. When the phase tracking loop was not locked, the input to the exciters were switched to fixed oscillators that operated at 19.125 MHz. The exciter multiplied the input signal frequency by a factor of four. A phase modulator at the output of the X4 multiplier accepted the telemetry modulation input and input from ranging modulation from the receiver. The phase modulated signal was multiplied in frequency by a factor of 30–2295 MHz and applied to one of two transmitters. The output signal level of the exciters was about 0.5 watts. Only one exciter unit was activated at a time. The coherent receiver output maintained a ratio of received frequency to transmitted frequency of 221/240.

The telecommunications subsystem contained two transmitters. Only one was used at a time and the output of either exciter could be switched to either transmitter. One transmitter was a triode vacuum tube coaxial cavity type and the other was a traveling wave tube (TWT) type. The triode cavity transmitter had output power of seven watts. The TWT transmitter had output power of 10 watts. The transmitters had individual power supplies. Power was applied to one or the other of the two transmitters. The outputs of the two transmitters were applied to a voltage-controlled circulator that allowed selection of one or the other transmitter output to be applied to the antennas. A second voltage-controlled circulator allowed selecting whether the transmitter output was applied to the high-gain or low-gain antenna.

Command Detector and Decoder

Command signals from Earth and processed by the receiver were in the form of biphase modulation of a sinusoidal subcarrier. The subcarrier was demodulated and the resulting signals were applied to a command decoder. The command decoder decoded the command messages and based on their address sorted them into commands that were to be acted on in real time and those that were to be stored in the CC&S for execution later. The real time commands were prefixed by "DC" and the stored commands were prefixed by "QC." A tabulation of commands available for the Mariner 5 mission and their function is given in the table below (Table 4.6). Some of the commands were back-up to commands issued by the CC&S.

Data Encoder

The data encoder processed 90 analog and digital engineering data items. It accepted spacecraft measurements in analog and digital form and converted the data to 7-bit non-return-to-zero serial words. Two analog-to-digital converter channels were provided for redundancy. The selection of one or the other could be toggled by a command from Earth. A synchronization signal was formed from a 63-bit pseudorandom noise (PRN) code generator clocked at 1200 Hz. The biphase modulated subcarrier was combined with the PRN synchronization signal. The resulting signal was

Table 4.6 Available commands from Earth for Mariner 5

Designation	Command
QC-V1–1	Pitch turn +55.27 degrees, duration 304 seconds.
QC–V1–2	Roll turn +70.95 degrees, duration 380 seconds
QC-V1–3	Motor burn time 17.66 seconds, $\Delta V = 16.13$ m/s
DC-V1	Switch encoder to data mode 1
DC-V2	Switch encoder to data mode 2
DC-V3	
DC-V4	Initiate tape playback
DC-V5	Toggle between 33½ and 8½ bits per second data rate
DC-V6	Toggle between analog-to-digital converters
DC-V7	Toggle between power amplifiers in transmitter
DC-V8	Toggle between exciters in transmitter
DC-V9	Turn ranging receiver on or off
DC-V10	Transmit through high-gain antenna and receive through low-gain antenna
DC-V11	Transmit and receive through high-gain antenna
DC-V12	Switch transmit and receive to low-gain antenna
DC-V13	Disable attitude control by CC&S
DC-V14	Reverts attitude control back to CC&S
DC-V15	Canopus sensor gate inhibit override
DC-V16	Activate back-up clock A
DC-V17	Start encounter backup clock
DC-V18	Turn on gyros
DC-V19	Turn off gyros
DC-V20	Remove roll control
DC-V21	Roll override
DC-V22	Change antenna pointing angle
DC-V23	Allow rocket motor burn the second time
DC-V24	Begin data acquisition system encounter mode. Activate planet sensor
DC-V25	Activate science
DC-V26	Turn off science
DC-V27	Initiate midcourse maneuver
DC-V28	Switch battery charger
DC-V29	Arm propulsion system

amplified and sent to the biphase modulator in the exciter units of the transmitter and on to Earth.

The digital words were applied to a data selector function in the data encoder. Science data gathered and formatted by the data automation subsystem was also applied to the data selector function for encoding and transmission to Earth. The format of data output from the data encoder depended on the particular mission mode of the spacecraft.

Data Automation Subsystem

The Data Automation Subsystem (DAS) controlled the science experiments and converted the various forms of data from the experiments to a standard form to be transmitted to Earth or to be recorded on tape. It provided synchronization signals to the various experiment elements and it controlled the sampling, data conversion, formatting, and encoding of data from the experiments.

The DAS accepted various inputs from the experiments and transformed them into 10-bit data words. The data words were assembled and fed to the data encoder subassembly where they were inserted into a time slot along with engineering data and telemetered to Earth. There were four telemetry data modes. Each master data frame of telemetry contained 420 bits.

Mode 1 (Maneuvers) – Only engineering data transmitted
Mode 2 (Cruise) – 280 bits of science and 140 bits of engineering
Mode 3 (Encounter) – 406 bits of science and 14 bits of engineering
Mode 4 (Playback) – Playback of science data recorded during encounter

Tape Recorder Subsystem

The telemetry subsystem used a low data rate of 8 $\frac{1}{3}$ bits per second to achieve useable signal-to-noise ratio in the receiver on Earth when the spacecraft was in the vicinity of Venus. The distance to Earth was about 79.5 million km during encounter with Venus. This low data rate would not accommodate the data rate of the experiments in real time so a tape recorder was used to accept two data streams, each at a rate of 66 $\frac{2}{3}$ bits per second during encounter. The data was played back to through the data encoder at a data rate of 8 $\frac{1}{3}$ bits per after the spacecraft had passed by the planet.

The tape recorder recorded on two tracks. It had two record heads and two playback heads. The tape speed during the record mode was 4.8 inches per minute giving a record time of 127 minutes. The length of tape was 16 meters. An end-of-tape sensor stopped the recorder when the end of the tape was reached. The tape speed during the playback mode was 1.5 cm per minute and the playback time was 17 hours 48 minutes per track.

Science Experiments Carried by Mariner 5

The Mariner 5 spacecraft carried five science experiments: trapped radiation detector, solar plasma probe, helium magnetometer, ultraviolet photometer, and dual frequency receiver. The dual frequency receiver experiment was new for the Mariner 5 mission. The other experiments used modifications of instruments carried on the Mariner 4 spacecraft that explored Mars in 1964.

Trapped Radiation Detector

The Trapped Radiation Detector (TRD) had multiple objectives. First, detect and measure magnetically trapped particles in the vicinity of Venus. Second, detect and measure solar cosmic rays and electrons in interplanetary space. Third, monitor soft X-rays from the sun throughout the flight.

The TRD contained four separate radiation detectors. Three of the detectors, labeled A, B, and C, were Geiger-Mueller tubes. The fourth detector, labeled D, was a solid-state silicon surface-barrier diode type of radiation detector. The three Geiger-Mueller tubes were configured as end window counters with different end-window material to accept particles of different energies.

Detector A had a field of view of 10 degrees circular centered along the sun-probe line (Z axis of the spacecraft). Its mica end-window restricted electrons to energies greater than 95 keV and protons to energies greater than 2.7 MeV. It detected X-rays with wavelengths less than 10 angstroms. Detectors B, C, and D had their look angles centered 70 degrees from the sun-probe line. The fields of view of detectors B, C, and D were 60, 60, and 80 degrees respectively. The end-window of detector B restricted electrons to energies greater than 45 KeV and protons to energies greater than 0.65 MeV. Detector C had an end- window that restricted electrons to greater than 150 keV and protons to energies greater than 3.1 MeV.

The solid-state detector D had four channels, D1, D2, D3, and D4. The detector was not sensitive to electrons. Channels D1, D2, and D3 were sensitive to protons with energies, E_P, as indicated below.

Channel	Sensitivity
D1	$0.30 < E_P < 12$ MeV
D2	$0.50 < E_P < 5.3$ MeV
D3	$0.98 < E_P < 2.2$ MeV

Channel D4 was sensitive to alpha particles between ranges of 2.1 MeV to 18 MeV.

Each output of the TRD was sampled for 37.2 seconds out of 403.2 seconds by a commutator. Counters recorded the number of pulses at the outputs of the detectors during the sample periods. These counts became the data outputs of the trapped radiation detector.

Several solar flares occurred during the cruise phase of the mission. During five of these both protons and electrons were detected. During another five flare events only protons were detected and on two flare events only electrons were detected. Several hundred solar X-ray events were observed during cruise to Venus. No increase in energetic particles above interplanetary background levels were observed in the vicinity of Venus.

Solar Plasma Probe

The solar plasma probe was designed to measure the energies of protons and alpha particles contained in the solar wind. The plasma probe sorted particles as a function of energy. It consisted of a cup-like structure 15.3 cm in diameter and 11.4 cm long with a series of grids and collector elements on the inside. The grids were made from tungsten wire mesh. A sketch of the instrument is given below (Fig. 4.10).

The cup was aligned with its axis along the Z-axis of the spacecraft and hence the opening faced the sun during most of the mission. Protons and alpha particles are positively charged and that property was used to sort them as a function of velocity (energy). The first screen in the probe was at the electrical potential of the spacecraft. The second screen was the modulation screen that was connected to a positive voltage of varying amplitude. Incoming positive particles encounter a repelling positive field between the first grid and the modulating grid. Only particles with sufficient velocity (energy) passed through the modulating grid and would strike the collector array.

The modulation voltage was a square wave at a frequency of 2 kHz superimposed on a positive dc voltage. There were 32 different values of the square wave plus dc offset modulating voltage. The modulating waveform ranged from a low of 30 volts and high of 40 volts to a low of 7700 volts and high of 10,000 volts. Each of the 32 different voltage windows allowed a select velocity range of particles to pass through to the collector. The 32 voltage intervals covered energy intervals from 40 to 9400 electron volts. The voltage pattern and hence energy window was controlled by the data automation subsystem (DAS). The DAS caused the voltage waveform to step to a new level each 2.1 seconds at the high data rate of the DAS and each 8.4 seconds at the low data rate.

The collector at the bottom of the cup was segmented into three pie-shaped segments. Each segment fed an independent signal processing channel. The direction of the particles could be determined by the relative amplitudes of the signals in the three collector channels. During the cruise mode of Mariner 5, the sum of the three segments was sampled first and then the three channels were sampled separately.

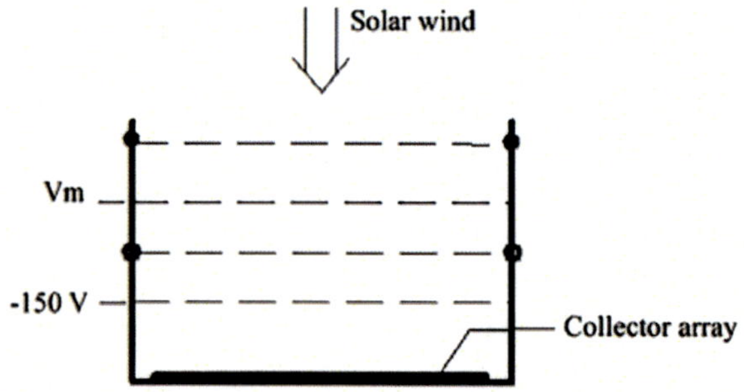

Fig. 4.10 Sketch of solar plasma probe. (Adapted from JPL/NASA drawing)

During the encounter mode, only the sum of the three signals was sampled. A calibration signal was periodically gated into the three signal processing channels.

The plasma probe performed well during both the cruise and encounter modes of the Mariner 5 mission. The energy spectrums obtained allowed computing positive ion density and bulk velocity. Data taken about 210 hours before encounter indicated positive ion number density of about 4 per cubic centimeter and bulk velocity of about 600 km/sec. Fluctuation in the density as the spacecraft approached and receded from the encounter point (point of closest approach) allowed scientists to conclude that Venus generates a bow shock as it passes through the solar wind similar to the bow shock of Earth passing through the solar wind.

Helium Magnetometer

The purpose of the magnetometer was to determine if Venus has a very weak magnetic field and if so, measure its intensity and direction. It was also intended to measure the magnetic field originating in the sun transported out by the solar wind. The magnetometer was a helium vapor type that made use of the magnetic properties of two electrons in the helium atom. That type magnetometer was sensitive to weak magnetic fields and it had been successfully used on the Mariner 4 spacecraft to Mars in 1964.

The operation of the magnetometer was based on the transparency of a plasma of helium gas to radiation at the resonant wavelength of 1.083 microns. The pumping frequency of 1.083 microns, which is in the infrared wavelength region, ionized the helium creating a plasma. The helium plasma was contained in an enclosure referred to as the absorption cell. The transparency of the plasma was dependent on magnetic fields enveloping the plasma.

A tri-axial set of Helmholtz coils was used to generate rotating magnetic field vectors in the plasma. A Helmholtz coil consists of two coils of wire separated by about the radius of the coils. A nearly uniform magnetic field is produced along the axis between the coils when the coils are energized by a dc electrical current.

A beam of infrared radiation at 1.083 microns was transmitted through the absorption cell and an infrared detector was used to sense modulation of the infrared beam by the magnetic fields. In the absence of an ambient magnetic field, the modulation of the infrared beam was at twice the frequency of the rotating magnetic field. The presence of an external magnetic field generated a frequency component of the infrared beam at the fundamental frequency of the rotating field.

The presence of a fundamental frequency component was synchronously detected. The resulting dc signal was amplified and applied to a servo amplifier that applied a dc current to a particular Helmholtz coil set to null out the fundamental frequency component. This nulled out one component of the external magnetic field. The output of that axis of the magnetometer was derived from the amount of current applied to the coil to null the fundamental component. The amount of current was proportional to the strength of the magnetic field. The current was converted to a dc analog voltage that in turn was applied to an analog-to-pulsewidth converter. The pulsewidth data was sent to the data automation subsystem. The other two components of the ambient magnetic field were similarly processed.

The sensitivity of the magnetometer was set by a noise level of 0.1 nanotesla (nT) where Tesla is a unit of magnetic field strength. The digital uncertainty of the data was 0.2 nT. When near the encounter with Venus at a distance of 5768 km, the magnetometer measured total magnetic field strength of about 8 nT. It measured field strength of about 20 nT when the spacecraft reached the vicinity of the bow shock front. For comparison, the magnetic field of the Earth over the equator at the prime meridian at a distance of 5768 km is about 3976 μT. At the earth's surface at that location, the magnetic field strength is 31,957 μT.

Ultraviolet Photometer

The purpose of the ultraviolet photometer was to measure the density of atomic hydrogen and atomic oxygen in interplanetary space and in the outer atmosphere of Venus. Atomic hydrogen resonates at a wavelength of 1218 Å and atomic oxygen resonates at a wavelength of 1304 Å. The symbol Å denotes angstroms, which is a unit of length equal to 0.1 nanometers. The ultraviolet band is usually referred to as lying between wavelengths of 100 Å and 4000 Å.

The ultraviolet photometer contained three ultraviolet sensors labeled A, B, and C. Each sensor consisted of a photomultiplier tube with a cesium iodide photocathode. The windows of the tube were made from lithium fluoride. Each tube had a nearly flat response between 1050 angstroms and 1900 angstroms. Tube A was fitted with an optical filter that resulted in a sensor bandpass of 1250 Å to 1900 Å. Tube B was fitted with an optical filter that resulted in a sensor bandpass of 1350 Å to 1900 Å. Tube C had no filter and had a bandpass of 1050 Å to 1900 Å. The intensity of atomic hydrogen resonance was determined by taking the difference between the amplitudes between tubes A and C and the intensity of atomic oxygen resonance was determined by taking the difference between tubes A and B.

The fields of view of tubes A and B was a cone with half angle of 0.5 degrees at the half amplitude point. Tube C had a conical field of view with half angle of 1.25 degrees at the half amplitude point. The three tubes were pointed in the same direction. The instrument was capable of measuring brightness between 500 Rayleighs and 100,000 Rayleighs. A Raleigh is the standard unit of brightness for emission lines. The outputs of the photomultiplier tubes were a dc current. The output of each tube was connected to a dc amplifier that had a voltage output proportional to current input. The output of the amplifier for each of the three channels was sent to the data automation subsystem.

The results of this experiment on Mariner 5 indicated that the amount of atomic hydrogen in the upper atmosphere of Venus was about the same as in the upper atmosphere of Earth. No evidence of atomic oxygen was found in the upper atmosphere of Venus.

Dual–Frequency Receiver

The dual-frequency receiver experiment was designed to measure electron number density in the vicinity of Venus. It did so by viewing RF signals at two different frequencies sent from earth as the spacecraft passed by Venus. The experiment viewed the RF signals through the atmosphere just before the signals were occluded by the planet. The path passed tangent to the planet and therefore signals from earth traveled a significant distance through the Venusian atmosphere. A second look through the atmosphere was provided as the spacecraft emerged from total occlusion on the other side. Occlusion lasted for about 26 minutes. Mariner 5 approached Venus from the dark side of the planet and followed a curved path due to gravity of Venus and passed around to the sunlit side. That allowed electron concentrations in both the dark side and sunlit side of Venus to be measured.

Two continuous wave RF carriers, one at a frequency of 49.8 MHz and the other at 423.3 MHz, were transmitted to the spacecraft from a 46-meter (150-foot) diameter parabolic antenna at Stanford University. The two frequencies were harmonically related by a ratio of 2/17. The RF power transmitted was 350 kW for the 49.8 MHz signal and 30 kW for the 423.3 MHz signal. The two carriers were phase modulated by either an 8692 Hz tone or a 7692 Hz tone.

The receive antenna on the spacecraft for the 423.3 MHz signal was a stub type with two reflector elements and four ground plane radials. It was mounted at the end of the Bay 3 solar panel. The receive antenna for the 49.8 MHz signal was formed by feeding the spars for Bay 1 and Bay 3 solar arrays in parallel. The lengths of cables feeding the two spars were adjusted to cause the peak of the antenna pattern to point in the direction of Earth during the time of spacecraft encounter with Venus. The peak gain of the 423.3 MHz antenna was also near the direction of Earth at spacecraft encounter with Venus.

The dual receiver and associated signal processing was implemented to measure signal amplitudes, signal frequencies, differential Doppler frequency, and differential group path velocities. The receiver data was fed to a tape recorder from a time before encounter to a time after encounter. The tape was played back and data transmitted to earth after leaving Venus. The electron number density results of the experiment were primarily obtained by analyzing differential Doppler measurements and signal amplitude measurements in the two RF channels. The scientists who analyzed data from the experiment did not consider the frequency data and the group path data to yield as accurate measure of electron number density as did the differential Doppler and amplitude data.

The electron number density was found to be about an average of 300 electrons per cubic cm at a radial distance of 9500 km to 7500 km during the approach to Venus on the night side. The electron number density increased to about 10,000 electrons per cubic cm at a radial distance of about 6200 km on the sunlit side.

Flight of Mariner 5

Mariner 5 was launched from Cape Kennedy on 14 June 1967 by an Atlas/Agena-D launch vehicle. The Atlas was a SLV-3 model, a standardized version of the Atlas-D vehicle. Previous versions of the Atlas-D used for space launches had been Intercontinental Ballistic Missiles (ICBMs) modified for space vehicle launch. The SLV-3 model also had a new improved autopilot. The Agena-D upper stage was a standardized version of Agena-B that had been used for previous space launches. Agena-D was a restartable stage that burned the hypergolic combination of unsymmetrical dimethylhydrazine (UDMH) as fuel and inhibited red fuming nitric acid (IRFNA) as oxidizer. It developed thrust of 71 kN (67,000 pounds). Here, N is the symbol for newton, the International System of *Units* (SI) for *unit* of force.

The Atlas SLV-3 used a Rocketdyne MA-3 propulsion system that included two booster engines, a sustainer engine, and two vernier engines. The total thrust at sea level of the booster engines, sustainer engine, and the two vernier engines firing together was 1725 kN (387,780 pounds). The launch vehicle easily lifted the total weight of the Mariner 5 spacecraft and launch vehicle of about (127,273 kg (280,000 pounds) off of the ground.

Atlas performed well with significant events of booster engines cutoff and jettison, sustainer engine cutoff, vernier engine cutoff, and separation of Atlas from Agena all occurring as planned. Separation of Atlas from Agena occurred 322 seconds after launch. Following separation, Agena was fired for 142.5 seconds to place Agena/Mariner 5 into a circular parking orbit 185 km above the earth. After coasting for about 13.3 minutes Agena was fired again. The second Agena firing lasted for 94.4 seconds and increased the speed to 11.4 km per second which injected Agena/Mariner 5 into a geocentric escape hyperbola in a direction to intercept Venus.

Firing of Agena was controlled from the ground using trajectory information from tracking stations that were part of the Deep Space Instrumentation Facility (DSIF). Firing information for Agena was transmitted up to the spacecraft from the DSIF. At injection towards Venus, Agena/Mariner 5 was at an altitude of 193 km above the Earth and traveling at a speed of 11.4 km per second.

The DSIF included a series of spacecraft tracking stations located around the world so that there would be at least one tracking station in view of the spacecraft as the Earth rotated. There were five such stations at the time of the flight of Mariner 5. The number designators of the Deep Space Stations (DSS) and locations of the stations were: DSS 41 Goldstone in the Mojave Desert of California, DSS 41 Woomera at Island Lagoon, Australia, DSS 42 Tidbinbilla near Canberra, Australia, DSS 51 Johannesburg near Johannesburg South Africa, and DSS 61 Madrid near Madrid, Spain. These stations were equipped with parabolic dish tracking antennas 26 meters (85 feet) in diameter and associated transmitter and receiver equipment. Goldstone also had a second antenna, DSS 14, that was 70 meters (230 feet) in diameter. That antenna was used during a portion of the Mariner 5 mission. The Goldstone and Johannesburg stations had the ability to send commands to the spacecraft.

Mariner 5 was separated from Agena 26 minutes, 17 seconds (0:26:17) after launch and Mariner 5 began its long journey to Venus. Because of tolerances on performance of the launch vehicle, the trajectory of the spacecraft was deliberately biased to assure that it would not impact and contaminate the planet. Agena was also maneuvered so it would not impact Venus.

The solar panels were deployed 0:29:11 after launch. The numeric notation represents hours: minutes: seconds. The spacecraft used sun sensors to drive pitch and roll angles of the spacecraft to acquire the sun 0:42:25 after launch. The Z axis of the spacecraft was pointed at the sun under control of the primary sun sensors. Roll search for the star Canopus was initiated 16:37:00 after launch. The Canopus sensor first acquired the Earth and it was forced to resume search. Canopus was acquired at a time 19:08:12 after launch. The spacecraft then cruised towards Venus in a stable orientation with its Z axis pointed at the sun and roll angle set by the Canopus sensor locked to Canopus.

Midcourse correction – After several days of tracking by the DSIF, the trajectory of the spacecraft was accurately determined. It was projected to pass 69,693 km from the leading edge of Venus. A midcourse correction had been planned to adjust the trajectory so that the spacecraft would pass a desired distance from Venus. Computations made on the ground included changes in attitude of the spacecraft in pitch and roll from the stable cruise conditions to point the thrust of the rocket motor in the direction to achieve the needed trajectory correction. The computations also included burn time of the motor to bring about the correction.

Midcourse correction information was sent to the spacecraft by the DSIF and stored in the central computer and sequencer (CC&S). The computed data included a pitch adjustment of 55.27 degrees, roll adjustment of 70.95 degrees, and firing of the propulsion system for 17.66 seconds. The firing was expected to change spacecraft velocity by 16.13 meters per second. The midcourse correction procedure was initiated 5.6 days after launch and it was completed about 1.4 hours later. Trajectory data after midcourse correction indicated that a velocity change of 15.39 meters per second was attained, slightly less than that expected.

Cruise Mode – Following midcourse correction, the sun and Canopus were reacquired and the spacecraft continued cruise towards Venus. Early trajectory information indicated that the spacecraft would have a closest approach to Venus of 3.99 km during the flyby.

The central computer and sequencer (CC&S) was programmed to make a series of adjustments to the spacecraft as the flight progressed. Four changes to the cone angle of the Canopus sensor were made between 24 August and 10 October 1967. The transmitter was switched to 10 watts output on 27 June. The bit rate of the telemetry was switched from 33 $\frac{1}{3}$ to 8½ bits per second on 24 July. The transmitter was switched to the high-gain antenna on 1 October 1967.

Encounter Mode – An isometric projection of the trajectory of Mariner 5 at encounter with Venus from NASA document SP-190 is shown below (Fig. 4.11). The numbers on the graphic represent time in minutes from encounter (E). Encounter occurred near the leading edge of the planet in its orbit around the sun. The notation WIND indicates solar wind from the sun. Mariner 5 approached from above the

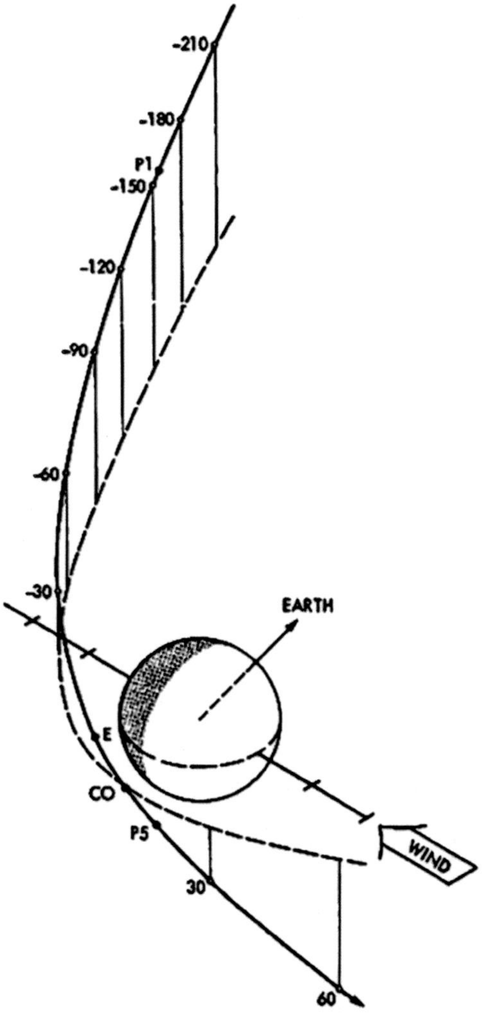

Fig. 4.11 Trajectory of Mariner 5 at encounter with Venus. (NASA graphic)

planet on the dark side and emerged from the encounter below the planet on the sunlit side.

The CC&S was capable of orchestrating the encounter sequence automatically but operators on the ground decided to preempt some commands and initiate them from the ground via the Deep Space Instrumentation Facility (DSIF). The command DC-V25 was sent to Mariner 5 at 02:49:00 Greenish Mean Time (GMT) on 19 October to initiate the encounter sequence. That turned on science instruments, turned on the tape recorder, and turned on the terminator sensor. The command DC-V24 was sent to the spacecraft at 10:50 to activate the planet sensor and begin the encounter mode of the Data Acquisition System (DAS). The command DC-16 was sent at 15:01:43 to start the encounter backup clock.

The planet sensor detected the planet at 16:33:57 on 19 October and that started DCS timer B, started the tape recorder motor, and started recording on track 1. The tape recorder started recording data on track 2 at 16:46:46. The S-Band signal was lost by the Goldstone DSS at 17:39:08 signifying occlusion by the planet. The terminator sensor signal occurred at 17:46:51. That signal initiated changing the pointing angle of the S-Band high-gain antenna by 17.6 degrees. Closest approach to Venus occurred at 17:34:55 at an altitude above the surface of 4094 km. The S-Band signal was regained by the Goldstone DSS at 17:59:59 at the exit of the occlusion.

The 49.8 MHz channel of the dual frequency receiver lost the signal from Earth at 17:33:10 due to occlusion by the planet and reacquired it at 18:03:13 at the end of occlusion. The 423.3 MHz channel lost the signal at 17:39:00 due to occlusion and reacquired it at 18:00:31 after occlusion.

Data transfer to tape recorder track 2 was ended by DCS timer B at 18:34:18. The tape recorder was stopped when an end of tape indicator was detected at 18:41:51 on 19 October 1967. Playback of data from the tape recorder was initiated at 07:25:23 on 20 October by a signal from the CC&S. Playback had been completed and the data encoder was switched to Mode 2 (cruise mode) at 07:55:00 on 23 October by a DC-V2 command from the ground.

Mariner 5 continued to operate well in the post encounter period. Communications subsystem performance tests were made along with additional calibration data for the UV photometer. The spacecraft was set up for long term cruise. Contact with the spacecraft was lost on 4 December 1967.

Pioneer Venus Orbiter

The Pioneer Venus program was designed to gather in-depth information about the planet Venus. The program included two spacecraft: the Pioneer Venus Orbiter and the Pioneer Venus Multiprobe. The orbiter collected information about the atmosphere and surface of Venus from orbit. Its mapping radar penetrated the clouds to reveal the topography of the surface. The Multiprobe spacecraft consisted of a bus, three small probes, and one large probe. The probes were released from the bus when near Venus and gathered data on the clouds and atmosphere while descending to the surface.

An artist's rendering of the Pioneer Venus Orbiter in its orbit around Venus is shown on the next page (Fig. 4.12).

Background of Pioneer Venus Mission

Studies of a potential expanded Venus mission and a report by the National Science board in 1970, *Venus—Strategy for Exploration*, led to formulation of a plan to develop new spacecraft to explore Venus. NASA issued an Announcement of

Fig. 4.12 Artist's rendering of Pioneer Venus Orbiter. (NASA graphic)

Opportunity for scientists to participate in defining the Venus exploration program in July 1971. NASA Ames was assigned responsibility for the program in January 1972.

Two spacecraft, an Orbiter and a Multiprobe bus were planned. The Hughes Aircraft Company was awarded a contract to develop the two spacecraft in February 1974. Final selections of scientific instruments for both spacecraft were made in June 1974.

Pioneer Venus Orbiter was launched from Cape Canaveral on 20 May 1978. It fired its orbit insertion rocket motor on 4 December 1978 and entered into a near polar orbit around Venus. The inclination of the orbit was 106 degrees and the period was about 24 hours. Periapsis of the orbit was at latitude of 16 degrees north. Altitude at periapsis was maintained between 140 and 250 km to favor radar altimeter operation. Apoapsis of the orbit was about 66,900 km. The Orbiter remained in orbit around Venus until fuel for the maneuvering jets ran out in 1992.

The Multiprobe bus was launched on 8 August 1978. A large probe was released as the bus approached Venus on 16 November 1978. Three small probes were released on 20 November 1978. The probes transmitted measured data on the clouds and atmosphere as they descended towards the surface.

Mechanical Configuration of Pioneer Venus Orbiter

A photograph of the Pioneer Venus Orbiter during testing is shown below (Fig. 4.13). The Pioneer Venus Multiprobe is in the background of the photograph.

Fig. 4.13 Pioneer Venus Orbiter during testing. (NASA HEASARC photograph)

The cylindrical main body of the Orbiter spacecraft was 2.5 meters in diameter and 1.2 meters high. A cylindrical solar array was installed around the circumference of the cylinder. The reason for the cylindrical array was that the spacecraft was spin stabilized with the spin axis perpendicular to the ecliptic plane. The nominal spin rate of the spacecraft was 15 RPM during cruise to Venus and 5 RPM while in orbit.

A labeled sketch showing the mechanical configuration of the Pioneer Venus Orbiter is shown on the next page (Fig. 4.14). The primary structural element of the Orbiter was referred to as a thrust tube. The thrust tube had a conical frustum shape, meaning that it was a cone with the top portion sliced off parallel to the base. The thrust tube is the structure within the cylindrical shell with support tubes attached in the sketch. The larger diameter end of the thrust tube was towards the bottom of the spacecraft. That end interfaced with the spacecraft attach fitting that held the spacecraft during launch.

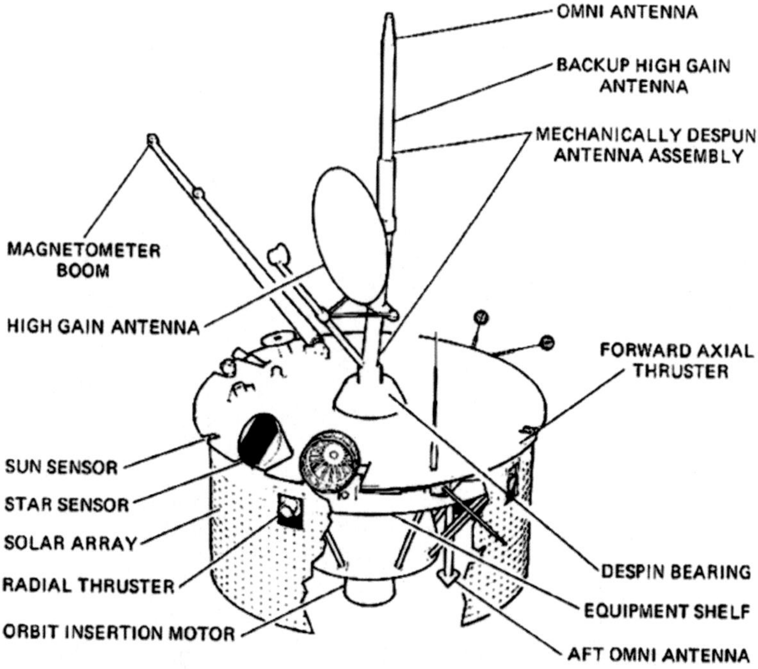

Fig. 4.14 Sketch of mechanical configuration of Pioneer Venus Orbiter. (NASA graphic)

The upper end of the thrust tube supported the center of a circular equipment shelf by a series of 12 struts. Electronics for spacecraft systems and the experiments were mounted to the top of the shelf. The shelf was a circular honeycomb plate two inches thick with aluminum faceplates. A series of 15 louver modules were mounted to the bottom of the shelf to control thermal radiation out of the shelf. Electronic packages with the highest heat dissipation were located directly over louvers.

The solar array elements were attached to a cylindrical structural shell of aluminum honeycomb 1.5 meters in diameter and 1.2 meters high. The solar array formed the outer cylindrical portion of the spacecraft. The cylindrical structure was attached to the spacecraft by 24 fittings around the periphery of the equipment shelf.

An antenna mast at the top of the spacecraft held a high-gain parabolic antenna and two omni directional antennas. The mast was supported by a bearing that allowed the high-gain antenna to be pointed at Earth as the spacecraft spun under it. The bearing assembly for the mast was attached to the upper end of the thrust tube.

A solid fuel rocket motor, used for insertion into orbit around Venus, was mounted to the inside of the thrust tube. The nozzle of the rocket motor was pointed in the opposite direction to the antenna mast. Two tanks holding hydrazine fuel for attitude control thrusters were mounted to the inside of the thrust tube.

A 4.7-meter long boom holding three magnetometer elements is shown on the left side of the sketch of Fig. 4.16. The boom was folded during launch. Two magnetometers were located near the end of the boom, one parallel to the spin axis and the other perpendicular to it. A third magnetometer was located one third of the distance down the boom and pointed 45 degrees to the spin axis. The object at the top front of the spacecraft with a circular cage is the antenna for the surface mapping radar.

Thermal control was an important consideration for the spacecraft since the solar irradiance near Venus is 2601 watts per square meter compared with 1361 watts per square meter near Earth. A thermal blanket consisting of 15 layers of aluminized Kapton was installed above the equipment shelf to protect the interior from heat. The Kapton sheets were aluminized on one side and crinkled to provide small spaces between layers. Heaters were provided on several components to maintain acceptable temperature range for the equipment.

The spacecraft weighed 582 kg when launched from Earth. The weight included 43 kg of scientific instruments.

Pioneer Venus Orbiter Systems

Orbit Insertion Rocket Motor

The orbit insertion rocket motor was mounted on the inside of the thrust tube structure with the exhaust directed along the –Z axis of the spacecraft. The rocket motor was a Thiokol Model TE-M-604 solid fuel motor. Propellant for the motor was TP-H-3062, which was aluminized ammonium perchlorate (AP) using carboxyl terminated polybutadiene (CTPB) as binder. The propellant was contained within a spherical titanium alloy combustion chamber 0.62 meters in diameter. An expansion nozzle with exit diameter of 38 cm directed the exhaust of the motor along the –Z axis of the spacecraft. The total length of the rocket motor to the end of the nozzle was 1.03 meters. The weight of the motor with propellant was 218 kg.

The propellant was cast in an eight-point star configuration on the inside of the chamber to provide a large burn area. The propellant, which burned from the inside out, was ignited by a pyrogen igniter with dual initiators. The rocket engine had an average thrust of 18,600 newtons (4181 pounds) and specific impulse of 286 seconds. The 200 kg of propellant in the combustion chamber provided a burn time of 30 seconds. The engine imparted a velocity change of about 1059 meters per second to the spacecraft.

Spacecraft Power

Solar cells located on the circumference of the cylindrical orbiter provided the primary source of power for the spacecraft. A total of 15,988 solar cells were used, each cell being 2.0 cm by 2.0 cm in size. Power generated by the solar array decreased at higher temperatures and it could be reduced by damage due to solar flares. However, power increased as the spacecraft traveled towards Venus because of increased solar constant. Under worst case assumptions of a massive solar flare and warmest temperature, the solar array was expected to generate 232 watts of power at 28 volts DC near Earth and 336 watts of power at 28 volts DC near Venus.

The output voltage of the solar array could increase to over 30 volts under favorable conditions but it was limited to 30.1 volts DC by shunt limiters. Seven limiters in parallel with resistive loads were used to dissipate excess power. The first 132 watts of excess power was shunted to resistors on the back of the equipment shelf as part of thermal control of the shelf. The bus voltage range with solar power alone was 27.8–30.1 volts.

Two nickel-cadmium batteries, each with capacity of 7.5 ampere hours, powered the spacecraft when the solar array did not provide adequate power, as when in the shadow of Venus. Batteries were automatically connected to the load whenever the bus voltage dropped below 27.8 volts. The longest time in the shadow of Venus during orbiting was three hours and ten minutes. The maximum depth of discharge of the batteries was expected to be about 70%. Each battery was controlled by a charge/discharge controller. The two batteries were connected in parallel across the power bus. The charge/discharge controllers managed charge of the batteries and applied battery power to the bus when the bus voltage dropped below 27.8 volts DC.

DC power at 28 ± 10% was provided to spacecraft systems and to each of the experiments. The electronic units for these systems and experiments contained voltage converters and regulators to provide appropriate voltages to their circuitry.

Power to the spacecraft was apportioned in four different power busses: Essential Bus, Science Bus, Switched Load bus, and RF Transmitter bus. The essential bus, which powered the command subsystem, RF subsystem drivers and critical heaters, could not be switched off. If a fault increased spacecraft current load beyond 16.5 amperes, the three unessential busses were switched off sequentially. If the current decreased to acceptable values after switching off a particular bus, the switching process was stopped. Likewise, if a fault occurred that lowered battery voltage below 27.55 volts, the nonessential busses would be switched off sequentially. If bus current or battery voltage recovered to the decision levels, the nonessential busses were switched back on. The switching function could be reset or overridden by commands from Earth.

Power was fed to individual experiments through overload control units. There were two such units, each with six individual outputs. Each overload control unit had an overload trip sensor that opened the circuit if the total current drawn by that unit exceeded a preset value. The six outputs could be individually switched on/off by command.

Telecommunications

Communications between the spacecraft and ground stations on Earth were carried out in S-band. Command information from the Deep Space Network on Earth was received at the spacecraft by low-gain, broad beamwidth antennas with near spherical total angular coverage. High data rate information from the spacecraft was sent to Earth by a steerable high-gain parabolic antenna or a backup medium-gain antenna.

Antennas – The spacecraft contained two omnidirectional (omni) antennas, one high-gain parabolic antenna, and a backup high-gain antenna. One omni antenna was mounted on top of the despun mast that held the high-gain parabolic antenna. The other omni antenna was mounted on a short mast that extended on the opposite side of the spacecraft. NASA documentation referred to the two antennas as the forward omni and the aft omni.

The forward omni was a log conical spiral type that provided antenna coverage from zero degrees (along the +Z axis) to 95 degrees. The aft omni antenna was a turnstile element mounted over a conical ground plane. The antenna and ground plane were constructed from aluminum to withstand heat from the plume of the orbit insertion rocket motor. It provided omnidirectional coverage in the azimuth plane and coverage from 95 degrees to 180 degrees from the +Z axis in the elevation plane.

The high-gain antenna was a parabolic reflector type 1.09 meters in diameter. It was mounted on the despun mast. It had a gain of 25 dB and beamwidth of 7.6 degrees at the transmitted frequency of 2295 MHz. The high-gain antenna could also be used to receive commands from Earth at a frequency of about 2112 MHz. The antenna beam could be directed ±15 degrees in the elevation plane from normal to the spin axis by tilting the reflector by a jackscrew driven by a motor in response to commands from Earth. It could be steered through the full azimuth plane. Both the azimuth and elevation angles of the high-gain antenna were periodically adjusted during the mission to keep Earth within the beamwidth of the antenna.

The high-gain parabolic antenna contained an X-Band feed horn as well as an S-Band feed structure. The beamwidth of the antenna at X-Band was 2.2 degrees. The X-band signal was unmodulated and it was only transmitted from the spacecraft during occultation experiments along with the S-Band signal.

The backup high-gain antenna was a sleeve dipole type mounted on the mast above the parabolic antenna. Its purpose was to provide communications with Earth in case of failure of the parabolic antenna or failure of the despin mechanism. It had an omnidirectional pattern in the plane perpendicular to the spin axis (azimuth plane) and a narrow beamwidth in the plane of the spin axis (elevation plane). The peak gain was about 7.5 dB and the beamwidth in the elevation plane was about 10 degrees.

A series of switches allowed any antenna to be connected to either of two redundant receivers. Switches also allowed the transmitter to be connected to any antenna.

Commands from Earth – Commands from Earth were transmitted to the spacecraft by the very large parabolic antennas of the Deep Space Network stations. Pioneer Venus Orbiter contained two redundant transponders each containing a receiver and an exciter. One receiver was tuned to an input frequency of 2112.289 MHz and the other was tuned to a frequency of 2112.630 MHz. Operators on the ground could select one or the other receiver by transmitting the frequency for that receiver.

The receiver included a phase lock loop and detectors that demodulated the uplink command data. The receiver generated a coherent signal offset from the received carrier frequency to drive the exciter. The downlink transmitted signal was phase coherent with the uplink carrier and that allowed Doppler shift of the communication path to be measured. The two-way Doppler shift provided a very accurate measure of relative velocity between the spacecraft and the antenna on earth. The velocity accuracy was reported to have been about three meters per hour (0.00083 meters per second).

Commands were transmitted to the spacecraft in PCM/FSK/PM format where PCM denotes a digital signal form, FSK is frequency shift keying, and PM is phase modulation. The command word was transmitted to the spacecraft at a rate of four bits per second. A subcarrier was encoded as phase modulation and it contained ones and zeros of binary signals in the form of frequency tones. The output data stream from the receiver was PCM/FSK. Data one was a tone at a frequency of 250 Hz and data zero was a frequency of 100 Hz.

The PCM/FSK data stream at the output of each receiver was applied to a Command Processor that demodulated the FSK sequence and verified a 13-bit sync code. The FSK demodulator was paired to a receiver. The subsequent processing elements of the Command Processor were redundant between the two Command Processors and either could be connected to either FSK demodulator. Each of the Command Processors contained a 128-bit memory to store commands that were to be acted on later.

The selected Command Processor routed the commands to one of six Command Output Modules. The Command Output Modules, which interfaced with the element in the spacecraft to be controlled, could each disperse 64 discrete (on/off) commands and four quantitative commands. The quantitative commands, which represented magnitude, were in the form of 16-bit words. A bit in the command word signified whether the command was a discrete command or a quantitative command.

Prior to transmitting a command or a contiguous series of commands, a 48-bit preamble consisting of alternating zeros and ones was transmitted to the spacecraft. The command words that followed were 48 bits in length. Of those, the first 13 bits were used for synchronization and the next two bits were used to identify whether the data was for the Orbiter or the Multiprobe. The Orbiter code was 1–0. The next three single bits commanded discrete functions: Command Processor select, stored command or real time command, and time delay or no time delay. The next four bits addressed a particular Command Output Module. The next bit in the command word identified whether the data was discrete or quantitative. The next two bits represented address of the particular quantitative command. A series of 16 bits followed that represented the magnitude of the quantitative command. The last seven bits were parity bits to check validity of the information bits (13 through 40) of the command word.

The command structure included a total of 381 assigned commands including both discrete (on/off) commands and magnitude commands. The word size for magnitude commands was 16 bits.

Data transmission to Earth – The two redundant transponders in the spacecraft contained exciter units to develop and format signals to be transmitted to Earth. The redundant receivers in the transponders provided carrier signals to the exciters. The output of either receiver could be switched to either exciter. The carrier signals to the exciters were coherent with the uplinked carrier and at a frequency that was a set ratio of the uplink frequency. The coherency of the transmitted signal with the uplink signal allowed equipment on the ground to measure the two-way Doppler component of the signal received from the spacecraft.

The selected exciter was phase shift keyed (PSK) modulated by telemetry data from the experiments and from spacecraft subsystems. The output signal from the exciter was at a power level of about 126 milliwatts at the transmitted frequency of about 2294 MHz. Since the uplink frequencies were slightly different in frequency, the transmitter frequency was also slightly different, depending on which receiver was used.

A switch selected the output of one or the other transponder to drive the transmitter. The transponder signal was applied to a four-way power splitter. Each of the four outputs of the power splitter was applied to a driver amplifier that drove a power amplifier. There were four power amplifiers. Each generated a minimum of 9.5 watts at the transmitted frequency. The nominal power output was 10 watts.

The four power amplifier outputs were grouped into two pairs for redundancy. Switches allowed selecting one of the two amplifiers from each pair. A selected amplifier from each pair was applied to a power combiner that resulted in a nominal power level of about 20 watts. Switches were arranged so that one power amplifier output of 10 watts could be applied to any antenna or the summed output of 20 watts could be applied to any antenna.

Telemetry data – Data from the experiments and engineering data from spacecraft systems were conditioned and assembled into data words by the Data Handling Subsystem. The processed and assembled telemetry data was fed to the communications subsystem for transmission to Earth.

The Data Handling Subsystem included eight Digital Input Modules (DIMs) that gathered engineering data from the spacecraft systems and from the experiments. Each DIM contained a 32-channel commutator to sample data from the various sources. The DIMs accepted analog, serial digital, and on/off data from experiments and spacecraft subsystems and multiplexed the data onto a single output line for transfer to two PCM Encoders. The DIMs provided a clock signal to read out serial digital data sources and a precision one milliampere current source for resistive transducers. The output from any DIM could be switched to either of two PCM Encoders

The PCM Encoders converted analog data to eight-bit digital form and reformatted digital data. The digital bit stream at the output of the PCM Encoders was applied to two redundant Telemetry Processors. The Telemetry Processor generated clocks and timing signals for the Data Handling System and it organized data from the PCM Encoders into minor frames and major frames. The minor frames consisted of 64 eight-bit words. There were 64 minor frames in a major frame.

There was a total of 13 preprogrammed minor frame formats associated with particular segments of the mission. For example, there were five formats known as Periapsis A through D that could be commanded at various times when in the vicinity of periapsis of the orbit. Each format assembled data from a selected set of scientific experiments. The Periapsis E frame, for example, included 28 words from the radar mapper, 18 words from the ultraviolet spectrometer, four words from the infrared radiometer, and four words from the magnetometer. Engineering data was given priority in the Launch and Cruise minor frame where 34 words were allocated to engineering data. A total of 21 words were allocated to experiments that were active during cruise. In addition to the 13 preprogrammed minor frame formats, there was one programmable minor frame that could be assembled during the mission.

The Telemetry Processor generated a subcarrier at 16,384 kHz and applied telemetry data to the subcarrier in the form of biphase modulation. The biphase modulated carrier was sent to transponders in the communications subsystem for transmission to Earth.

The Data Handling Subsystem contained two storage units capable of 524,288 bits of storage each for a total of 1,048,576 bits. Data could be put into the storage units by the Telemetry Processor when the spacecraft was behind Venus and communication with Earth was blocked. The data was read out of memory by the Telemetry Processor and transmitted to Earth at a later time.

Attitude and Maneuvering Control

Thrusters – The Pioneer Venus Orbiter was maneuvered by controlling the firing of selected thrusters out of a group of seven small thrusters mounted to the spacecraft. The thrusters were used to control orientation of the spin axis, control the spin rate, and impart small velocity increments to the spacecraft. The thrusters, which generated about 22 newtons (5 pounds) of thrust, were developed by Hughes Aircraft Company.

The thruster assemblies consisted of a thruster valve, cylindrical combustion chamber, and a conical expansion nozzle. Scaling from photographs, the cylindrical reaction chamber was about 5 cm in diameter and 9 cm long. The conical expansion nozzle that attached to the reaction chamber was about 7.5 cm long with an exit diameter of about 5.6 cm.

The thrusters used hydrazine as propellant. Hydrazine is a monopropellant liquid that spontaneously decomposes when in contact with a catalyst. A bed of catalyst was placed inside the reaction chamber of each thruster. The gasses resulting from the decomposition exhausted through a conical nozzle generating thrust.

Hydrazine was contained in two tanks, each holding 15.9 kg of hydrazine. The tanks were pressurized with nitrogen gas to force the propellant out into the thrusters when selected thruster valves were opened. Temperatures of the hydrazine tanks and propellant flow paths were maintained between 4 and 60 degrees centigrade to keep the hydrazine from freezing. The temperature of the thruster catalytic bed and nozzle was maintained above −6.6 degrees centigrade.

There were four thrusters known as radial thrusters and three thrusters known as axial thrusters. The thrusters could be fired independently or in pairs. They could be fired continuously or pulsed with ON times of either 128 milliseconds or 512 milliseconds.

The three axial thrusters were offset from the spin axis and used to change the orientation of the spin axis (precession) and to impart axial velocity increments to the spacecraft. Orientation of the spin axis could be controlled by pulsing the axial thrusters individually at the spin frequency or pairs of radial and axial thrusters at the spin frequency.

The four radial thrusters were used to control the spin rate and to impart radial velocity increments to the spacecraft. When fired individually, the spin rate changed by 0.09 rpm per second of firing. When fired as a pair, the spin rate changed by 0.19 rpm per second of firing.

Attitude sensing – The Orbiter spacecraft used gyroscopic properties of the spinning mass to maintain attitude stability about a preselected spin axis. The orientation of the spin axis was controlled by using sun and star sensors to determine spacecraft orientation relative to the sun and stars. The sun was the primary celestial reference source to set the orientation of the spin axis. Output from the sun sensor was used to set up an acceptance angle gate for viewing a selected star.

The sun sensor had a narrow slit aligned with the spin axis such that a brief view of the sun occurred once each spin cycle. The light pulse was detected and the associated electronics generated an electrical pulse. The sensor contained a second narrow slit canted relative to the first slit as shown in the sketch below (Fig. 4.15). Slit "A" was aligned with the spin axis and the slit "B" was canted from it. The angle of the sun from the spin axis was determined by measuring the time delay, Δt, between detected light pulses from the two viewing slits.

Two sun sensor assemblies, a Mid-Range assembly and an Extended Range assembly, were used to provide complete angular coverage. The two assemblies were mounted on top of the spacecraft near the periphery to give them a wide elevation viewing angle. The mid-range assembly provided angular coverage of 55–125 degrees from the +Z axis. The Extended Range assembly provided upper range coverage of 15–90 degrees and the lower range provided coverage from 90 to 165 degrees from the +Z axis. Two redundant sun sensor elements were provided

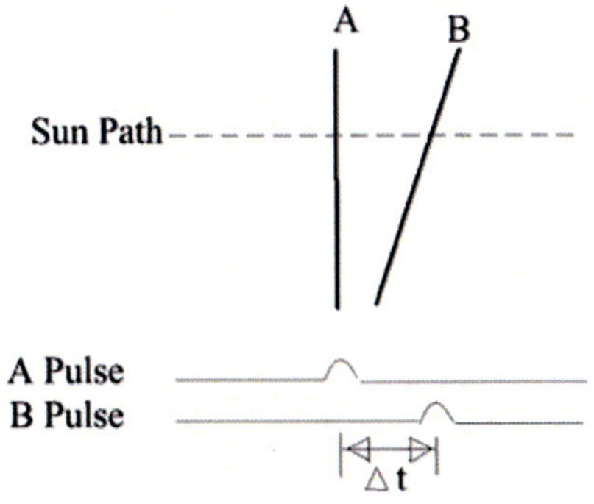

Fig. 4.15 Viewing angles of sun sensor

for each angular coverage range. One set of three sun sensor outputs was applied to one Attitude Data Processors and the other set was applied to a second Attitude Data Processor.

The star sensor was also a dual slit instrument. It had a 24-degree elevation field of view centered 56 degrees from the positive spin axis. The azimuth field of view of the slits was 0.5 degrees. The principle of operation of the dual slit star sensor was the same as the sun sensor. The star sensor had two redundant sensor channels. Either channel could be applied to either Attitude Data Processor.

Outputs from the sun sensors and star sensor were applied to redundant Attitude Data Processors along with command data from the ground. The Attitude Data Processors developed signals to control the thrusters to maneuver the spacecraft as required.

Science Experiments Carried by Pioneer Venus Orbiter

The Pioneer Venus Orbiter carried 12 science instruments. They can be broadly grouped into those experiments that operated in the near environment of the spacecraft and those that gathered information by remote sensing.

Spacecraft Environment Experiments

- Neutral mass spectrometer
- Ion mass spectrometer
- Electron temperature
- Retarding potential analyzer
- Magnetometer
- Electric field detector
- Solar wind plasma analyzer

Remote Sensing Experiments

- Surface Radar mapper
- Infrared radiometer
- Ultraviolet spectrometer
- Cloud photopolarimeter
- Gamma ray burst detector

In addition to the 12 experiments conducted by these instruments, five additional experiments were conducted using transmitted signals from the spacecraft. Those experiments were:

- Dual frequency radio occultation
- Density distribution
- Orbital celestial mechanics
- Atmospheric turbulence

A description of each of the experiments and several photographs of the instruments is given in NASA report SP-461, *Pioneer Venus*, by Fimmel, Colin, and Burgess. The report was assembled in book form by NASA Scientific and Technical

Information Branch and the report is available on the internet. In recognition of that work, this book will only cover experiments in the remote sensing category and provide a little more engineering detail than given in SP-461. The magnetometer was described in SP-451 but a few more details are provided here.

Magnetometer Experiment

Three fluxgate type magnetometers were mounted on a boom to distance the magnetic sensors from the spacecraft. The boom was a three-element hinged structure that was folded on top of the Orbiter to fit within the launch shroud. The boom assembly was released by pyrotechnic pin pullers a few hours after launch and the boom extended by centrifugal force of the spinning spacecraft. The boom, which was built from non-magnetic graphite-epoxy material, was about 4.7 meters (15.5 feet) long when fully extended. It tapered from 10.5 cm in diameter at the base to 3.8 cm diameter at the tip.

Two magnetometers were located at the end of the boom. One measured fields parallel to the spin axis and the other perpendicular to it. A third magnetometer was located about 3.3 meters from the base of the boom and it was aligned to measure fields 45 degrees from the spin axis. Measurements from the third magnetometer were used to cancel some of the residual magnetic field from the spacecraft.

Previous spacecraft to Venus determined that its magnetic field was very low. Consequently, the upper measurement range of the magnetometer for Pioneer Venus was set at 128 nanotesla. The resolution of the magnetometers ranged from 0.06 to 0.5 nanotesla, depending on field strength. For reference, the strength of Earth's magnetic field is 25,000–65,000 nanotesla. Magnetometer measurements by the Pioneer Venus spacecraft indicated that the intrinsic magnetic field of Venus was less than 10^{-5} of the magnetic field of Earth.

Surface Mapping Radar

The surface mapping radar had two modes of operation: a radar altimeter mode and a side-looking imaging mode. The radar altimeter mode was used to generate topographical data from the surface at the suborbital point. The side-looking imaging mode was used to obtain low-resolution radar maps of the terrain at angles to the side of the suborbital track.

Radar Altimeter mode – The radar altimeter mode employed a pulsed radar to measure variations in surface elevations at the suborbital point during spacecraft orbit around Venus. Pioneer Venus Orbiter was in a nearly polar orbit inclined 106 degrees from the equator. The orbit extended to apoapsis at 67,000 km altitude and descended to about 150 km at periapsis. The period of the orbit was about 24 hours. The suborbital point at periapsis was at latitude 16 degrees north. The altitude of the spacecraft with respect to the mean elevation of the surface of Venus was accurately known from tracking data from Earth. The radar measurements gave the surface elevation at the suborbital point with respect to the mean elevation.

The eccentricity of the orbit took the spacecraft down to 150 km above the surface at periapsis but the range to the surface increased rapidly as the spacecraft continued in its orbit. The radar needed to be within about 4700 km from the surface to have sufficient signal-to-noise ratio for accurate altitude measurements. That constraint, and the inclination of the orbit, limited the latitude of measurements by the radar to between 74 degrees north and 61 degrees south. About 92% of the surface of Venus is between these latitude limits and was mapped by the radar altimeter.

The Pioneer Venus Orbiter spacecraft rotated about its spin axis at a rate of five revolutions per minute or one revolution each 12 seconds during its orbit around Venus. The radar antenna beam was positioned to swing by the suborbital point on the surface once each rotation of the spacecraft. The orbital velocity was about 10 km per second near periapsis. The distance covered on the surface between radar altimeter measurements was about 120 km near periapsis of the orbit.

The near polar orbit of the spacecraft about Venus had a period of 24 hours (one Earth day). Venus rotates with a sidereal period of 243 Earth days. Thus, the surface of Venus was displaced in longitude by 1.48 degrees each successive orbit. The equivalent distance on the surface at the equator was about 150 km.

The radar used a coherent, pulse compression process in the radar altimeter mode. Pulse compression allowed accurate range measurement to be made using a relatively wide pulsewidth and a transmitted power of only 20 watts. The transmitted pulse was modulated by phase shift keying in the form of a 55-bit pseudorandom code. The bit length was 4 microseconds (μs) and the 55-bit sequence length was 220 μs at altitudes below 1515 km. At altitudes above 1515 km the bit length was 6 μs and the sequence length was 330 μs. The pulsewidth was equal to the sequence length. Resolution of the radar was related to the bit length. The relatively long pulsewidth provided a significant amount of effective energy per pulse with relatively low peak transmitted power.

The pulse repetition interval was 5280 μs at altitudes below 700 km, 10,780 μs at altitudes between 700 km and 1515 km, 9240 μs at altitudes between 1515 km and 2575 km, and 5280 μs at altitudes above 2575 km. An ephemeris generator informed the radar of the approximate altitude of the spacecraft at all times and that allowed resolving range ambiguities with the short pulse repetition interval at higher altitudes.

The antenna was a backfire reflector type mounted at the edge of the cylindrical spacecraft on the end that contained the high-gain antenna. The diameter of the reflector was 35 centimeters. A photograph of the antenna and its positioner is shown below (Fig. 4.16). The beamwidth of the antenna was 25 degrees in a plane parallel to the spin axis and 29 degrees in a plane perpendicular to the spin axis. The gain of the antenna was 15 dB. The antenna could be positioned in the plane of the spin axis from the direction of the spin axis, which was the zero-degree position, to 140 degrees from the spin axis. The antenna position was moved during the orbit so that the antenna beam was normal to the surface of Venus at the suborbital point. For example, the antenna position was 90 degrees from the spin axis when the spacecraft was overflying the equator of Venus.

Fig. 4.16 Antenna and positioner for radar mapper. (NASA photograph)

In the radar altimeter mode, the received biphase modulated pulse was decoded and applied to a correlator. Velocity data from the ephemeris generator was used to adjust the frequency of the receiver to accommodate the Doppler shift of the return signal. Altitude data from the ephemeris generator was used in the signal processor to set up an acceptance gate for the echo pulse. The acceptance gate was in the form of trial delays of a reference 55-bit pseudorandom code. Correlation of the received signal with delayed reference pseudorandom code was made for 64 different delays of the reference code. Each trial reference code was separated in timing by the duration of one bit of the code. The delay setting that gave the highest value of correlation corresponded to the time delay of the echo pulse and that gave a measure of altitude of the spacecraft at that time. The amplitude of the received signal was also measured and used to compute the reflectivity of the surface at normal incidence.

Radar altimeter data from the Pioneer Venus Orbiter was assembled in Mercator projection color map form by NASA and the Massachusetts Institute of Technology. The map is shown on the next page (Fig. 4.17). The highest mountain measured on Venus was Maxwell Montes at 10,500 meters (34,440 feet) above the mean elevation of the planet. Maxwell Montes is located at about zero degrees longitude and 65 degrees north latitude on the map. The lowest point measured was 2500 meters (8200 feet) below mean elevation.

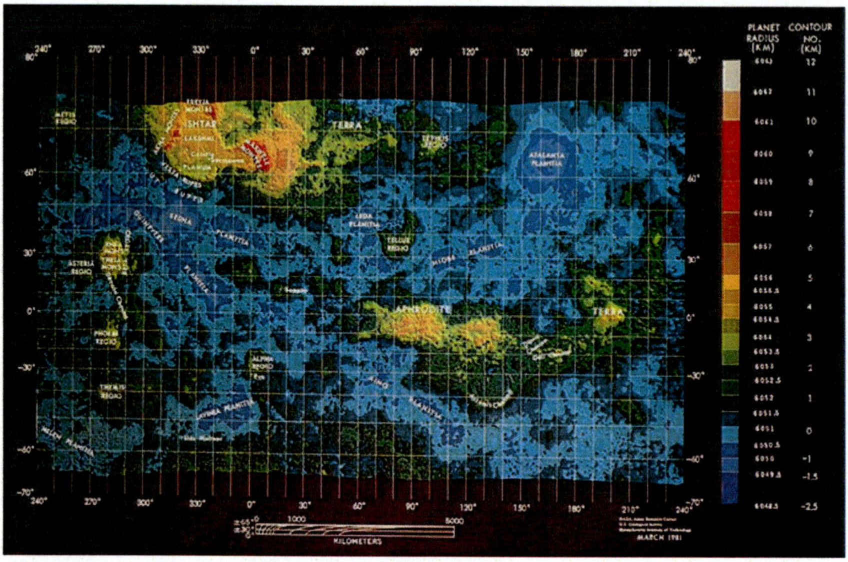

Fig. 4.17 Topographical map of Venus from Pioneer Venus Orbiter radar altimeter data. (NASA graphic)

The accuracy of the altitude measurements was better than 200 meters. The footprint size of the altitude measurement at the suborbital point was about 10 km at a spacecraft altitude of 200 km and about 50 km at a spacecraft altitude of 4700 km.

Side looking imaging mode – The side-looking imaging mode could only be used at altitudes below 550 km where adequate signal-to-noise ratio was obtained. The transmitted pulse was unmodulated in this mode with a pulsewidth of 128 μs and pulse repetition interval of 5280 μs. The side-looking mode was active during the interval of the spacecraft spin just before the antenna beam reached the suborbital point and again just after it swung by the suborbital point.

The imaging mode could be commanded from the ground or from stored commands. Imaging consisted of establishing an eight-by-eight cell formation on the surface of Venus consisting of eight range intervals and eight Doppler intervals. The Doppler intervals were along the direction of travel of the spacecraft. The center of the range intervals was set by the ephemeris generator so that the center of the imaged area on the surface was offset from the suborbital point track by about 260 km.

The output of the receiver in the imaging mode was a quadrature pair (usually referred to as I and Q channels). Each channel was applied to an analog-to-digital converter and the resulting digital signals were separated into eight successive range intervals. The quadrature signals for each range interval were processed by an eight-point Fast Fourier Transform (FFT) to extract amplitude information in eight Doppler frequency bins. The resulting amplitudes of the 64 range/Doppler cells were telemetered to Earth and used to generate low resolution radar images of the surface. The resolution of the images was about 20 km at the lowest spacecraft altitudes and about 40 km at higher altitudes.

Orbiter Infrared Radiometer

The Orbiter Infrared Radiometer (OIR) measured the thermal emission from clouds of Venus and reflected solar radiation from the clouds. The goal of the OIR was to measure thermal emission from the atmosphere above the clouds at different pressure levels. The OIR was an eight-channel instrument that observed the clouds through a telescope equipped with a 48 mm parabolic mirror. The channels observed in different wavelength bands. The field of view was 1.25 degrees for channels 2 through 8. The field of view of Chanel 1, which used a volume of pressure modulated CO_2 gas as a selective filter, was about 5 degrees.

The instrument was mounted on the spacecraft with its field of view at an angle of 45 degrees from the spin axis. The field of view scanned across the clouds as the spacecraft rotated at five revolutions per minute. Radiometer measurements were made when the spacecraft was near periapsis for durations of about 90 minutes each orbit.

Thermal measurements concentrated on the wavelengths near the absorption lines of carbon dioxide at 2 microns and at 15 microns. The emission characteristics of CO_2 change with pressure and since the pressure of the atmosphere changes with altitude above the surface of Venus, close measurements of emission as a function of wavelength allowed sounding of temperature as a function of altitude.

The OIR used a grating monochromator to select specific wavelengths in the 15-micron band for channels 2 through 5. One of the channels was adjustable by commands from Earth to three different wavelengths. The seven selectable wavelengths allowed temperatures to be measured at seven different altitudes in the range of 60 km to 150 km.

Channel 1 of the instrument used a pressure modulated sample of carbon dioxide gas as a spectrally selective filter in the 15-atmosphere above the micron CO_2 band. The pressure in the CO_2 gas cell was modulated at a frequency of 40 Hz by a resonant piston. Incoming radiation was passed through the gas cell before being applied to detectors. That channel was used to measure temperatures in the altitude range of about 100 km up to the altitude of the exosphere above 220 km.

Channel 6 used a narrow filter centered at the CO_2 band at 2 microns. Channel 7 was a wide band channel that covered the range of 0.2–4.5 microns. It was used to measure the radiance from reflected solar radiation from the clouds. Channel 8 used a filter centered at the water vapor band around 45 microns. Data from that channel was used to map water vapor in the clouds.

Detectors at the output of each channel generated output proportional to the amplitude of the radiation. These outputs were applied to the Data Handling Subsystem for formatting for transmission to Earth.

A paper by Taylor et al. (1979), reported some of the results of the OIR during orbiting of Venus. A series of curves in the paper indicated very little temperature difference between the nightside and dayside of the planet in the altitude range of 63 km to 100 km. The temperature on the nightside was about 175 K from altitude of 100 km down to about 85 km. Below 85 km, the temperature increased approximately linearly to about 252 K at 63 km altitude.

The size of the OIR instrument was 248 × 175 × 251 mm. It weighed 5.9 kg.

Ultraviolet Spectrometer

The Orbiter Ultraviolet Spectrometer (OUVS) measured ultraviolet radiation emitted by the clouds and from sunlight scattered and reflected by the clouds and haze layers. The instrument was mounted so that the field of view was centered 60 degrees from the +Z axis of the orbiter. The spacecraft orbited with the +Z axis in the direction of the south pole of the planet. Reference pulses were generated by the spinning space craft at selected points in its rotation. These pulses were used by the OUVS to time when to take data. A RAM pulse was generated when the X-axis of the spacecraft was along the velocity vector and a NAIDIR pulse was generated when the X-axis was directed at the planet.

The instrument used a Cassegrain telescope with an aperture diameter of 250 mm. The output of the telescope was focused on a monochromator with a 6.0 mm by 0.6 mm entrance slit that resulted in a field of view of 1.4 degrees by 0.14 degrees. The output of the slit was directed at a diffraction grating. The diffraction grating was rotated by a stepping motor though a 27-degree range in 512 steps. Each step corresponds to a wavelength shift of 4.4 Å and the total range covered was 2253 Å. The symbol Å, called an angstrom, denotes wavelength of 10^{-10} meters. The stepping motor could be commanded to step through the full range or to remain fixed at one setting.

The output of the monochromator was applied to two photomultiplier tube detectors. One tube was sensitive to wavelengths in the band 1100 Å to 1800 Å and the other was sensitive to wavelengths of 1600 Å to 3300 Å. The 1100 to 1800 Å wavelength range was referred to as the F channel and the 1600 to 3300 Å wavelength range was referred to as the G channel. Each detector was followed by a pulse amplifier. The output of the either the F channel or the G channel was selected and the output pulses were applied to a 15-bit counter with selectable integration time. The most significant four bits was shifted out of the counter and 4 bits were added to indicate the number of shifts. The resulting 8-bit word was applied to a 256-word data buffer. When the data buffer was full it was read out into the telemetry system.

A paper by Stewart et al. (1979), gives some results of the OUVS experiment on the orbiter. A scan of the field of view across the lighted disk of Venus at a wavelength of 2662 Å indicated reflectivity agreeing closely with a model consisting of a homogeneous layer of sulfuric acid droplets under a clear layer of carbon dioxide. Absorption features of sulfur dioxide were observed in the atmosphere above the clouds. Emissions found in the ultraviolet spectrum of the airglow indicated presence of atomic hydrogen and atomic oxygen.

The size of the OUVS instrument was about 35 cm by 14.2 cm by 14.2 cm. It weighed 3.1 kg.

Cloud Photopolarimeter

The Orbital Cloud Photopolarimeter (OCPP) was used to generate images of the clouds around Venus at ultraviolet wavelength and to determine the size, shape, and refractive index of cloud particles by observing amplitudes and polarization of reflected sunlight at different wavelengths.

The central instrument for the OCPP was a telescope having an aperture of 3.7 cm and focal length of 17.5 cm that collected scattered sunlight from the clouds. The field of view of the telescope could be steered in the plane of the spin axis from the +Z axis (zero degrees) to 145 degrees from the +Z axis of the spacecraft. The spin stabilized spacecraft orbited with its +Z axis along the south polar axis of Venus. There were 63 commandable look angles within a 100-degree range that could be selected from the ground. A particular look angle was selected to direct the field of view at the planet depending on the location of the spacecraft in its orbit.

A 16-position filter wheel at the output of the telescope allowed selection of particular wavelength for observation. The cloud imaging mode used a filter centered at an ultraviolet wavelength of 365 nm. The notation nm is nanometers or 10^{-9} meters. The polarimetry mode could observe within four spectral bands: 255–285 nm (far ultraviolet), 255–380 nm (ultraviolet), 540–555 nm (visible), and 930–945 nm (infrared). There were three filters available at different wavelengths in each band. Silicon photodetectors converted the amplitude of the ultraviolet signals from the telescope to analog electrical form.

Imaging mode – The field of view of the telescope in the imaging mode was 0.4 by 0.45 milliradians (about 0.025 degrees) at 365 nm. The resolution of the image on the clouds was about 30 km at the nadir when viewed from apoapsis (the high point of the orbit) at 66,000 km. The rotating spacecraft swept the field of view of the telescope across the planet while motion of the spacecraft advanced the scan along the surface of the clouds between the rotation scans to form a two-dimensional image. The amplitude of the signal from the photodetector was sampled each 0.488 milliseconds and digitized by an 8-bit analog-to-digital converter. Each scan line consisted of 1016 8-bit words. The digital data plus 64 status bits was placed in an 8192-bit memory. The memory was read out into the telemetry system for transfer to Earth.

Each image consisted of about four hours of scanning. At the spacecraft spin rate of five RPM, 300 scan lines would be generated each hour and 1200 lines in a four-hour period. Each line consisted of 1016 data points. Pairs of images separated by about four hours were taken to determine motion of the clouds. Small features within the clouds allowed tracking. Two ultraviolet images of Venus taken a day apart are shown below (Fig. 4.18). A marking referred to as a "horizontal Y" is apparent in the images.

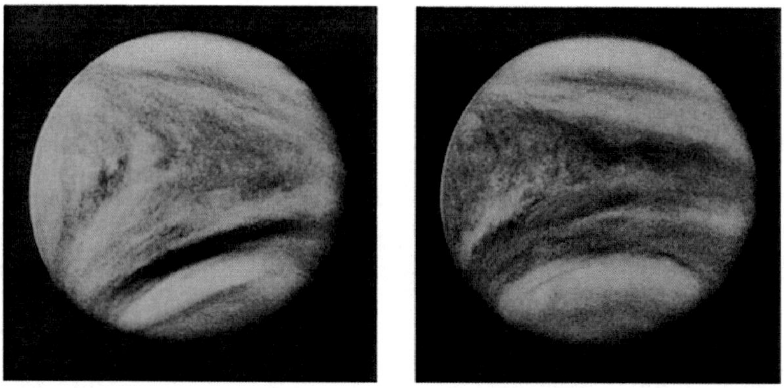

Fig. 4.18 Images of clouds of Venus one two consecutive days taken at ultraviolet wavelength. (NASA images)

Analysis of the ultraviolet images over time indicated that the clouds are pushed by high velocity winds in the same direction as the rotation of the planet. The average zonal period of rotation of the clouds around the planet was found to be about five days. This translates to an average wind speed of 320 km per hour (89 meters per second) at the upper altitude of the clouds of 70 km. The winds were found to have complex wave modes. In addition to the zonal rotation, there was an equatorial wave with a period of four days.

Polarimetry mode – The polarimetry mode was used to determine the size, shape, and refractive index of cloud particles around Venus. The intensity and the polarization of the scattered signal were measured in four wavelength bands from the ultraviolet to infrared: 255–285 nm (far ultraviolet), 255–380 nm (ultraviolet), 540–555 nm (visible), and 930–945 nm (infrared). The size of the particles was determined quite accurately from the intensity of scattering at the different frequencies. The polarization angle was determined by first passing the optical signal through a retarder plate sequentially set at an angle of 0°, 22½°, and 45° with respect to a reference plane in the instrument. The signal at each retarder setting was split into two orthogonally polarized optical signals. The amplitudes of the two signals were measured simultaneously by two photodetectors for each position of the retarding plate. The two signals were sampled simultaneously at 9.52 milliseconds intervals and the sampled voltages were digitized by 11-bit analog-to-digital converters. These data were used to determine the magnitude and the polarization angle of the scattered sunlight signals.

From measurements made in the polarimetry mode, scientists were able to determine that the clouds contain droplets of sulfuric acid (H_2SO_4). The diameter of the droplets was determined to be two to three microns. A haze layer existed above the clouds.

Gamma Ray Burst Detector

The Orbiter Gamma Ray Burst Detector (OGBD) experiment was part of an experiment involving other spacecraft to determine the origins of gamma ray bursts. Triangulation of the temporal gamma ray burst characteristics with data from similar instruments on other spacecraft provided a long baseline sensor path to determine direction of arrival of the bursts. Other spacecraft involved in the experiment included the International Sun-Earth Explorer 3 (ISEE-3), Helios-2, Vela, and the Soviet Union Venera 11 and Venera 12 spacecraft.

The OGBD recorded temporal and spectral characteristics of gamma ray bursts. The experiment included two sensors mounted on the upper periphery of the spacecraft diametrically opposite one another. The two sensors provided nearly complete angular coverage of space around the spacecraft. The sensors consisted of Sodium Iodide (NaI) scintillator detectors followed by photomultiplier tubes. The NaI crystal absorbed energy from the gamma ray and responded by emitting a flash of light. The wavelength of the light was about 415 nanometers (violet light). The intensity of the light was proportional to the energy of the gamma ray. The flash of light was optically coupled to a photomultiplier tube that generated an electrical pulse proportional to the strength of the burst.

The first gamma ray burst location to be determined with precision from data from several spacecraft was a burst received on 5 March 1979. That event produced a gamma ray burst of historic energy. Triangulation located the origin of the burst to be in a supernova remnant called N49 in the large Magellanic Cloud. The Large Magellanic Cloud is a small galaxy that orbits our home galaxy at a distance of about 160,000 light years.

Pioneer Venus Multiprobe

The Multiprobe spacecraft consisted of a bus, three small probes, and one large probe. The probes were released from the bus when near Venus and gathered data on the clouds and atmosphere while descending to the surface. A photograph of the Pioneer Venus Multiprobe spacecraft is shown on the next page (Fig. 4.19). The spacecraft is being inspected by Charlie Hall, Program Manager for Pioneer Venus at NASA Ames.

The Pioneer Venus Multiprobe was launched from Cape Canaveral on 8 August 1978. After 123 days of travel, the large probe was released from the bus on 16 November at a distance of 11.1 million km from Venus. The three small probes were released on 20 November 1978 at 9.3 million km from the planet. All four probes began their descent through the Venusian atmosphere on 9 December 1978. The bus also entered the atmosphere on 9 December. It burned up at an altitude of about 110 km, shortly after entering the atmosphere.

The four probes were targeted to descend through the atmosphere in different regions of the planet. One small probe, called the North Probe, descended at a latitude

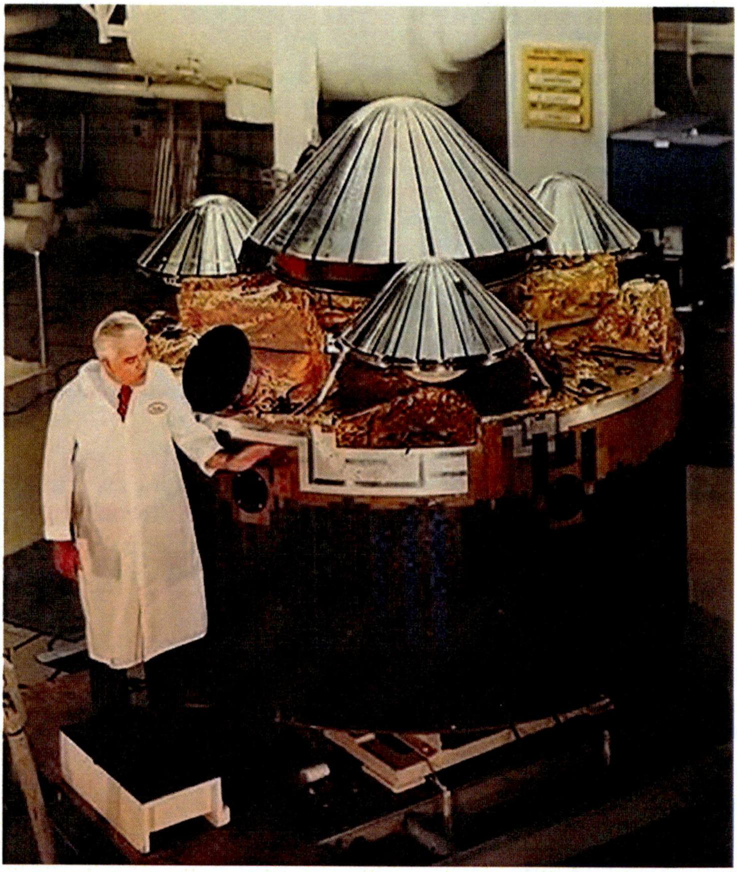

Fig. 4.19 Charlie Hill inspecting Pioneer Venus Multiprobe. (NASA photograph, cropped by author)

of 59 degrees North on the night side of the planet. Another small probe, called the Night Probe, descended at a latitude of 26.7 degrees South on the night side. The third small probe, called the Day Probe, descended at a latitude of 31.3 degrees South on the day side of the planet. The Large Probe descended near the equator at a latitude of 4.4 degrees North on the day side of the planet. A sketch showing the regions of the planet sounded by the probes is shown on the next page (Fig. 4.20).

Mechanical Configuration of Pioneer Venus Multiprobe

The general arrangement of the Multiprobe spacecraft is shown in the photograph in Fig. 4.19. It consisted of a cylindrical bus and four probes that descended independently to the surface of Venus. The large probe was mounted in the center of the

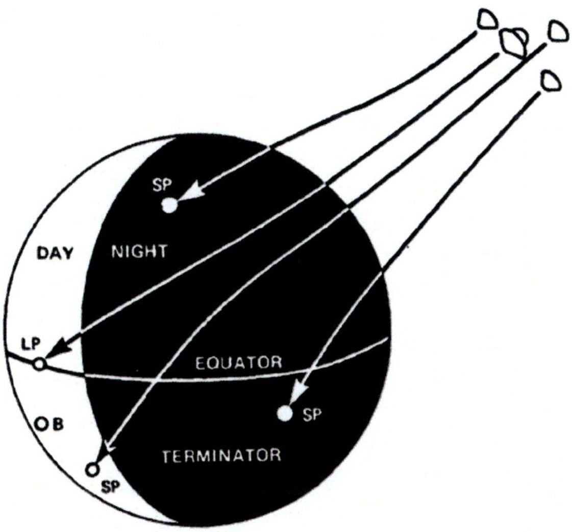

Fig. 4.20 Entry points of probes into atmosphere of Venus. (NASA graphic)

bus and the three spall probes were mounted with angular separation of 120 degrees with their centers at a radius of 99 cm from the Z-axis of the bus.

A drawing of the Multiprobe spacecraft with structural parts labeled is shown on the next page (Fig. 4.21).

The spacecraft was spin stabilized with the spin axis nearly perpendicular to the ecliptic plane until approaching Venus. The nominal spin rate was 15 RPM during cruise to Venus. The spin axis was adjusted to point along the trajectory of the large probe when near Venus.

The Pioneer Venus Orbiter and the Multiprobe bus were designed to have a significant amount of commonality. The cylindrical main structure of the Multiprobe bus was basically the same as that of the Orbiter. The thrust tube, equipment shelf structure, and shelf support struts were also essentially the same. Many of the subsystems including power, communications, data handling, and command subsystems were also essentially the same in the two spacecraft.

The cylindrical main body of the Multiprobe bus was 2.5 meters in diameter and the height from the bottom of the bus to the top of the large probe was 2.9 meters. A cylindrical solar array was installed around the circumference of the bus. The weight of the Multiprobe spacecraft, including probes, was 907 kg. The total weight of the probes was 585 kg and the weight of the bus including 32 kg of hydrazine fuel was 322 kg.

The thrust tube, labeled in Fig. 4.21, was the primary structural element of the spacecraft. The equipment shelf and support struts for the shelf were mounted to it. It also was used to connect the spacecraft to the launch vehicle. The circular equipment shelf, which was 248 cm in diameter, contained electronic packages for

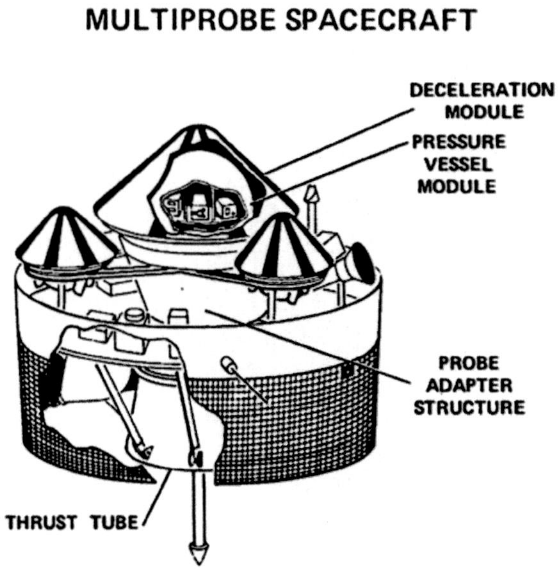

Fig. 4.21 Drawing of Multiprobe spacecraft. (NASA graphic)

spacecraft systems and the experiments. A series louver modules mounted to the bottom of the equipment shelf controlled thermal radiation out of the shelf. Two tanks holding hydrazine fuel for the thrusters were located below the equipment shelf and supported by structure connected to the thrust tube.

Each of the four probes was contained in a pressure vessel protected by a conical heat shield. The large probe was supported on the spacecraft by an inverted conical frustum that was attached to the thrust tube. The small probes were supported by structure connected to the inverted conical frustum.

The large probe was launched on 16 November 1978 when about 11.1 million km from Venus. It was launched by a spring separation system released by pyrotechnics. The three small probes were launched towards Venus on 20 November when the bus was about 9.3 million km from the planet. The small probes were retained by clamps until time for launch. When it became time to launch the small probes, the spin rate of the Multiprobe spacecraft was increased to 48 RPM, the clamps were released by pyrotechnics, and the small probes were spun off the bus by centrifugal force. The probes were spun off tangentially from the bus and that established their different trajectories to Venus. All four probes entered the atmosphere of Venus on 9 December 1948. The large probe entered the atmosphere a few minutes before the first of the small probes.

Multiprobe Bus Systems

Power Subsystem

Electrical power for the Multiprobe spacecraft was furnished by the cylindrical solar array during most of the mission. The solar array contained 13,060 solar cells, each two cm by two cm in size. The height of the array was 67 cm. This was somewhat less than the 79 cm height of the array on the Orbiter since not as much power was required for the Multiprobe bus. The solar array generated a minimum of 230 watts in the vicinity of Earth and 297 watts in the vicinity of Venus The output voltage of the solar array was limited to 30 volts DC by shunt limiters with resistive loads. Five limiters in parallel were used to dissipate excess power. Two of the limiters had their load resistors mounted on the back side of the equipment shelf. When the temperature of the shelf fell below about 1.7 °C, thermal switches lowered the set points of the two limiters and up to 320 watts of excess power from the solar arrays in 33-watt increments was applied to heat the equipment shelf.

Two nickel-cadmium batteries, each with capacity of 7.5 ampere hours, powered the spacecraft when the solar array did not provide adequate power. Each battery was controlled by a charge/discharge controller. The two batteries were connected in parallel across the power bus. The charge/discharge controllers managed charge of the batteries and applied battery power to the bus whenever the bus voltage dropped below 27.8 volts DC. Each battery was maintained in a fully charged state when it was not being used by a small boost charge solar array. A separate boost charge array consisting of 230 solar cells was provided for each battery.

DC power in the voltage range of 30.1–27.5 Volts (28 volts nominal) was provided to spacecraft systems and the experiments. The electrical load was divided up into four separate busses: Essential bus, Science bus, Switched Load bus, and RF Transmitter bus. The Essential bus could not be turned off. In case of overcurrent, the other three busses could be turned off sequentially to isolate the bus drawing excessive current.

Communications

Antennas – The bus contained two omnidirectional (omni) antennas and a medium-gain conical horn antenna. The omnidirectional antennas, referred to as forward omni and aft omni, were the same as used on the Orbiter. The conical horn antenna was located near the bottom of the bus and pointed parallel to the –Z spin axis. The diameter of the aperture of the horn was 45.6 cm. The horn had a peak gain of 13.8 dB and a beamwidth of 40 degrees. It was used for transmission during a time from just before launch of the probes when the spin axis of the bus was pointed towards Venus and Earth was within the field of view of the antenna.

Transmissions from Earth were received from one or the other omnidirectional antennas depending on which antenna had the better signal strength. Transmission to Earth also used one of the two omnidirectional antennas until the spacecraft was oriented such that the conical horn antenna could be used.

Commands from Earth – Commands from Earth were transmitted to the Pioneer Venus Multiprobe spacecraft by the very large parabolic antennas of the Deep Space Network stations. The Multiprobe spacecraft contained two redundant transponders each containing a receiver and an exciter. One receiver was tuned to an input frequency of 2111.266 MHz and the other was tuned to a frequency of 2110.584 MHz. Both receivers were always turned on. Operators on the ground could select one or the other receiver by transmitting the frequency for that receiver. The two frequencies were different from the two receive frequencies of the Orbiter.

The receiver included a phase lock loop and detectors that demodulated the uplink command bits. The receiver also generated a coherent signal offset from the received carrier frequency to drive the exciter. The downlink transmitted signal was phase coherent with the uplink carrier and that allowed Doppler shift of the communication path to be measured. The two-way Doppler shift provided a very accurate measure of relative velocity of the spacecraft and the antenna on earth.

Commands were transmitted to the spacecraft in PCM/FSK/PM format at a rate of four bits per second. A subcarrier was encoded as phase modulation and it contained ones and zeros of binary signals in the form of frequency tones. The output data stream from the receiver was PCM/FSK. Data one was a tone at a frequency of 250 Hz and data zero was a frequency of 100 Hz.

The command processing in the Multiprobe bus was essentially the same as described previously for the Orbiter. The command words were 48 bits in length. Of those, the first 13 bits were used for synchronization and the next two bits were used to identify whether the data was for the Orbiter or the Multiprobe. The Multiprobe code was 0–1.

The command subsystem provided commands to the probes as well as to the bus. Commands to the probes were used to conduct checkout of the probes before launch.

Data transmission to Earth – The two redundant transponders in the Multiprobe spacecraft contained exciter units to develop and format signals to be amplified for transmitted to Earth. The redundant receivers in the transponders provided carrier signals to the exciters. The output of either receiver could be switched to either exciter. The carrier signals to the exciters from the receivers were coherent with the uplink carrier and at a frequency that was a set ratio of the uplink frequency. The coherency of the transmitted signal with the uplink signal allowed equipment on the ground to measure the two-way Doppler component of the signal received from the spacecraft.

The selected exciter was phase shift keyed (PSK) modulated by telemetry data from spacecraft subsystems and from the experiments. The output signal from the exciter was at a power level of about 126 milliwatts at the transmitted frequency of about 2293 MHz. Since the uplink frequencies were slightly different in frequency, the transmitter frequency was also slightly different, depending on which receiver was used.

A switch selected the output of one or the other transponder to drive the transmitter. The transponder signal was applied to a four-way power splitter. Each of the four outputs of the power splitter was applied to a driver amplifier that drove a power amplifier. There were four power amplifiers. Each generated a minimum of 9.5 watts at the transmitted frequency. The nominal power output was 10 watts.

The four power amplifier outputs were grouped into two pairs for redundancy. Switches allowed selecting one of the two amplifiers from each pair. A selected amplifier from each pair was applied to a power combiner that resulted in a nominal power level of about 20 watts. One power output from one pair could be selected to be applied to the antennas. Switches were arranged so that one power amplifier output of 10 watts could be applied to any antenna or the summed output of 20 watts could be applied to any antenna.

Telemetry data – Data from the experiments and engineering data from bus subsystems were conditioned and assembled into data words by the Data Handling Subsystem. The processed and assembled telemetry data was fed to the communications subsystem for transmission to Earth. Checkout data from the probes before they were launched was included in the data stream.

The Data Handling Subsystem included eight Digital Input Modules (DIMs) that gathered data from the spacecraft systems (engineering data) and from the experiments. Each DIM contained a 32-channel commutator to sample data from the various sources. The DIMs accepted analog, serial digital, and on/off data from experiments and spacecraft subsystems and multiplexed the data onto a single output line for transfer to two PCM Encoders. The output from any DIM could be switched to either of two PCM Encoders

The PCM Encoders converted analog data to eight-bit digital form and reformatted digital data. The digital bit stream at the output of the PCM Encoders was applied to two redundant Telemetry Processors. The Telemetry Processor generated clocks and timing signals for the Data Handling System and it organized data from the PCM Encoders into minor frames and major frames. The minor frames consisted of 64 eight-bit words. There were 64 minor frames in a major frame.

There was a total of four preprogrammed minor frame formats: Bus Engineering, Bus Entry, ACS, and Command Memory Readout. In addition, there was one programmable minor frame that could be assembled during the mission and four subcommutated formats. The subcommutated format contained only one word in a minor frame.

The Telemetry Processor generated a subcarrier at 16,384 kHz and applied telemetry data to the subcarrier in the form of biphase modulation. The biphase modulated carrier was sent to transponders in the communications subsystem for transmission to Earth.

Attitude and Maneuvering Control

Thrusters – The Pioneer Venus Multiprobe bus was maneuvered by controlling the firing of selected thrusters out of a group of six small thrusters mounted to the spacecraft. The thrusters were used to control the spin rate of the spacecraft, control orientation of the spin axis, and impart a velocity increment to the spacecraft. The operation and characteristics of the thrusters were the same as described for the thrusters in the Pioneer Venus Orbiter described in the previous section.

Attitude sensing – The Multiprobe spacecraft used gyros to maintain attitude stability about a preselected spin axis. The orientation of the spin axis was controlled by using sun and star sensors to determine spacecraft orientation relative to the sun and stars. The sun was used as the primary celestial reference source to set the orientation of the spin axis. Output from the sun sensor was used to set up an acceptance angle gate for viewing a selected star.

The sun and star sensors were the same as used on the Orbiter spacecraft. Outputs from the sun sensors and star sensor were applied to redundant Attitude Data Processors along with command data from the ground. The Attitude Data Processors developed signals to control the thrusters to maneuver the spacecraft as required.

Science Experiments Carried by Multiprobe Bus

Only two experiments were carried on the Multiprobe bus since the bus did not have a protective aeroshell and it was expected to burn up upon entering the atmosphere of Venus. The two experiments were: a neutral mass spectrometer and an ion mass spectrometer. Data was collected from about 700 km altitude to about 110 km where the unprotected bus burned up.

Neutral Mass Spectrometer

The neutral mass spectrometer was designed to measure the density of components of the upper atmosphere of Venus. The instrument consisted of an ionization box where a sample of the atmosphere was bombarded by electrons from a tungsten filament to ionize the components. The resulting ions were accelerated, passed through a slit, and the ion beam was applied to the mass/charge analyzer.

The ion beam was first directed to an electric analyzer consisting of two curved plates with an electric field between them. Only ions with suitable mass/charge ratios pass through gap in the pair of curved plates. The emerging ion beam was directed to the magnetic analyzer where a magnetic field caused the ion beam to deflect in a curved path. The magnetic analyzer structure was curved with two exit slits. Heavier components curved around farther in the device and exited through a high-mass slit. Lighter components were bent more sharply by the magnetic field and exited through a low-mass slit.

Ions exiting through the low mass and high mass slits were applied to ion counters. The low-mass collector system covered the range of 1–18 Atomic Mass Units (AMUs) and the high-mass collector covered the range of 12–46 AMUs. An atomic mass unit is essentially equal to the number of protons and neutrons in an atomic nucleus. Mass spectra were developed by varying the ion accelerating voltage and varying the electric field across the electric analyzer.

Measurements made by the neutral mass spectrometer in the Multiprobe bus indicated densities of carbon dioxide and helium of 6×10^6 and 5×10^6 per cubic centimeter respectively at an altitude of 150 km. A helium homopause was found at an altitude of about 137 km. Homopause is the altitude above which gasses do not mix but become stratified. The bus burned up in the atmosphere at an altitude of about 110 km.

Ion Mass Spectrometer

The ion mass spectrometer was designed to measure the composition and density of positively charged ions in the ionosphere of Venus. It had useful output at altitudes from the upper boundary of the ionosphere down to about 120 km where aerodynamic heating precluded meaningful measurements. The measurement range of the instrument was one AMU (a proton) to 56 AMUs (an iron ion).

The experiment used a Bennett spectrometer tube. That device inputted ions at one end of the tube and it had an ion collector at the other end. A series of grids set up electric fields within the tube. Incoming ions were accelerated by a grid with a negative DC voltage. The accelerating voltage was stepped to a series of discrete voltages that selected particular ions to be measured. The accelerated ions passed into another set of grids that had a radio frequency (RF) signal applied. Ions whose velocity matched the accelerating portion of the RF cycle gained sufficient energy to transit through a third grid region that had a positive or repelling DC potential. Those electrons with sufficient energy to penetrate the repelling field struck the collector plate and the resulting current was measured. The magnitude of the ion current was proportional to the density of the particular ion. The magnitude of the accelerating voltage determined which ion would be selected.

The instrument operated in a two-mode cycle. During the explore mode, which lasted for 1.8 seconds, a search for up to 16 different ions was made by stepping the acceleration voltage. The adapt mode, which lasted for 4.5 seconds followed. During the adapt mode, repetitive measurements were made of up to eight of the most prominent ions identified in the explore mode. Each measurement cycle lasted for a total of 6.3 seconds.

Analysis of the data returned from the ion mass spectrometer indicated that a peak density of O^+ ions occurred at an altitude of 200 km. Lower abundances of C^+, N^+, H^+, and He^+ ions were found at that altitude. A peak density of O_2^+ molecular ions was found at altitudes near 150 km.

Pioneer Venus Large Probe

The Pioneer Venus Large Probe was a significant spacecraft in itself. It followed a choreographed descent through the inhospitable atmosphere of Venus using an aeroshell and then a parachute to slow its descent. The probe conducted seven different experiments during its descent and sent data back to Earth. It impacted the surface at a speed of about 8.9 meters per second after descending 55 minutes from entry into the atmosphere. The probe was not designed for a soft landing or to survive impact with the surface.

Mechanical Configuration of Large Probe

The aeroshell of the large probe is the dominant conical structure on top of Multiprobe spacecraft in the photograph in Fig. 4.19. A drawing showing large probe nested in the aeroshell and separate from the aeroshell is given below (Fig. 4.22). An exploded view showing major portions of the probe is given on the next page (Fig. 4.23).

The aeroshell, referred to as the deceleration module in the exploded view, was a blunt, 90 degrees conical structure with base diameter of 1.42 meters. It was made of aluminum. It was covered with carbon phenolic material that provided thermal protection by ablating during entry into the atmosphere. A pilot parachute and a mortar, which was fired to deploy the pilot chute, were mounted at the top of the aeroshell.

The main element of the large probe was a spherical pressure vessel, 73.2 centimeters in diameter. It contained the probe operating systems and the experiments. It is referred to as the descent module in the exploded view. The pressure vessel held a communications antenna at its top, parachute attach points, and viewing ports for the experiments. The shell of the pressure vessel was made from titanium. A multilayer Kapton blanket 2.5 cm thick was fitted to the inside of the shell for insulation.

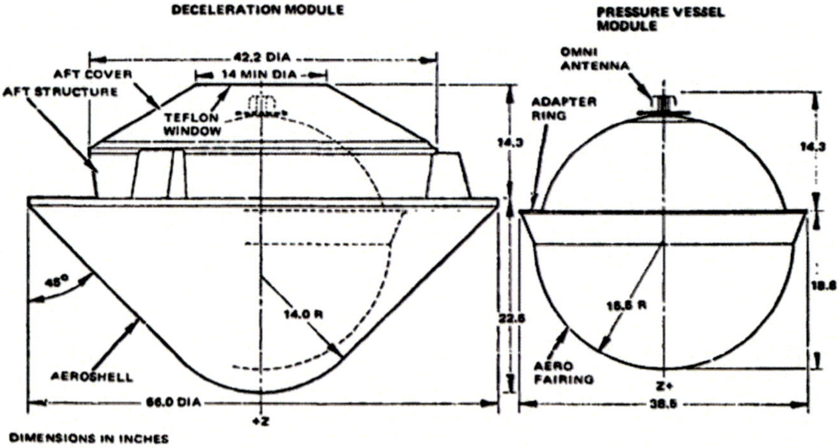

Fig. 4.22 Structure of large probe. (NASA graphic)

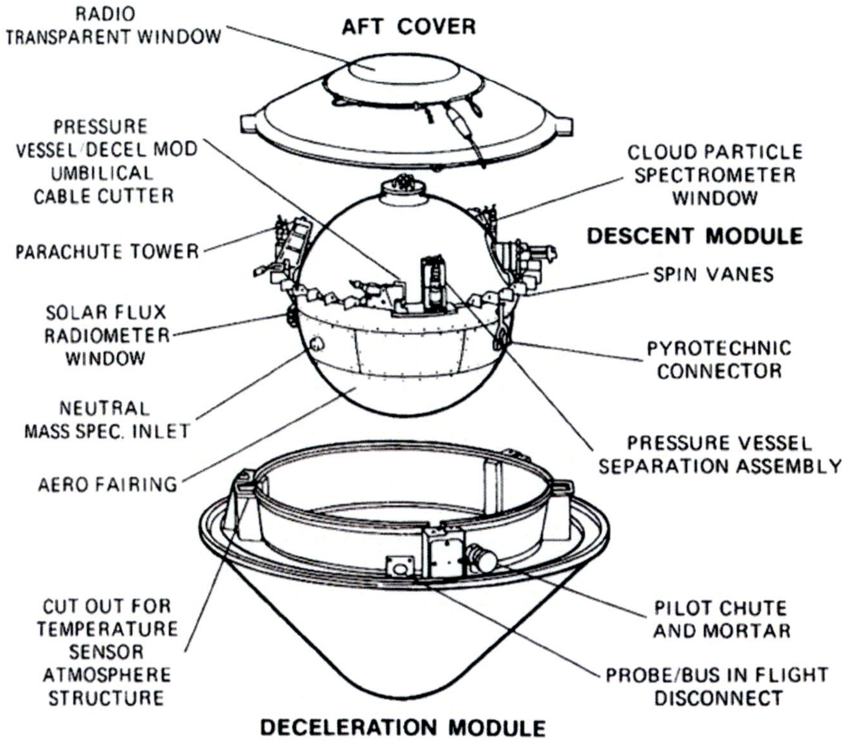

Fig. 4.23 Exploded view of large probe. (NASA graphic)

Components for spacecraft systems and experiments were mounted on two beryllium shelves within the pressure vessel. The interior was filled with nitrogen at a pressure of 102 kPa (15 psi).

An aero fairing was fitted over the bottom hemisphere of the pressure vessel to stabilize its orientation along the velocity vector during its descent through the atmosphere. The appendages labeled spin vanes in the exploded view were attached to the pressure vessel. The vanes caused the vessel to rotate at about one rpm for added stability and to provide a scanning platform for some of the experiments.

The aft cover, made from fiberglass, protected the pressure vessel from heat in the aft region during entry into the atmosphere. The cover had a Teflon window over the antenna area that was transparent to radio frequency energy. The antenna, shown at the top of the pressure vessel in Fig. 4.22, was a crossed dipole type.

Special windows were provided in the pressure vessel as view ports for some of the experiments. Eight sapphire windows ranging in size from 9.5 mm to 38.6 mm in diameter were provided along with one diamond window 15.4 mm in diameter.

Descent of the Large Probe

The descent sequence of the large probe is illustrated below (Fig. 4.24). The large probe entered the upper atmosphere of Venus at an altitude of about 200 km at a speed of 11.6 km per second. The deceleration peaked at about 320 Earth gs at an altitude of about 78 km. The aeroshell slowed the probe to about 188 meters per second by the time it had descended to an altitude of 68 km. A mortar at the top of the aeroshell was fired at that point to deploy the pilot parachute. Explosive nuts holding the aft cover were fired and the pilot parachute pulled off the aft cover and extracted the main parachute. The main parachute, which was a ribbon type 4.9 meters in diameter, was attached to the pressure vessel by three attach points.

Shortly after the probe stabilized on the main parachute, wiring between the aeroshell and pressure vessel was severed, explosive nuts holding the aeroshell were fired, and the aeroshell dropped away. The probe descended on the parachute to an altitude of about 47 km. Explosive nuts that held the parachute to the pressure vessel were fired at that point and the parachute was jettisoned. The pressure vessel then fell 36 minutes through the dense atmosphere to the surface. Spin vanes attached to the pressure vessel caused it to rotate at about one rpm for stability and to provide a scanning platform for some of the experiments. Velocity at the time of impact was 8.9 meters per second. Transmissions from the probe ceased at impact.

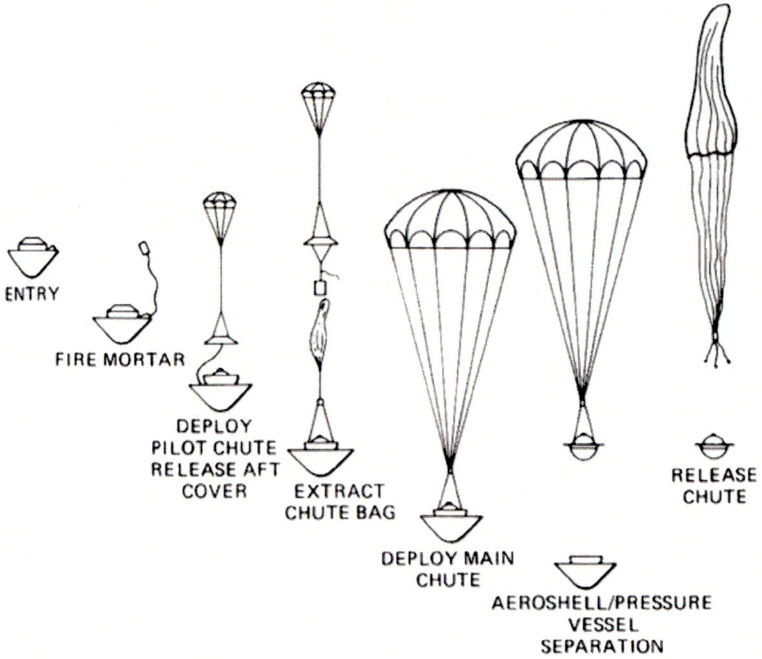

Fig. 4.24 Descent sequence of large probe. (NASA graphic)

Large Probe Systems

The large probe was controlled by its internal systems after release from the Multiprobe bus. A coast timer was started at the time of separation from the bus. The coast timer initiated turn-on of the receiver and a battery heater 2.5 hours before entry into the atmosphere. It reset the descent timer in the Entry Sequence Programmer (ESP) to zero 22 minutes before entry into the atmosphere. The ESP controlled the probe systems and the experiments throughout remainder of the descent. In the event that the ESP was not activated by the coast timer, a deceleration sensor sensed when deceleration decreased below its peak and this signal was used as a backup to activate the ESP.

Power Subsystem

Power to the large probe was supplied by the Multiprobe bus during checkout before deployment of the probe. After deployment from the bus, power was supplied by a 28-volt silver zinc battery with a rated capacity of 40 ampere-hours. Silver zinc batteries offer relatively high energy density and they have a very low self-discharge rate. The battery was charged on Earth before the flight and it was not charged during the mission.

The predicted total current load on the spacecraft with all systems and the experiments operating was 16.8 amperes at 28 volts. The experiments required a total average current of 3.2 amperes at 28 volts. The computed total current drain for the mission was about 24 ampere-hours. The expected voltage range on the battery bus during the powered portion of the mission was 28.2–26.2 volts. A latching relay connected the battery to the probe power bus just before deployment from the Multiprobe bus. Additional relays, activated by the internal sequencer, turned power on to individual portions of the probe and to the experiments.

Communications Subsystem

The large probe did not receive commands from Earth. It received an unmodulated carrier from Earth to generate a coherent reference signal to drive the transmitter. The received frequency was 2110.317 MHz and the transmit frequency was 2291.747 MHz The coherent transmitter signal allowed two-way Doppler measurements to be made and that provided very accurate velocity tracking of the large probe from the time of its release from the Multiprobe bus. The communications system telemetered engineering data and measurements from the experiments back to Earth.

The antenna, which can be seen at the top of the pressure vessel in Fig. 4.2, was a crossed dipole type. The transmitter contained four 10-watt power amplifiers of the same type as used in the Multiprobe bus. The outputs of the amplifier were summed by a 6-dB hybrid to feed the antenna. The RF power level at the input to the antenna was reported to be 28.2 watts after experiencing losses in the 6-dB hybrid, diplexer, and feed lines.

Command Subsystem

The Entry Sequence Programmer was programmed to provide 53 command functions. The sequence of commands, which was permanently programmed into the ESP, was begun by timeout of the coast timer. For example, Command No. 36 fired the pilot parachute mortar, Command No. 41 fired squibs for the three explosive nuts that separated the aft cover, and Command 38 fired squibs for the three explosive nuts that jettisoned the aeroshell. The command numbers were not in the order that they were used.

The ESP turned on the receiver, transmitter, and the gas chromatograph experiments within the first second after its timer was started from zero. Three minutes later the ESP turned on the rest of the experiments for warmup and calibration. Entry into the atmosphere occurred about 22 minutes after the ESP timer started from zero. Times planned for some events after entry into the atmosphere are given below.

Event	Time after entry
Fire pilot parachute mortar	36.6 seconds
Fire squibs for release of aft cover	37.1 seconds
Fires squibs to jettison aeroshell	39.9 seconds
Sequentially power on 7 experiments	90.3 to 91.4 sec
Fire squibs to jettison parachute	18 minutes

The ESP also generated clock and timing signals for probe subsystems and for the science experiments.

Data Transmission to Earth

The data handling subsystem of the large probe processed analog and digital signals from spacecraft systems and from the experiments and formatted the signals in a form to be transmitted to Earth. There was a total of 36 analog signals, 12 serial digital signals, and 24 discrete (on/off) signals processed. Of these, the science instruments contributed nine analog, seven serial digital, and seven on/off signals.

The data handling subsystem included separate multiplexers for the analog and digital signals. The analog multiplexer fed a 10-bit analog-to-digital converter, the output of which was applied to a data formatter. The digital multiplexer accepted both serial digital and on/off digital signals. The output of the digital multiplexer also fed the data formatter.

The data formatter arranged the data into major frames and minor frames. The major frames held 16 minor frames. Each minor frame contained 64 eight-bit words. Data was transmitted to Earth at a rate of 128 bits per second. The digital telemetry data was passed through a convolutional error correcting encoder. The output of the encoder was used to biphase modulate an 8.192 kHz square wave subcarrier. The biphase modulated subcarrier was applied to the exciter that drove the transmitter.

Data transmission to Earth was expected to be blocked due to ionization of the atmosphere during entry into the atmosphere. Similar "blackout" conditions exist when a spacecraft enter the Earth's atmosphere. A storage capability of 3072 bits was provided to store data during the blackout period and transmit it to Earth later.

Science Experiments Carried by Large Probe

The large probe carried seven experiments designed to determine properties of the clouds and atmosphere of Venus. The Experiments, their acronyms, and brief objectives are shown in Table 4.7.

Neutral Mass Spectrometer

The neutral mass spectrometer in the large probe operated from the time that the aeroshell was jettisoned at about 62 km to the surface. Principles of operation of the spectrometer were the same as described previously for the neutral mass spectrometer in the Multiprobe bus. The range of the instrument was 1–208 Atomic Mass Units (AMU).

Table 4.7 Experiments carried by large probe

Experiment	Acronym	Objective
Neutral mass spectrometer	LNMS	Measure composition of neutral atmosphere
Gas chromatograph	LGC	Measure gases that are likely constituents of the atmosphere
Solar flux radiometer	LSFR	Measure the deposition of solar energy during descent through atmosphere
Infrared radiometer	LIR	Determine distribution of thermal flux and determine infrared opacity of clouds
Cloud particle size spectrometer	LCPS	Measure particle size and density in the clouds and lower atmosphere
Nephelometer	LN/SN	Used on both the large probe and small probe to determine vertical structure of clouds
Atmosphere structure	LAS/ SAS	Used on both the large probe and small probe to measure temperature, pressure. And acceleration as a function of altitude

Table 4.8 Measured abundance of gasses relative to CO_2

Gas	Abundance, parts per million
Neon	10
Argon	40–120
Krypton	<0.2
Oxygen	<30
Chlorine	<10
Sulfur dioxide	<300
Water vapor	<1000

A sample of the atmosphere was obtained by two crimped tubes that extended through the pressure vessel and past the boundary layer of the probe as it descended. A cap over the tubes was ejected prior to entering the atmosphere. One of the tubes was closed when the atmospheric pressure reached 1.5 bars to limit the pressure of the sample in the spectrometer.

During descent of the probe, blockage of the inlet tubes occurred at an altitude of 50 km and new samples of the atmosphere could not be obtained. The blockage was thought to be caused by droplets of sulfuric acid from the dense clouds. The droplet apparently evaporated by the time the probe descended to 30 km altitude and operation of the instrument resumed.

Carbon dioxide (CO_2) was known to be the dominant gas in the atmosphere of Venus. It makes up about 96% of the atmosphere. Measurements of abundances of other gasses were made as a fraction of the CO_2. An exception was the case of nitrogen gas that was determined to be about four percent of the atmosphere. Measurements by the large probe of the abundance of other gasses relative to CO_2 in parts per million are listed in Table 4.8.

Gas Chromatograph

The gas chromatograph in the large probe measured the relative abundances of gasses in the atmosphere of Venus. Each measurement sequence lasted several minutes and so only three measurement sequences were made during the descent of the probe. Altitudes of the measurements were: 51.6 km, 41.7 km, and 21.6 km.

The gas chromatograph separated gasses by virtue of their transit times through a long narrow column filled with special mesh material. Separation of the gasses was based on different strengths of interaction of each gas with the mesh material. The stronger the interaction the longer time it took that particular gas to move through the column to a detector at the end. An inert helium carrier gas was used to move the atmospheric sample through the column. A detector at the output of the column produced an electrical signal displaced in time by the transit time through the column with amplitude proportional the abundance of the gas.

The gas chromatograph in the large probe used two sets of columns.

A long column, 15.85 meters long, was used to resolve gasses from neon to carbon dioxide. A short column, 2.13 meters long, was used to resolve gasses from carbon dioxide to sulfur dioxide. The inside diameter of the columns was 1.1 mm. The columns were wound in a coil to make a compact instrument.

Two columns of each length were used. The first was the sensing column containing a sample of the atmosphere. The other was a reference column that carried the same flow rate of helium as the sensing column. Separate detectors were used at the output of the sensor column and the reference column and the difference between the two gave a measure of the amount of a particular gas. It was necessary to control the temperature of the columns quite accurately for accurate operation of the gas chromatograph. The long column set was maintained at a temperature of 18.3 °C and the short column set was maintained at a temperature of 62 °C.

Raw data from the first measurement at 51.6 km altitude indicated a strong peak of nitrogen at a delay time of 300 seconds. Peaks of lesser amplitude occurred at about 312 seconds for oxygen, 325 seconds for argon and 337 seconds for carbon dioxide.

Analysis by Oyama et al. (1980), indicated some differences in the relative abundances for the three altitudes. At the middle altitude of 41.7 km the relative abundances in terms of per cent of the total were: carbon dioxide 95.9%, nitrogen 3.54%, and water vapor 0.519%. Other gasses were less abundant and given in parts per million (ppm). They were: oxygen at 16.0 ppm, argon at 63.8 ppm, carbon monoxide at 30.2, neon at 10.6 ppm, and sulfur dioxide at 176 ppm.

Solar Flux Radiometer

The purpose of the solar flux radiometer was to measure the temperature of the atmosphere above and below the probe to obtain a measure of the contribution of solar radiation to the heat of the atmosphere as a function of altitude. The instrument operated from an altitude of 67 km to the surface.

The radiometer measured solar flux along five narrow beams. The beams were formed by quartz lenses set behind five sapphire windows in the hull. The beamwidth of each of the narrow beams was about five degrees. The beams were directed at angles from the zenith of 27.5, 60, 83.5, 102.2 and 142.1 degrees. Three beams looked upwards and two beams looked down. Data was taken at six different azimuth angles relative to the azimuth of the sun as the probe rotated. Those angles from the sun were 0, 12, 24, 60, 120, and 138 degrees. The difference between the flux in the upward and downward direction was used to determine the solar energy absorbed by various layers of the atmosphere.

Energy gathered in each of the five beams was measured in two spectral bands selected by filters: 0.4–1.0 μm and 1.0–1.8 μm. Visible light is contained in the 0.4–1.0 μm band and near infrared is contained in the 1.0–1.8 μm band. The filter outputs were applied to photodetectors followed by logarithmic amplifiers.

A portion of the energy gathered by the 60-degree to zenith beam and the 142.1-degree to zenith beam were applied to narrowband filters with passbands of 0.59–0.67 μm. Measurements in this narrow band were used to determine the scattering albedos of clouds. The output of the narrowband filters was applied to photodetectors followed by logarithmic amplifiers.

The instrument had a total of 12 channels. The outputs of the twelve logarithmic amplifiers were sampled by a 12-channel multiplexer. The selected output was applied to an 11-bit analog-to-digital converter. The output data stream was transmitted to earth at a rate of 20 bits per second.

Analysis of the data telemetered to earth indicated that there were three different cloud regions with bottoms at altitudes of 57.5 km, 49.7 km, and 47.9 km. The atmosphere was clear beneath the clouds. Absorption of solar radiation was found to occur in the upper cloud layer and in the atmosphere below about 35 km. There was little absorption in the lower two cloud layers. It was found that about 17 watts per square meter was absorbed by the ground on the surface of Venus. That is about 2.5% of the solar energy incident on Venus.

Infrared Radiometer

The infrared radiometer was designed to measure infrared radiation in the atmosphere of Venus during descent of the large probe. The instrument viewed the atmosphere through a 15.4 mm diameter diamond window in the shell of the probe. A rotating angled light pipe was used to generate a conical viewing beam with a field of view of 25 degrees. The angled light pipe was rotated to direct the beam 45 degrees upwards and then 45 degrees downward. Two black body calibration sources at different temperatures were located at angles of 90 degrees and 270 degrees of rotation of the light pipe. The radiometer was commanded into the calibration mode about six percent of the operating time.

The radiometer measured radiation in a spectral range of 3–50 μm. The spectral range was examined by six channels established by infrared filters. The spectral width of each channel is given in Table 4.9.

Channel A measured the total infrared energy in the 3.0–50 μm wavelength band. Channels B and C were used to search for water vapor. Channel D was used to determine the opacity of clouds. Channel E encompassed a strong CO_2 band and it was used to determine if the outside of the window had become contaminated. Channel F was used to determine temperature uniformity of the diamond window as the beam was directed upwards and then downwards.

Table 4.9 Infrared radiometer filter channels

Channel	Spectral extent, microns
A	3.0–50
B	6.0–7.0
C	7.0–8.0
D	9.0–9.0
E	14.5–15.5
F	4.0–5.0

Analysis of the data returned from the experiment indicated problems with the reasonableness of some data. A paper by Revercomb et al. (1984), *Net Thermal Radiation in the Atmosphere of Venus,* identified likely reasons for the problems and provided corrections. Their corrected plots of net flux (upwards infrared flux minus downward flux) indicated that between altitudes of 50 km and the surface the average net infrared flux was about 10 watts per square meter (W/m^2). The infrared flux increased rapidly with higher altitudes to about 40 W/m^2 at 60 km and about 70 W/m^2 at 63 km altitude. The water vapor mixing ratio beneath the clouds was $>5 \times 10^{-4}$ relative to the dominant CO_2 atmosphere gas.

Cloud Particle Size Spectrometer (LCPS)

The purpose of the cloud particle size spectrometer was to measure the particle size and density in the clouds of Venus. A vertical profile of particulate size and density was determined as the probe descended through the clouds.

The instrument used a laser source to generate a beam of light that passed through a sapphire window in the probe. Optics focused the light onto one-half of a double prism supported 15 cm from the outer surface of the probe. The other half of the double prism reflected the light back through the window to a set of optics that split the beam into three paths. Each path was focused on an array of ten photodiodes. Each of the three paths was implemented to accommodate a different range of particle sizes. The three ranges were: 5–50 μm, 20–200 μm, and 50–500 μm. When a particle entered the field-of-view of the instrument, it cast a shadow on the photo detector arrays. The size of the shadow was measured by the response of individual photodiodes in the array. Data from all three arrays was telemetered to Earth. One of the three channels was best suited for a particular particle size.

The LCPS also included a light scattering method of determining particle size. This method was effective for particles 0.5–5.0 μm in size.

Results of the large probe cloud particle size spectrometer experiment have been analyzed by Knollenberg et al. (1980), *The clouds of Venus: A Synthesis Report.* Using data from the LCPS and other instruments of the Pioneer Venus program, they identified specific regions of the vertical extent of the atmosphere of Venus. Those regions and their characteristics are summarized in Table 4.10.

Table 4.10 Vertical extent of particles in atmosphere of Venus

Region	Altitude range, km	Mean particle size, μm	Ave. density, N/cm^{-3}
Upper haze	70–90	0.4	500
Upper cloud	56.5–70	0.4 and 2.0	50
Middle cloud	50.5–56.5	0.3, 2.5, 7.0	50
Lower cloud	47.5–50.5	0.4, 2.0, 8.0	50
Lower haze	31–47.5	0.2	2–20

Nephelometer (LN/SN)

The nephelometer instrument was designed to gather information about particles in the clouds of Venus. The nephelometer was carried by the three small probes as well as by the large probe. The instrument focused a beam of light at a distance beyond the aerodynamically disturbed region and measured the intensity of the light reflected back from particles. The reflected energy was a function of backscattering cross section of the particles and the number of particles in the illuminated path.

The light transmitter consisted of a pulsed light emitting diode (LED) operating at a wavelength of 9000 Å and a lens that focused the light pulses through a sapphire window to a point outside of the probe beyond the turbulent wake. Light backscattered from particles passed through a sapphire window in the receive path and to a lens that focused the signal onto a detector. In addition, sunlight scattered by the particles was received and applied to filters centered at wavelengths of 3550 Å and 5200 Å. The filter outputs were detected and processed.

The nephelometer data confirmed the existence of three different cloud layers, each with distinctive particle characteristics. The upper layer extended from 70 km to 56.5 km. The middle layer extended from 56.5 to 50.5 km and the lower level extended from 50.5 to 47.5 km. Data from the nephelometer allowed researchers to conclude that the particles in the clouds of Venus are spherical and consist of a solution of sulfuric acid (H_2SO_4) with 75–85 percent concentration.

Atmosphere Structure Experiments (LAS/SAS)

The large probe and the small probes carried instruments to measure temperature, pressure, and acceleration as the probes descended through the atmosphere. There were two temperature sensors. One, referred to as "free wire sensor," consisted of an array of 0.1 mm diameter platinum wires supported by a frame but insulated from it. The wire array was exposed to the atmosphere. The resistance of the wire was a function of temperature. A constant current source was connected to the array and the voltage drop across the array was measured and compared to calibration data for temperature reading. The other temperature sensor was similar but it was bonded to the support frame. The temperature measurement range of the sensors was 200 K to 800 K. Temperature measurements by the probes indicated temperature of −83 °C at 90 km altitude, −48 °C at 70 km, 13 °C at 57 km, 72 °C at 50 km, 105 °C at 46 km, and 209 °C at 31 km. The temperature readings became unreliable on all four probes at an altitude of 12.5 km during descent.

The pressure sensors were diaphragm types with one side of the diaphragm facing a sealed vacuum chamber. A total of 12 sensors were used to cover a pressure range of 30 millibars to 100 bars. Each pressure sensor covered a pressure range of about a factor of two. Two strain gauge resistors were mounted at the center of the diaphragm and two reference resistors were mounted to the frame. The four resistors formed a sensitive bridge circuit to allow measurement of the deflection of the diaphragm. Measurement of atmospheric pressure by the four probes indicated pressures of about 0.03 bars at altitude of 75 km, 0.4 bars at 56 km, 1.5 bars at 47 km, 8.2 bars at 31 km and 99 bars at the surface.

The accelerometers were pendulous mass type where the mass was maintained in a null position by current in a coil attached to the mass in the presence of a permanent magnetic field. The accelerometers covered a range of 0.4 micro g to 600 g in eight ranges. Four ranges were used during entry and two ranges were used during terminal descent. The accuracy of the accelerometers allowed velocity to be tracked to an accuracy of about two meters per second. The large probe carried four accelerometers, two in the axial direction and one along each of the two other axes of the spacecraft. Accelerometer measurements indicated that deceleration of the large probe peaked at about 320 Earth gs at an altitude of 78 km.

Pioneer Venus Small Probes

The three Pioneer Venus Small Probes descended independently to the surface of Venus after being released from the Multiprobe Bus. Release took place on 20 November 1978 about 9.3 million km from Venus. The small probes were retained by clamps until time for launch. When it became time to launch the small probes, the spin rate of the Multiprobe spacecraft was increased to 48 RPM, the clamps were released by pyrotechnics, and the small probes were spun off the bus by centrifugal force. The probes were spun off tangentially from the bus and that established their different trajectories to Venus.

The names given to the three small probes, their landing coordinates, illumination at the landing site, and descent time from entry into atmosphere at 200 km altitude are indicated in Table 4.11.

The probes conducted three types of scientific experiments during descent and transmitted data back to Earth. The probes were not designed for a soft landing or to survive impact with the surface. The predicted velocity at impact was about eight meters per second. The Day Probe survived impact and transmitted data for 67 minutes after landing. The Night Probe transmitted data for two seconds after impact.

Table 4.11 Probe landing coordinates, illumination, and descent times

Probe Name	Landing Latitude	Landing Longitude	Illumination	Descent time from 200 km
North Probe	59.3° N	4.8° East	Night	53 minutes
Day Probe	31.3° S	317° East	Day	56 minutes
Night Probe	28.7° S	56.7° East	Night	56 minutes

Mechanical Configuration of Small Probes

A photograph of the top of the Multiprobe bus showing the aeroshells (decent modules) of the three small probes was given previously in Fig. 4.19. An exploded view of the small probe with the pressure vessel module and the descent module separated is given (Fig. 4.25). The lower portion of the pressure vessel fitted inside of the descent module and the two remained joined together during the mission.

The aeroshell, referred to as the deceleration module in the exploded view, was a blunt, 90 degrees conical structure with base diameter of 0.76 meters. It was made of titanium to survive high temperatures during the descent. The aeroshell was covered with carbon phenolic material that provided thermal protection by ablating during entry into the atmosphere.

The main element of the small probe was a spherical pressure vessel, 46 centimeters in diameter made from titanium. It contained the probe operating systems

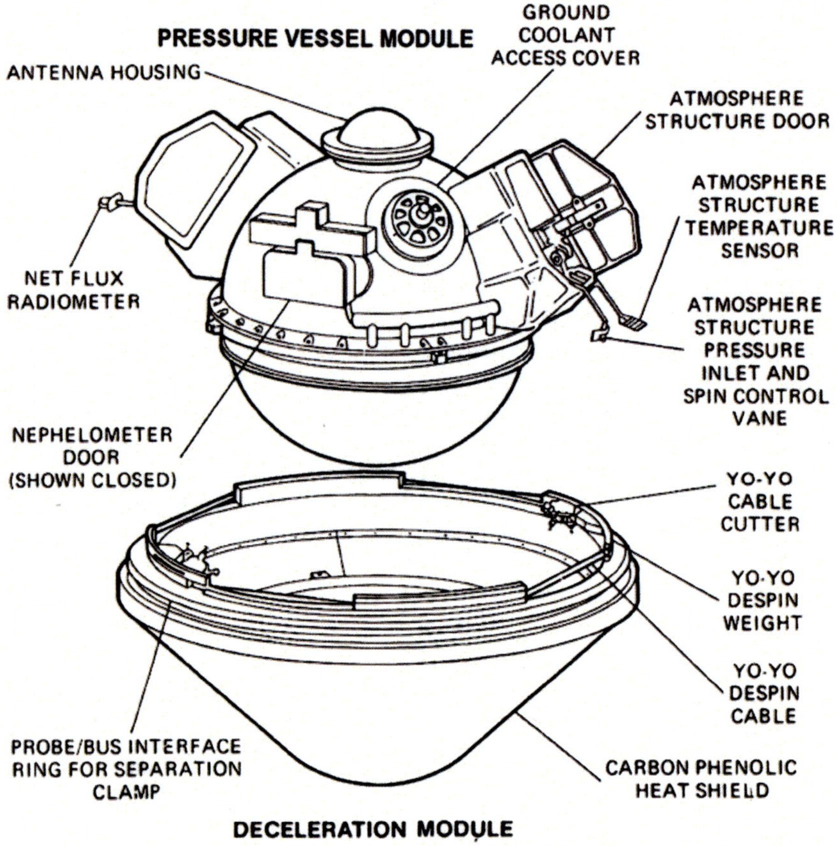

Fig. 4.25 Exploded view of small probe. (NASA graphic)

and the experiments. The pressure vessel held a communications antenna at its top and viewing ports for the experiments. A multilayer Kapton blanket was fitted to the inside of the shell for insulation. Components for spacecraft systems and experiments were mounted on two beryllium shelves within the pressure vessel. The interior was filled with Xenon at a pressure of 102 kPa (15 psi).

An aft cover, made from aluminum, protected the pressure vessel from heat in the aft region during entry into the atmosphere. The cover was hemispherical in shape with protrusions to match protrusions on the pressure vessel. The antenna, shown at the top of the pressure vessel, was a crossed dipole type with a hemispherical antenna pattern.

The completed assembly of the pressure vessel mounted within the descent module had an envelope 0.76 meters in diameter and 0.60 meters high. The probe weighed 75 kg.

Descent of the Small Probes

The small probes entered the atmosphere of Venus at an altitude of 200 km with a velocity of about 11 km per second. The probes were still spinning at a rate of about 48 RPM that had been imparted at launch from the bus. The spin rate was slowed to about 15 RPM by deploying two weights at the ends of cables 2.4 meters long. The weights were released and centrifugal force extended the weights on the cables. The outlying weights on the cables caused the spin rate of the probes to decrease to preserve angular momentum. The cable and weight arrangement was fittingly referred to as a yo-yo. The weights and cables were jettisoned after a spin rate of 15 RPM had been obtained. The lower spin rate made it easier for the probe to align aerodynamically with the flight path. The conical aeroshell provided aerodynamic braking to slow the probe in the thick atmosphere of Venus. The shape of the aeroshell aligned the axis of the probe with the flight path.

Covers over the view ports of the scientific instruments were opened when below the high heat region of entry into the atmosphere. An acceleration switch was triggered when deceleration decreased after passing peak deceleration. Covers over the view ports were opened five seconds after receiving the reset signal from the accelerometer switch. The altitude of the probe at this time was about 70 km.

The covers over the viewports of the atmospheric sensors (temperature and pressure) and the net flux radiometer are shown in the open position in the drawing of the small probe in Fig. 4.25. The cover over the nephelometer is shown in the closed position. Opening the cover of the atmospheric sensors allowed a strut holding the temperature sensor and a strut containing the pressure inlet for the pressure gauge to be deployed as shown in the drawing. Likewise, opening the cover of the net flux radiometer allowed the sensor to be deployed. There were two sapphire windows in the pressure vessel under the cover of the nephelometer. One window was used to transmit an optical pulse and the other window was used to receive signals reflected off of particles.

Small Probe Systems

Checkout of the small probe systems were conducted by hardwired connections to the Multiprobe Bus prior to release of the probes from the bus. After release from the bus the small probe was controlled by its internal systems. Those internal elements included a coast timer, Entry Sequence Programmer (ESP), and acceleration switch.

The coast timer was started at the time of separation from the bus. It initiated turn-on of heaters and it reset the descent timer in the ESP to zero before entry into the atmosphere. The ESP controlled the probe systems and the experiments throughout remainder of the descent. A total of 128 commands were issued by the ESP. These were organized into eight sequences of 16 commands each. In the event that the ESP was not activated by the coast timer, a deceleration sensor sensed when deceleration decreased below its peak and this signal was used as a backup to activate the ESP.

Power Subsystem

Power for the small probes was supplied by a 28-volt silver-zinc battery with a rated capacity of 11 ampere-hours. The battery was charged on Earth before the flight and it was not charged during the mission. The expected voltage range on the battery bus during the powered portion of the mission was 27.6–29.7 volts.

A latching relay connected the battery to the probe power bus just before deployment from the Multiprobe bus. Additional relays, activated by the internal sequencer, turned power on to individual portions of the probe and to the experiments.

Communications Subsystem

The small probe did not receive commands from Earth and it did not have a receiver as did the large probe. The oscillator driving the transmitter was precise enough that reasonably accurate one-way Doppler measurements could be made on Earth to determine spacecraft velocity. The communications system telemetered engineering data and measurements from the experiments back to Earth.

The antenna, which was located on top of the pressure vessel, was a crossed dipole type. The transmitter contained a single 10-watt power amplifier of the same type as the four 10-watt amplifiers used in the large probe. The transmitted frequencies were different for the three small probes. The frequencies were: 2292.282 MHz, 2292.437 MHz, and 2291.239 MHz.

Command Subsystem

The Entry Sequence Programmer was programmed to provide command functions for the small probe mission. The sequence of commands, which was permanently programmed into the ESP, was begun by timeout of the coast timer. There was a total of 63 commands programmed into the ESP. Of these, two were spare. The ESP also generated clock and timing signals for probe subsystems and for the science experiments.

The sequence of commands from the ESP started at 1319 seconds before entry into the atmosphere with turn-on of the transmitter. Entry was defined a passing through an altitude of 200 km. The science primary power relay was activated at 1020 seconds before entry and the squibs that released the despin device (yo-yo) were fired at 299 seconds before entry. The ESP generated a series of commands from the time of entry to just before impact. It used other signals as reference at times. For example, it generated signals to deploy the covers over the scientific instruments five seconds after receiving the acceleration switch reset indication.

Data Transmission to Earth

The data handling subsystem of the small probe processed analog and digital signals from spacecraft systems and from the experiments and formatted the signals in a form to be transmitted to Earth. The analog signals were converted to digital form by analog-to-digital converters. There were 16 data channels allocated to science and 20 channels allocated to housekeeping for a total of 36 channels. Each channel contained 16 minor frames and each minor frame contained sixty-four 8-bit words. The digital telemetry data was passed through a convolutional error correcting encoder. The output of the encoder was used to biphase modulate an 8.192 kHz square wave subcarrier. The biphase modulated subcarrier was applied to the exciter that drove the transmitter. The data rate was 64 bits per second when the probe was above 30 km altitude and at a rate of 16 bits per second when at 30 km and below.

Data transmission to Earth was expected to be blocked due to ionization of the atmosphere during entry into the atmosphere. A storage capability of 3072 bits was provided to store data during the blackout period and transmit it to Earth later. The memory was also used to store data for a short time when changing the data rate at 30 km altitude.

Science Experiments Carried by the Small Probe

The small probe carried three experiments to determine properties of the clouds and atmosphere of Venus. The Experiments, their acronyms, and brief objectives are shown in Table 4.12 below.

Table 4.12 Experiments carried by small probe

Experiment	Acronym	Objective
Nephelometer	LN/SN	Used on both the large probe and small probe to determine vertical structure of clouds
Net flux radiometer	SNFR	Measure distribution of radiative heat transfer
Atmosphere structure	LAS/SAS	Used on both the large probe and small probe to measure temperature, pressure, and acceleration as a function of altitude

The nephelometer and the atmosphere structure experiments were the same as carried on the large probe and described previously. The atmospheric structure results of temperature and pressure for all four probes was given previously in the Large Probe section of this chapter.

Net Flux Radiometer

The net flux radiometer measured the radiative flux above and below the probe and used the difference between the two to arrive at net flux as a function of altitude. The differential radiative heat sensor was a flat plate mounted on an extension shaft that protruded from the pressure vessel. The shaft deployed when the cover for the instrument was opened at an altitude of about 70 km.

The heat sensor plate was oriented parallel to the planet surface. The plate was protected by diamond windows fitted above and below the plate. The plate sensed the difference in radiant flux striking the plate from above and below. This net flux resulted in a small electric current that was processed by an electronics module. The extension shaft and plate were rotated by 180 degrees at a one second rate to cancel any offsets in the instrument. The instrument was composed of a sensor assembly and an electronics assembly. The total weight of the instrument was 1.1 kg and it required 3.8 watts of power at 28 volts DC.

Analysis of the net flux measurement data from the small probes disclosed significant errors that resulted in unrealistic values of net flux at lower altitudes. The measurements and corrections were discussed and in a paper by Revercomb et al. (1984), *Net Thermal Radiation in the Atmosphere of Venus.*

The sensors apparently failed on all three small probes at an altitude of about 13 km. The paper referenced above gives uncertainty bounds for corrected net flux as a function of altitude and the corrected net flux from a model that applied corrections. Separate graphs are given in the reference for each of the small probes and for the large probe. A few data points from the day probe graph are listed below.

Altitude, km	**Net flux, W/m²**
62	80
60	50
55	25

(continued)

51	44
40	18
20	8
14	5

The presence of a cloud layer at about 55 km altitude is apparent in the data. The night probe data has the same general trend as the day probe data. The north probe data indicates a peak of about 90 watts per square meter at 45 km altitude and decreasing to about 20 watts per square meter at an altitude of 20 km.

Bibliography

Barath, F. T., et al., *Mariner 2 Microwave Radiometer Experiment and Results,* The Astronomical Journal Vol. 69, No. 1 Feb. 1964

Chase, S. C., et al., *The Mariner 2 Infrared Radiometer Experiment,* Journal of Geophysical Research, Vol. 68, No. 22 Nov. 1963

Colin, L., et al., Pioneer Venus Experiment Descriptions, Space Science Reviews, Vol 20, July 1977

Davies, Merton E., et al., *Atlas of Mercury*, NASA History Office SP-423

Driscoll, Walter G., et al., *Handbook of Optics,* pp (1-13–1-15) McGraw-Hill Book Company, New York

Fimmel, Richard O., et al., *Pioneer Venus* NASA report SP-461, 1983

Hughes Aircraft Company, *Final Report System Design of the Pioneer Venus Spacecraft,* Vol 1, 3, 4, 6, 7, 8, 9, 10, Prepared for Ames Research Center, July 1973

Hughes Aircraft Company, PC-402, Pioneer Venus Orbiter Spacecraft Mission Operational Characteristics Document, Feb. 1978 Hughes Aircraft Company, *PC-403, Pioneer Venus Multiprobe Spacecraft Mission Operational Characteristics Document,* Vol. 1, 2, 3 May 1978

Jet Propulsion Laboratory, EPD – 46, Rev. 1, *Mariner R Spacecraft for Mission P-37/P38,* 1961

Jet Propulsion Laboratory, *Mariner Scientific Experiments,* JPL News July 19, 1962a

Jet Propulsion Laboratory, *Mariner-Venus* 1962 Final Project Report, NASA SP-59, 1965

Jet Propulsion Laboratory, *Mariner-Venus* 1967 Final Project Report, Volume I, JPL Technical Report 32-1203, 1968

Jet Propulsion Laboratory, *Mariner-Venus* 1967 Final Project Report, Volume II, JPL Technical Report 32-1203, 1969a

Jet Propulsion Laboratory, *The Mariner V Flight Path and its Determination from Tracking Data, JPL Technical Report 32-1363,* 1969b

Jet Propulsion Laboratory, *Venus Trajectory,* JPL News July 19, 1962b

Jones, Douglas Emron, *The Mariner II Microwave Radiometer Experiment,* NASA/JPL Technical Report No. 32-722, 1966

Kasprzak, W. T., *The Pioneer Venus Orbiter: 11 Years of Data* NASA Technical Memorandum 100761, 1980

Knollenberg, R. L., et al., *The clouds of Venus: a synthesis report,* Journal of Geophysical Research, Vol. 95, 1980

LePage, Andrew J. *The Return to Venus: The Mission of Mariner 5,* https://www.drewexmachina.com 2017

Murray, Bruce and Burgess, Eric, *Flight to Mercury*, Columbia University Press, New York 1977

NASA News, Press Kit Pioneer Venus 2, July 1978a

NASA News, Press Kit, Pioneer Venus, May 1978b

NASA, *Mariner-Venus* 1967 Final Project Report, NASA Report SP-190, 1971

NASA-NSSDCA, *Mariner 2*, NSSDCA/COSPAR ID: 1962-041A

NASA-NSSDCA, *Mariner 5*, NSSDCA/COSPAR ID: 1967-060A

O'Donnell, Franklin, *The Venus Mission*, https://www.jpl.nasa.gov/mariner2/

Oyama, V.L., et al., *Pioneer Venus Gas Chromatography of the Lower Atmosphere of Venus,* Journal of Geophysical Research, Vol 85, No. A13, Dec. 1980

Pettengill, G. H., et al., *Pioneer Venus Orbiter Radar Mapper: Design and Operation* IEEE Transactions on Geoscience and Remote Sensing, Vol GE-18, No, 1, January 1980

Revercomb, R. E., et al., *Net Thermal Radiation in the Atmosphere of Venus*. 1984, NASA-CR-204917

Siddiqi, Asif, A., *Beyond Earth A Chronicle of deep Space Exploration, 1958-2016*, NASA SP-2018-4041, 2018

Sonnett, Charles P., *A summary Review of the Scientific Findings of the Mariner Venus Mission*, Space Science reviews 2 (1963), D. Reidel Publishing Company, Dordrecht, Holland

Staff of Lewis Research Center, *Preliminary Launch Vehicle flight Evaluation Report NASA Applications Technological Satellite (ATS) Program Flight 1, NASA Atlas-Agena No. 19,* NASA TM-52348, 1967

Stewart, A. I., et al., *Ultraviolet Spectroscopy of Venus: Initial Results from the Pioneer Venus Orbiter*, Science, Vol.203, Issue 4382, 1979

Taylor, F. W., Infrared remote Sounding of the Middle Atmosphere of Venus from the Pioneer Orbiter, Science, Vol. 203, Issue 4382, 1979

The Mission of Mariner II: Preliminary Observations Profile of Events, Science, Vol. 138, Issue 3545, December 1962 (No author listed)

Chapter 5
Soviet Union Spacecraft that Explored Venus 1980–1990

The Soviet Union continued vigorous exploration of Venus throughout the 1980s. That exploration was put on hold in 1991 when the Soviet Union broke up under pressure of people in satellite countries yearning to be free. The disarray that followed curtailed further missions to Venus. The Soviet space programs were picked up by Russia, the remaining large block of the Soviet Union. Design of space hardware and operation of the space program had been a Russian endeavor since the beginning. A summary of spacecraft that were launched towards Venus by the Soviet Union during the 1980s is given in the table on the next page (Table 5.1). Those spacecraft were all successful and provided expanded and new information about Venus.

The fact that all of those spacecraft were successful speaks to the maturity of the Soviet space programs in the 1980s. Their last offering, Vega 1 and Vega 2, were remarkable spacecraft. Each spacecraft dropped off a lander at Venus and its bus went on to photograph and make measurements of the comet Halley during a close encounter. The successful landers deployed instrumented balloons in the clouds of Venus. Driven by high winds, the balloons traversed about a third of the way around the planet.

4 V-1 Spacecraft (Venera 13 and Venera 14)

The Venera 13 and Venera 14 spacecraft were last in the 4 V-1 series of spacecraft. The two spacecraft had basically the same bus and lander configuration as Venera 11 and 12. The spacecraft were prepared for the 1981 launch window to Venus by Design Bureau NPO Lavochkin. Descriptions of Venera 11 and Venera 12 were given in Chap. 3 of this book.

Venera 13 and Venera 14 were both very successful spacecraft and good data was returned from all instruments on the bus and on the lander. In particular, the landers returned the first color photographs of the surface and first results of a soil sample

T. Lund, *Spacecraft that Explored the Inner Planets Venus and Mercury*,
Springer Praxis Books, https://doi.org/10.1007/978-3-031-29838-7_5

Table 5.1 Soviet Union spacecraft sent to Venus in the 1980s

Spacecraft/ type	Launch date	Objective	Results
Venera 13/ 4 V-1	30 Oct 1981	Fly by Venus and land a capsule	Bus flew by Venus and capsule descended to surface and returned meteorological data. Color photos of the surface returned
Venera 14/ 4 V-1	4 Nov 1981	Fly by Venus and land a capsule	Bus flew by Venus and capsule descended to surface and returned meteorological data. Color photos of the surface returned
Venera 15/ 4 V-2	2 June 1983	Orbit Venus and image surface by radar	The bus carried a synthetic aperture radar in a near polar orbit to create radar images of a portion of the planet above 30 degrees latitude.
Venera 16/ 4 V-2	7 June 1983	Orbit Venus and image surface by radar	The bus carried a synthetic aperture radar in a near polar orbit to create radar images of a portion of the planet above 30 degrees latitude.
Vega 1/ 5VK	15 Dec 1984	Deploy balloon, land capsule on Venus and explore comet Halley	Bus flew by Venus and released a Lander. Lander deployed an instrumented balloon that was blown by winds 11,000 km around the planet. Bus went on to reconnoiter comet Halley.
Vega 2/ 5VK	21 Dec 1984	Deploy balloon, land capsule on Venus and explore comet Halley	Bus flew by Venus and released a Lander. Lander deployed an instrumented balloon that was blown by winds 11,000 km around the planet. Bus went on to reconnoiter comet Halley.

by its chemical laboratory. The bus deployed the lander during a flyby and it relayed data from the lander to Earth as the lander descended to the surface.

An excellent photograph showing models of the overall Venera 13 spacecraft and of the lander is given in the NASA HEARSARC website https://heasarc.gsfc.nasa.gov/.../venera1314.html. The models are on display at the Exhibition of the Achievements of the National Economy in Moscow. The holder of the copyright for the photograph did not respond to the author's requests for permission to use in this book.

Main elements of the lander and its instruments were essentially the same as described for Venera 11 and Venera 12 in Chap. 3 of this book. The lander weighed 760 kg (1676 pounds).

Flight of Venera 13 and Venera 14

Venera 13 was launched from the Baikonur Cosmodrome on 30 October 1981 by a Proton K launch vehicle with a block D upper stage. The launch and the insertion into a trans-Venus trajectory were successful. A midcourse correction was made on 10 November 1981 and a second midcourse correction was made on 21 February 1982. The spacecraft approached the sunlit side of Venus on 27 February and the lander was released from the bus.

After deploying the lander, the rocket motor in the bus was ignited for short burn to cause the bus to fly by Venus at a distance of 36,000 km. The trajectory of the bus provided a relatively long time window to relay data from the lander to Earth by the high-gain antenna on the bus. The Venera 13 lander entered the atmosphere of Venus on 1 March 1982. The lander was slowed by aerodynamic braking by the lander enclosure. Small parachutes were deployed to stabilize the lander enclosure and to remove the upper portion of the enclosure. The main braking parachute was deployed at an altitude of 62 km. Systems in the lander were powered on after deployment of the parachute and the lander began transmitting data to the bus. At an altitude of 47.5 km, which was about the base of the clouds, the braking parachute was jettisoned.

The lander continued its descent being slowed by an aerodynamic braking disk. A successful landing was made at a velocity of about 7.5 meters per second after a descent lasting 62 minutes. The landing, which occurred on 1 March 1982, was at latitude of −7.55° and longitude of 303.69° in the Venus coordinate system. Data transmission from Venera 13 was received at the bus for 127 minutes after landing.

Venera 14 was launched from the Baikonur Cosmodrome on 4 November 1981 by a Proton K launch vehicle with a block D upper stage. The launch and the insertion into a trans-Venus trajectory were successful. Midcourse corrections were made on 23 November 1981 and 25 February 1982. The spacecraft approached the sunlit side of Venus on 3 March 1982 and the lander was released from the bus. The bus flew by Venus with a trajectory that allowed relaying data from the lander to Earth.

The Venera 14 lander entered the atmosphere of Venus on 5 March 1982 and performed the same landing sequence as Venera 13. It made a successful landing on 5 March at latitude of 13.05° and longitude of 310.19° in the Venus coordinate system. Data transmission from Venera 14 was received for 57 minutes after landing.

Scientific Experiments on Venera 13 and 14 Landers

The scientific instrument group on the landers included:

- Color cameras
- Temperature sensors
- Barometers
- Gas chromatograph (Sigma 2)
- Mass spectrometer (MKh-6411)
- Gamma ray spectrometer
- X-Ray fluorescence spectrometer
- Backscatter nephelometer
- Optical spectrometer
- Penetrometer (PROP-V)
- Low frequency radio sensor (GROZA 2)
- Soil Analysis unit and drill

Rather detailed description of the instruments on Venera 11 and Venera 12 landers were given in Chap. 3 and only a brief review with differences for Venera 13 and Venera 14 are presented here. A summary of information learned from the experiments on the Venera 13 and Venera 14 landers is given along with descriptions of the instruments.

The gas chromatograph, mass spectrometer, and low frequency radio sensor were improved versions of those used on the Venera 11 and Venera 12 landers. A very important improvement was made to the jettisoning mechanism for the lens caps on the cameras. The lens caps did not jettison on the two cameras of Venera 11 or the two cameras of Venera 12 when commanded. As a result, neither of those two landers was able to take photographs.

Color Cameras

The Venera 13 and Venera 14 landers each carried two cameras. The cameras, which scanned 180 degrees in azimuth, were located on opposite sides of the lander. The two cameras provided nearly full azimuth photographic coverage around the lander. The cameras were essentially the same as carried by Venera 11 and Venera 12. The cameras sequentially photographed through sections of a color wheel having clear, red, green, and blue segments. The resulting images were assembled on Earth into color photographs of the surface of Venus.

The angular coverage by each camera was 37 degrees vertical and 180 degrees horizontal. The panoramic image was represented by a field of 211 by 1024 pixels. The camera was mounted just below the aerodynamic braking disk and about 90 cm above the surface of Venus. The optics of the camera were arranged to direct the center of the vertical coverage at an angle of 45 degrees down from the horizontal.

The image was formed by causing a mirror to nod rapidly in the vertical plane over an included angle of about 37 degrees. A calibrated light source was fed into the optics during retrace of the vertical scan. The mirror assembly was rotated in the horizontal plane so that each vertical scan traced a new region of the surface. The mirror assembly panned 180 degrees in the horizontal plane in a series of 1024 vertical lines. The vertical scanning rate was one vertical line each 0.82 seconds. Each vertical line was mapped into 211 pixels. In addition, 41 pixels were required by the calibration signal during retrace giving a total of 252 pixels of images per line.

The external scene was reflected from the mirror down through the objective lens to a pinhole aperture at the image plane. The image from the pinhole, which represented a scanned spot on the Venusian surface, was applied to a photomultiplier tube and subsequently amplified and digitized at nine bits per pixel. A tenth bit was added for a parity check. The objective lens was focused at the hperfocal distance and images at distances from 0.8 meters to infinity were acceptably sharp.

The data rate applied to the transmitter during imaging was 3072 bits per second. This comes about from processing 252 pixels per vertical line at 10 bits per pixel in a time of 0.82 seconds. On the basis of a 37-degree vertical scan for 211 pixels, the vertical resolution was about 0.18 degrees. Likewise, on the basis of 180 degrees and 1024 pixels, the horizontal resolution was about 0.18 degrees. At the top of the vertical scan looking straight ahead, the distance to the surface was about 2.5 meters and the resolution was about 7.9 mm. At a distance to the surface of 10 meters from the camera the resolution would have been about 31 mm.

The scanning protocol was programmed for one of the two cameras to scan 180 degrees in azimuth through the clear filter and then scan a 60-degree sector sequentially through the red, green, and blue filters. The scan of 180 degrees would have taken about 840 seconds and the three 60-degree scans would have also taken 840 seconds for a total of 1680 seconds or 28 minutes. This protocol was likely chosen on the assumption that the lander would operate for at least 30 minutes on the surface.

The second camera was programmed to scan 180 degrees through the clear filter and then sequentially scan 180 degrees through the red, green, and blue filters for a total sequence time of 56 minutes. The two cameras operated simultaneously and video data was transmitted to the bus by two separate transmitters.

Venera 13 continued to operate on the surface for 127 minutes and between its two cameras a total of 11 panoramic images 180 degrees wide and 10 images spanning 60 degrees were taken. Panoramic images from the two cameras taken through the clear filter are shown below (Fig. 5.1). A synthesized color image from camera A is shown on the next page (Fig. 5.2). The ladder-shaped item to left of center in the photograph from camera B in Fig. 5.1 is the PROP-V penetration device. The moon- shaped object near the center is a lens cap and the rectangular item at the right is a color calibration panel. An estimate of size of rock fragments on the surface can be drawn from the size of the small circular structure at the end of the PROP-V penetration device that was 6.5 cm in diameter.

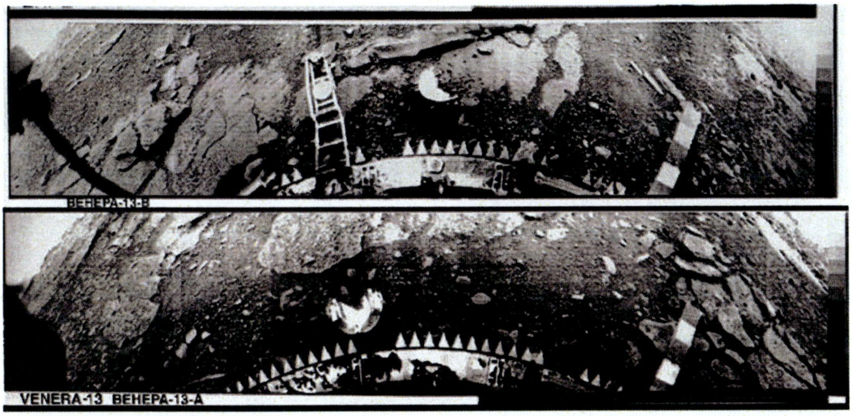

Fig. 5.1 Panoramic images taken by camera A and camera B of Venera 13 lander. (NSSDC Image Catalog, Credit USSR)

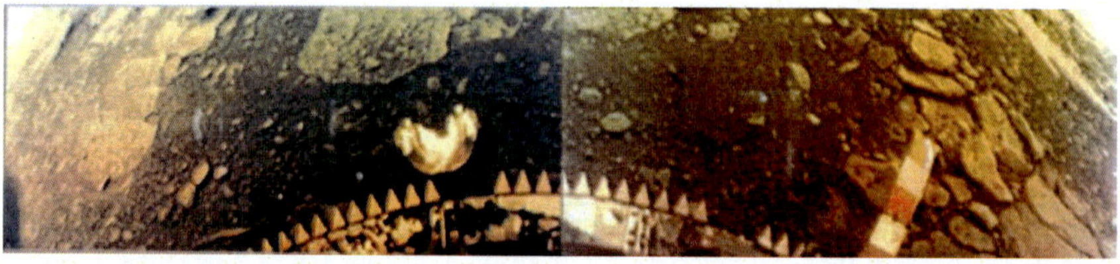

Fig. 5.2 Synthesized color image from camera A of Venera 13 lander. (NSSDC Image Catalog, Credit USSR)

Fig. 5.3 Panoramic images taken by camera A and camera B of Venera 14 lander. (NSSDC Image Catalog, Credit USSR)

The teeth on the landing ring in the foreground provided aerodynamic stability to the lander as it descended. The hue of the color image is due to the fact that very little blue light reaches the surface. The geometry of the azimuth scan and camera field of view allowed the horizon to be visible in the upper right and left corners of the images.

The panoramic views portrayed by Venera 13 show exposed layers of bedrock with flat layers of stratified rock. Dark soil-like material is present between exposed rocks. The Venera 14 lander also returned good quality panoramic images of the surface of Venus. The surface at its landing site was composed of stratified rocks with several stratified layers. There was no visible soil around the rocks.

Data was received from the Venera 14 lander for 57 minutes after landing. This was just enough time to scan a 170-degree azimuth sector through clear, red, green, and blue filters. Images from the two cameras of Venera 14 taken through the clear filters are shown above (Fig. 5.3).

Mass Spectrometer

The MX 6411 mass spectrometer installed on the Venera 13 and Venera 14 landers was an improved version of the mass spectrometer that operated well on the Venera 11 and Venera 12 missions. The improved version had higher sensitivity, improved electronics, and improved method of admitting a gas sample into the analyzer. The basic operation of the radio-frequency type mass spectrometer was the same as described for Venera 11 and Venera 12.

A paper by Istomin, et al., 1982, *Mass Spectrometry on the Venera 13 and Venera 14 Landers Preliminary Results*, gives results of the mass spectrometer experiment. The experiment concentrated on determining amounts of isotopes of the inert gasses, Neon (^{20}Ne and ^{22}Ne), Argon (^{36}Ar, ^{38}Ar, and ^{40}Ar), Krypton (^{83}Kr), and Xenon (^{131}Xe, ^{132}Xe, and ^{136}Xe). The following amounts were determined (Table 5.2).

Gas Chromatograph (Sigma 2)

A gas chromatograph was carried on the landers to complement the mass spectrometer in identifying abundances of different gasses in the atmosphere of Venus. The instrument was located inside of the spiral transmit antenna.

The Sigma 2 gas chromatograph was an improved version of the Sigma 1 instrument carried by Venera 11 and Venera 12 landers. The improvements included: increased sensitivity, increased selectivity of analysis, and identification of a wider range of constituents of the Venus atmosphere.

A basic gas chromatograph uses thin tubes filled with capillary material that gas molecules adhere to. An inert carrier gas such as neon is used to push the gas sample through the capillary tube. A gas constituent that has weaker adsorption to the walls has a shorter retention time and it is pushed through the capillary tube quickly by the neon carrier gas. A gas constituent that has stronger adsorption to the walls has a longer retention time and appeared later at the output of the tube. A detector at the output of the tube is used to determine retention time and amplitude of the constituents of the gas sample.

The instrument was calibrated beforehand so that the retention time of various gases was known. The output of the instrument was a plot of amplitude vs time from introduction of the sample. The time delay identifies a particular gas and the

Table 5.2 Amounts of inert gases in atmosphere of Venus

Element and isotope	Amount
Neon ^{20}Ne	~10 parts per million
Ratio ($^{20}Ne/^{22}Ne$)	11.0 ± 0.7
Argon (^{36}Ar, ^{38}Ar, and ^{40}Ar)	100 parts per million
Krypton (^{85}Kr)	10–100 parts per billion
Xenon ^{131}Xe and ^{132}Xe	10–100 parts per billion

amplitude of the detected signal is a measure of its relative abundance. The capillary tubes, which were several meters long, were coiled up to fit within the instrument.

The gas chromatograph installed in the Venera 13 and Venera 14 landers used three separate tubes referred to as columns to identify categories of gases. Two of the columns, each with a detector, were connected in series and the sample of gas passed through both columns before exiting the instrument. Neon gas was used to push the samples through the columns. The third column used nitrogen as a carrier gas. The three columns and detectors were enclosed in a temperature- controlled chamber while the analysis was being performed. The temperature was maintained at 70 ° C in the Venera 12 lander and this temperature was likely used for Venera 13 and Venera 14 as well.

In operation, a sample of gas about one cubic centimeter in volume was applied to the inlet of the first of two columns connected in series. The sample was pushed through the two columns by neon gas. The detectors at located at the exit of the columns responded to each constituent as it exited the column. The outputs of the detectors were applied to the telemetry system in analog form and also in digital form. Nitrogen was used as a carrier gas for the sample that was applied to the third column. The detector at the exit of that column was an electron capture type. That detector was very sensitive, particularly to sulfur hexafluoride (SF_6). The readings at the output of the detector were applied to the telemetry system in analog form.

The instrument in Venera 13 and Venera 14 were programmed to sample the gas in the clouds as the lander descended from an altitude of 58 km to 49 km. While one sample was being analyzed, a new sample was gathered. Analysis was performed each 400 seconds during the descent while a new sample was being gathered. A total of seven separate analyses were made during the descent of Venera 13. The Venera 14 lander performed six analyses using the neon detectors and nine analyses with the electron-capture detector.

A paper by L. M. Mukhin, et al., 1982, *Venera 13 and Venera 14 Gas Chromatograph Analysis of the Venus Atmosphere Composition,* gives results of the experiment for the Venera 13 and Venera 14 landers. The results for each of the constituents identified are given below (Table 5.3).

Table 5.3 Gas chromatography analysis of Venus atmosphere

Constituent	Percentage by volume	Sampling height, km
Hydrogen, H_2	$(2.2 \pm 1) \times 10^{-3}$	58–49
Oxygen, O_2	$(1.8 \pm 0.4) \times 10^{-3}$	58–35
Krypton, Kr	$(7 \pm 3) \times 10^{-5}$	49–37
Water, H_2O	$(7 \pm 3) \times 10^{-2}$	58–49
Hydrogen Sulfide, H_2S	$(8 \pm 4) \times 10^{-3}$	37–29
Carbonyl Sulfide, COS	$(4 \pm 2) \times 10^{-3}$	37–29
Sulfur hexafluoride, SF_6	$(2 \pm 1) \times 10^{-5}$	58–3.5

X-Ray Fluorescence Spectrometer Analysis of Clouds

The Venera 13 and 14 landers carried two different X-Ray fluorescence spectrometers. One was used to determine chemical properties of the clouds above the surface of Venus and the other was used to determine chemical properties of samples of soil on the surface.

The X-ray fluorescence spectrometer employed in the clouds used a fan to suck a sample of the atmospheric through a filter with fine pores about four microns in size. The sample was irradiated by gamma rays produced by a radio-isotope Fe-55 source. The gamma radiation induced fluorescence in the sample and emission of x-rays. The energy of the resulting X-rays was measured by gas-discharge proportional counters and associated analysis apparatus to give an indication of the various chemicals in the sample. A description of operation of the instrument is given in the next subsection on analysis of the soil and rocks of Venus.

The X-Ray fluorescence spectrometer on Venera 14 operated from an altitude of about 63 km down to an altitude of about 47 km. This altitude range encompasses the region of clouds. Don Mitchell in *Venera: The Soviet Exploration of Venus* gives the results of the measurements as follows:

Element	Density
Sulfur	1.1 ± 0.13 mg/m^3
Chlorine	0.16 ± 0.04 mg/m^3

The presence of iron had been measured by the X-Ray fluorescence spectrometer on Venera 12 but only sulfur and chlorine are listed in the results for Venera 14.

X-Ray Fluorescence Spectrometer Analysis of Soil and Rocks

The second X-Ray fluorescence spectrometer carried by the lander was designed to determine properties of the soil and rocks of Venus. The spectrometer was mounted on the inside of the pressure capsule and a series of airlocks was used to transfer a sample of soil gathered by a drill to the instrument.

The drill was a telescoping assembly that lowered a drill bit onto the surface. An electric motor powered the drill. The apparatus was capable of drilling up to three centimeters into rock and freeing a sample of fine rock chips about two cubic centimeters in volume. Based on the volume of the sample and depth of penetration, the drill bit was about 0.9 centimeters in diameter. Particles and dust from drilling were carried through tubes to airlocks. The tubes were initially at a much lower pressure than the atmosphere of Venus and when seals on the tube were breached by pyrotechnics the sample of rock was drawn through the tubes to airlocks and into the X-ray fluorescence chamber.

The X-ray florescence chamber contained one radioactive Plutonium isotope, Pu-238, and two radioactive Iron isotopes, Fe-55. The Pu-238 source had good efficiency in exciting the lighter elements of Mg, Al, and Si. The Fe-55 sources were effective in exciting the heavier elements. Gamma radiation from the radioactive sources induced fluorescence in the sample and emission of X-rays. The energy of the resulting X-rays was an indication of the chemistry of the sample.

The X-rays were detected by gas-discharge proportional counters. That type of detector consisted of an ionization tube to which a high voltage was applied through a load resistor. The ionization tube responded to X-ray radiation with a controlled avalanche of ion pairs. The amplitude of the current pulse resulting from conduction in the tube due to the ion pairs was proportional to the energy of the incoming radiation, hence its name "proportional counter." The current pulse through a load resistor resulted in a voltage pulse that was amplified and applied to a pulse height analyzer.

The analyzer was arranged with several channels, each with a narrow region of pulse height. The number of pulses per unit time was counted for each channel of the analyzer. The result was a spectrum of number of counts as a function of energy of the X-ray radiation. A paper by Surkov, et al., 1986, *Element Composition of Venus Rocks: Preliminary Results from Venera 13 and Venera 14,* shows a spectrum of counts vs channel number for Venera 14. The paper by Surkov, et al., gives results of the experiment on composition of Venus rocks as shown in Table 5.4 below.

The two most common elements found in the rocks were silicon and aluminum, their oxides being silicon dioxide and aluminum oxide. The paper by Surkov, et al., states that the composition of Venus rocks resembles basalts of the Earth's crust.

Optical Spectrometer

The Venera 13 and Venera 14 landers carried two optical spectrophotometers to determine the spectrum and angular distribution of sunlight scattered by the atmosphere of Venus. One of the optical instruments examined a 150-degree included angle cone of the upper hemisphere. The other optical instrument made directional

Table 5.4 Composition of Venus rocks, percentage by weight

Element (oxide)	Percentage by weight	
	Venera 13	Venera 14
MgO	11.4 ± 6.2	8.1 ± 3.3
Al_2O_2	15.8 ± 3.0	17.9 ± 2.6
SiO_2	45.1 ± 3.0	48.7 ± 3.6
K_2O	4.0 ± 0.63	0.2 ± 0.07
CaO	7.1 ± 0.96	10.3 ± 1.2
TiO_2	1.59 ± 0.45	1.25 ± 0.41
MnO	0.2 ± 0.1	0.16 ± 0.08
FeO	9.3 ± 2.2	8.8 ± 1.8
SO_3	1.62 ± 1.0	0.88 ± 0.77

assessment with a 20-degree beamwidth at angles of 0° (zenith), 45°, 135°, 180° (nadir), 225°, and 315°. The spectrometers scanned a wavelength band of 0.45 to 1.2 microns (μm). The wavelengths measured extend from blue light (0.45 to 0.485 μm) to red light (0.625 to 0.74 μm) and on to the near-infrared (0.75 to 1.4 μm). In addition, an ultraviolet (UV) photometric instrument operated at a wavelength of 0.37 microns. The ultraviolet instrument could be positioned at the same six angles as the second of the optical instruments.

A paper by V. I. Moroz, et al., 1982, *The Venera 13 and Venera 14 Spectrophotometry Experiments*, gives spectrums of intensity as a function of wavelength for several different altitudes from 61.3 km to 0.9 km. The spectrum at an altitude of 61.3 km had a broad peak of intensity at wavelengths around 0.55 microns of about 500 watts per square meter per micron per steradian. The intensity smoothly declined to about 120 watts per square meter per micron per steradian at a wavelength of 1.2 microns.

The spectrum at 0.9 km altitude had a broad intensity peak at wavelengths around 0.67 microns at about 32 watts per square meter per micron per steradian. The intensity declined with deep dips and peaks to about 3 watts per square meter per micron per steradian at 1.2 micron wavelength. The dips and peaks were due to absorption bands of water vapor and CO_2.

From the water vapor bands it was estimated that the water vapor abundance was about 0.1% (one part per thousand) in the altitude range of 40 to 60 km and 10 to 20 parts per million at the surface. The paper referenced above gives variation of normalized illuminance with altitude. The intensity was normalized to that that beyond the atmosphere of Venus. The measurements are given in the table below (Table 5.5).

The illuminance on the surface in terms of lux was 5000 lux for Venera 13 and 7000 lux for Venera 14. For reference, the illuminance on Earth in bright sunlight is about 111,000 lux and it is about 1000–2000 lux on an overcast day.

From illuminance data, the height of the underside of the clouds of Venus was determined to be 49 km for Venera 13 and 47.5 km for Venera 14. Three separate layers of clouds were observed between 62 km and 47 km. The ultraviolet photometric instrument, which operated at a wavelength of 0.37 microns, found that 90% of light at that wavelength is absorbed by the atmosphere above 58 km. The remainder is nearly completely absorbed in the altitude range 58 to 48 km.

Table 5.5 Illuminance as a function of altitude above Venus

Altitude, km	Normalized illuminance	
	Venera 13	Venera 14
60	0.54	0.80
48	0.36	0.50
0	0.025	0.035

Penetrometer (PROP-V)

The Venera 13 and 14 landers carried a penetration device with a rotating conical tip with a blade that was used to measure the bearing strength of the soil of Venus. The device was referred to as PROP-V, an acronym from the Russian term for an instrument for evaluating the characteristics of the surface of Venus. The rotating conical tip was mounted at the end of a ladder type assembly with two sections that was folded up on the lander prior to landing. After landing it unfolded and the end with the tip fell to the surface of Venus. The amount of penetration of the conical tip as a result of the fall was noted.

The penetrometer was about 60 cm long when unfolded. An image of the Venera 13 penetrometer extended on the surface is shown below (Fig. 5.4). The image was obtained by cropping a panoramic image taken by Venera 13.

The conical tip, which contained a blade, was rotated by a spring once it was imbedded in the surface. The amount of penetration and amount of rotation of the cone with the blade were observed by one of the cameras on the lander. A dial on the top of the cone indicated the amount of rotation. The mechanical properties of the rock were determined from the depth of penetration of the cone and the angle that the spring driven cone with blade cut through the rock. Electrodes embedded in the cone were used to measure the resistance of the rock.

Data from operation of the Venera 13 penetrometer on the surface of Venus indicated that the load bearing capacity of the rocks to be in the range of 2.6 to 10 kg per square centimeter. That is similar to compacted very fine sand on earth. The electrical resistance of the surface between electrodes of the penetrometer was 80 ohms. This value was lower than expected for basalt rocks. The cone of the

Fig. 5.4 Penetrometer extended on the surface of Venus (cropped from panoramic image)

penetrometer of Venera 14 fell on the jettisoned camera lens cover. Measurements were made, but the presence of the lens cap presented uncertainties.

Low Frequency Radio Sensor (GROZA 2)

The Venera 13 and Venera 14 landers carried a low frequency radio sensor, GROZA 2, which was an upgraded instrument from GROZA carried on Venera 11 and Venera 12. GROZA 2 used a loop antenna 25 cm in diameter, but the loop was exposed rather than being installed within a cylindrical cover as was done on Venera 11 and Venera 12.

GROZA 2 contained three low frequency receiving channels. The first was a wide-band channel, the width of which was not stated but may have been from 8 kHz to 42 kHz to encompass the other two channels. The second channel was centered at 10 kHz with 5 kHz bandwidth and the third channel was centered at 36 kHz with 12 kHz bandwidth. The wideband channel had a threshold detector set at 200 microvolts per meter per square root Hz. A counter with a 10 second count window recorded the number of pulses above the threshold.

The outputs of the three channels were recorded as the Venera 13 and Venera 14 landers descended to the surface. Both landers showed signals at 47 km altitude that likely was caused by jettisoning of the parachute. Both landers detected signals in the altitude interval 25 to 15 km and between 10 km and the surface. Analysis of data from Venera 11, Venera 12, Venera 13 and Venera 14 led researchers to conclude that the source of the low frequency electromagnetic signals was located below the cloud deck.

The GROZA 2 equipment included a microphone to record sounds of wind. The acoustical system accepted sound in the frequency region 400 to 2500 Hz. The signal processing equipment contained a high and a low sensitivity channel. The low sensitivity channel was used during descent where there would be considerable aerodynamic noise and the high sensitivity channel was used after landing. The response of the microphone to wind coming from various angles was established in the laboratory. Comparing the calibration data with data from the landers, the wind speed on the surface of Venus was estimated to be 0.35–0.39 meters per second for the Venera 13 landing site and 0.51–0.56 meters per second for the Venera 14 site.

Landing Sites for the Venera Spacecraft

The Russian Verena series of spacecraft made seven successful landings on the surface of Venus. Those spacecraft were Venera 8, 9, 10, 11, 12, 13, and 14. Locations of the landings in the Venus coordinate system in Mercator projection are shown on the next page (Fig. 5.5).

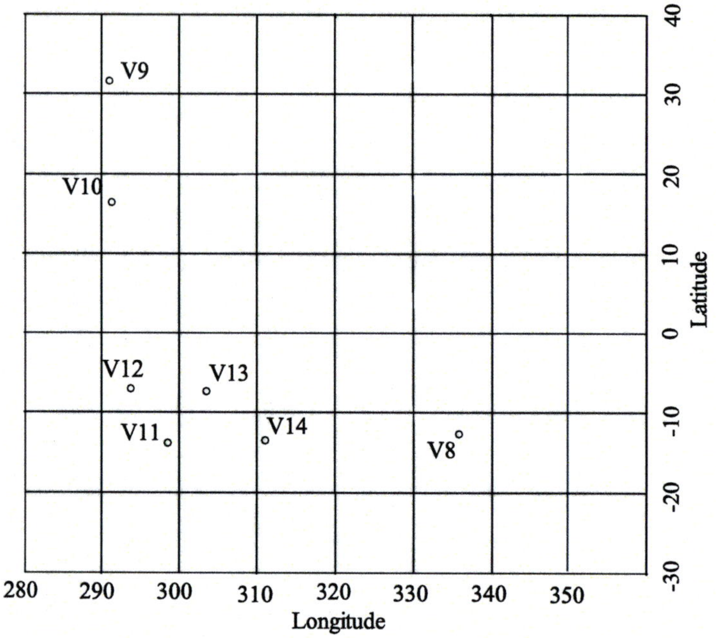

Fig. 5.5 Landing sites for Venera landers

4 V-2 Spacecraft (Venera 15 and Venera 16)

Venera 15 and Venera 16 were part of the 4 V-2 series of Russian spacecraft. The mission for Venera 15 and 16 was to image the surface of Venus through the clouds using a synthetic aperture radar and a radar altimeter. The two spacecraft were identical and both were placed into a near polar orbit of Venus. The spacecraft used a bus structure similar to the bus for Venera 9 through Venera 14. Rather than carry a landing capsule at the top of the bus, the structure was fitted with a large antenna and a toroidal equipment bay for the synthetic aperture radar. A photograph of a model of the spacecraft at the Cosmos Museum in Moscow is shown on the next page (Fig. 5.6).

Mechanical Configuration of Venera 15 and 16

The general configuration of the Venera 15 and 16 spacecraft is shown in Fig. 5.6. The basic structure was similar to that of Venera 14 but the lander on top of Venera 14 was replaced by the large antenna for the synthetic aperture radar in Venera 15 and Venera 16. The long cylindrical portion was lengthened to contain additional fuel for the rocket engine. The length of the main structure of the spacecraft, not

Fig. 5.6 Model of Venera 15 in Cosmos Museum, Moscow. (NASA image collection, Credit: Russian Space Research Institute)

counting the large antenna on top, was about 4.6 meters. The cylindrical core, containing propellant tanks, was 1.1 meters in diameter. The toroidal structure that made up the bottom of the main structure was 2.3 meters in diameter. It held the rocket motor and contained electronics for spacecraft systems and for the experiments.

There were several antennas on the spacecraft. The large circular parabolic antenna at the right side of the photograph was used for communication with Earth. It was 2.6 meters in diameter. Two conical spiral antennas mounted near the top of the main structure were omnidirectional antennas that were also used for communication with Earth. The large antenna on top of the spacecraft was used by the synthetic aperture radar. It was a parabolic segment six meters across the long dimension and 1.4 meters wide at the base. The antenna folded along its long dimension at about 40% of the distance from center on each side to fit within the shroud of the launch vehicle. The smaller parabolic antenna in front of the large antenna in the photograph was used by the radar altimeter. It was one meter in diameter.

Solar arrays extended from both sides of the spacecraft. Each side contained two panels. The panels are shown in the folded position in the photograph. The arrays were 2.1 meters high. The inner panel was 1.2 meters wide and the outer panel was one meter wide. The solar arrays spanned 8.9 meters when fully extended. Attitude control jets were mounted along the hinge area of the solar panels.

The several small spherical tanks around the bottom and top of the structure contained nitrogen gas for the attitude control jets. The tanks were about 35 centimeters in diameter. The amount of nitrogen carried limited the effective life of the spacecraft

The weight of Venera 15 with fuel was 5250 kg. The Weight of Venera 16, which contained one additional experiment, was 5300 kg.

Flight of Venera 15 and Venera 16

Venera 15 was launched from the Baikonur Cosmodrome on 2 June 1983 by a Proton K launch vehicle with a block D upper stage. The launch and the insertion into a trans-Venus trajectory were successful. A midcourse correction was made on 10 June 1983 and a second midcourse correction was made on 1 October 1983. The spacecraft successfully entered a near polar orbit around Venus on 10 October 1983. The final orbit had an inclination of 92.5 degrees and periapsis at an altitude of 1029 km and latitude of 62 degrees north. Apoapsis was at an altitude of about 66,000 km. Because of the eccentric orbit, good mapping data was restricted to latitudes greater than about 25 degrees north. A sketch of a portion of the orbit of the spacecraft in the vicinity of periapsis is given below (Fig. 5.7).

During mapping, the spacecraft was maneuvered to direct the radar altimeter beam along the local vertical and the synthetic aperture radar beam 90 degrees to the

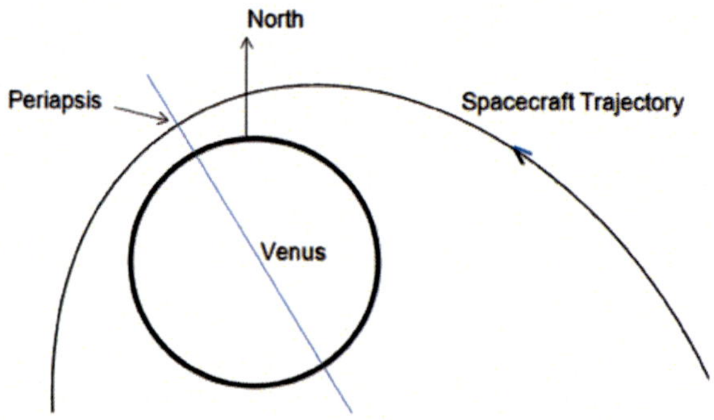

Fig. 5.7 Portion of orbit of Venera 15 and 16 near periapsis

side of the flight path and 10 degrees from the local vertical. Radar mapping data was gathered from the time the spacecraft passed over the north pole to reaching a latitude of about 25 degrees north in its orbit. Both Venera 15 and 16 orbited with their longitudinal axis (Z-axis) pointed towards the south polar axis of the planet.

The period of the orbits of Venera 15 and Venera 16 was set at 24 hours so that data could be transferred to deep-space communication facilities in the Soviet Union once each orbit. Data transfer took place during the time the spacecraft was in its long apoapsis loop. The primary mission of Venera 15 was scheduled to end on 10 March 1982 but the mission was extended to the end of 1994.

Venera 16 was launched from the Baikonur Cosmodrome on 7 June 1983 by a Proton K launch vehicle with a block D upper stage. The launch and the insertion into a trans-Venus trajectory were successful. Midcourse corrections were made on 15 June 1983 and 5 October 1983. The spacecraft was successfully placed in a near polar orbit around Venus on 14 October 1983. The final orbit was essentially the same as the orbit of Venera 15.

Venera 16 followed Venera 15 in its mapping operation around Venus. The orbit was displaced about 4.5 degrees in longitude from that of Venera 15 because of the time difference in entering orbit. The primary mission of Venera 16 was scheduled to end on 10 March 1982 but the mission was extended. The mission ended abruptly in June 1995 when Venera 16 responded to a command intended for the Vega 1 spacecraft that had just arrived at Venus.

Scientific Experiments on Venera 15 and 16

Experiments on Venera 15 and Venera 16 included:

- Synthetic aperture radar.
- Radar altimeter.
- Infrared spectrometer.

Venera 16 contained an additional experiment called cosmic ray detector.

Synthetic Aperture Radar

The synthetic aperture radar (SAR) was designed by the Institute for Radio Engineering and Electronics in Moscow. The antenna for the SAR was the large parabolic segment on top of the bus in Fig. 5.6. The spacecraft was oriented in flight so that the antenna was at the bottom so that the antenna beam could impinge on the surface. The center of the antenna beam was directed 90 degrees to the side of the flight path and 10 degrees from the local vertical of Venus. The radar used Doppler processing of the reflected signal from the surface to make use of motion of the spacecraft to create a long synthetic aperture in the plane of the flight path. The resulting resolution of the radar image was one to two km when the spacecraft was at an altitude of 1000 km near periapsis of the orbit.

The radar operated at a wavelength of eight centimeters (3.75 GHz). The 80-watt transmitter was biphase modulated by a code sequence 127 bits long. Each bit had duration of 1.54 microseconds (μs) and the code sequence was 195.58 μs long. Twenty sequences were transmitted in sequence for a total time of 3.9 milliseconds. The radar was then cycled to the receive mode for 3.9 milliseconds. Operation of the transmitter, receiver, and signal processing was controlled by an onboard digital computer. When processed, the effective pulsewidth for resolution purposes of the radar was 1.54 μs with effective power of that contained in the 195.58 μs sequence length.

The reflected signal from the surface contained both a time delay component and a frequency (Doppler shift) component. Each scattering element on the surface on a particular range contour experienced a positive Doppler shift as the spacecraft's path approached the element, zero shift as the spacecraft passed abreast of the element, and then a negative Doppler shift as the spacecraft's path moved away from the element. The Doppler information allowed the radar to achieve resolution in the along-track direction much better than based on antenna beamwidth alone.

The reflected signal was processed in the receiver by an in-phase channel and a quadrature-phase channel to preserve the phase properties of the signal. The in-phase and quadrature components of the receiver output were digitized and placed into RAM storage. Each radar look of 3.9 milliseconds produced 2540 bytes of data where each byte (8 bits) consisted of four in-phase and four quadrature phase components of the complex signal. The data stream was read out of RAM into tape recorders for later playback and transmission to ground stations on Earth. The data was cycled between two tape recorders in the spacecraft to accommodate the high data rate.

A new radar look was made each 0.3 seconds. The time required for the spacecraft to travel from the north polar region to latitude of 30 degrees north was about 16 minutes. In that time a total of 3200 radar looks were made. At 20,320 bits of data per look, a total of 65 megabits of data were generated each orbit.

Imaging began when the spacecraft approached 80 degrees north latitude on its way up and over the north pole. Imaging continued around the other side down to latitude of about 25 degrees north. The image path for each orbit was about 120 km wide and 7500 km long. Data was stored during the image taking process and then transmitted to Earth as the spacecraft continued its travel in the long loop around apoapsis.

The radar was switched between the synthetic aperture antenna and the radar altimeter antenna such that both a synthetic aperture and a radar altimeter sequence, including data processing, were made each 0.3 seconds.

Data from the spacecraft was transmitted to Earth by the 2.6-meter parabolic antenna at a frequency of 5 GHz. The data rate was 100 kilobits per second. Data transmitted included that from the experiments and engineering data as well as data from the synthetic aperture radar and radar altimeter. Transmission of 65 megabits of data per orbit from the synthetic aperture radar data took about 10.8 minutes. Data from the synthetic aperture radar was analyzed on Earth by capable computer programs. Each radar look generated an image after processing that was 31 pixels wide in the along-track direction and 127 pixels wide in the cross-track direction. The resolution of the resulting image was one to two km.

Radar imaging from Venera 15 and Venera 16 covered the upper portion of Venus from the north pole to latitudes of nominally 30 degrees north. The mapped area was about 25% of the surface area of the planet. The data from the spacecraft was processed on the ground to resolve each radar look of 2540 bytes into 31 cells along-track and 127 cells cross-track. The radar looks were stitched together along with data from adjacent orbits to form an image mosaic. A total of 27 image mosaics were used to cover the imaged area on the planet.

A small portion of mosaic number five that includes the region around Maxwell Montes is shown below (Fig. 5.8). Maxwell Montes is a mountain range with highest point 10,943 meters (39,904 feet) above the planet's mean radius. The image incorporates data from both Venera 15 and Venera 16. The radars were looking from the left at an incidence angle of about 12 degrees from an altitude of about 1000 km. The prominent circular item on top of the Maxwell Montes range is the impact crater Cleopatra. It is about 100 km in diameter. It is located at 65.8° N and 7.1° E in the Venusian coordinate system. The western slope of Maxwell Montes, which appears on the left side of the image, was found to be rough and very steep.

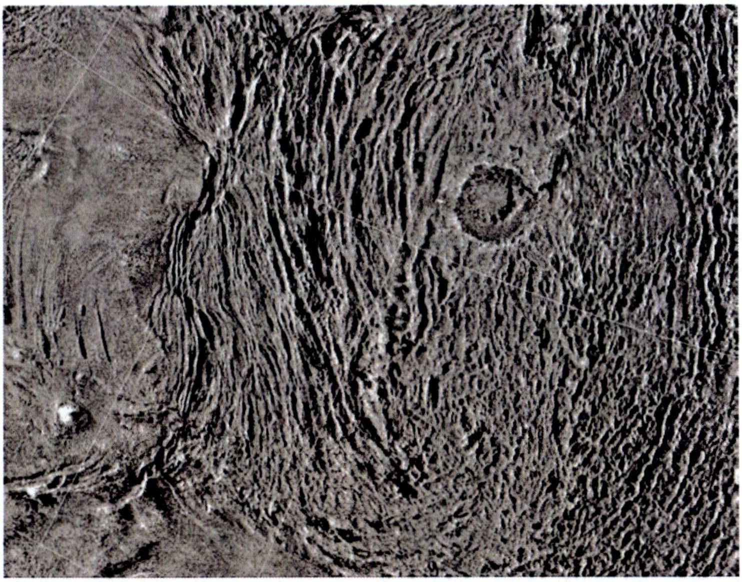

Fig. 5.8 Synthetic aperture image of Maxwell Montes from Venera 15 and Venera 16 measurements. (NASSA PDS Geoscience Node, Credit: Verndsky Institute and Moscow Power Institute)

Radar Altimeter

The radar altimeter used the same basic radar functions as the synthetic aperture radar. The transmitted power was about 80 watts at a wavelength of eight centimeters. The radar was switched from the large synthetic aperture radar antenna to a parabolic dish antenna one meter in diameter for radar altimeter measurements. The beamwidth of the radar altimeter antenna was about 5.6 degrees. The radar altimeter antenna beam was directed parallel to the Z-axis of the spacecraft. The spacecraft was continuously oriented with its Z-axis aligned with the local vertical.

The transmitted waveform in the radar altimeter mode was the same as for the synthetic aperture radar mode initially. That code sequence, which was 195.58 microseconds long, resulted in a range ambiguity of 29.33 km. The accuracy of tracking data from Earth allowed resolving that ambiguity and allowed altimeter measurements to be made. The code sequence was shortened to 31 bits once the ambiguity was resolved and initial radar altimeter measurements were made. The shorter 31-bit sequence with duration of 47.74 microseconds resulted in higher accuracy and it was used for the mapping mission. Fourteen sequences were transmitted that took a total of 0.67 milliseconds and then the radar switched to the receive mode for the same length of time. A new radar altimeter measurement was made each 0.3 seconds.

The accuracy of the altitude measurements was about 50 meters. The cross-track resolution was about 40 km at an altitude of 1000 km. The along-track resolution was enhanced by processing Doppler data in the return signal, just as in the case of the synthetic aperture radar. The resulting along-track resolution was about 10 km.

Topographical mapping began when the spacecraft approached 80 degrees north latitude on its way up and over the north pole. Mapping continued around the other side down to latitude of about 25 degrees north. Altitude measurements were made at 2.5 km intervals near periapsis along the flight path. Data was stored during the mapping process and then transmitted to Earth as the spacecraft continued its travel in the long loop around apoapsis.

A paper by V. A. Kotelnikov, et al., 1985, *Radar Study of Venus Surface by Venera-15 and -16 spacecraft*, presents elevation profiles of certain areas of Venus. A profile on the Maxwell Montes range that includes the impact crater Cleopatra indicated that the depth of the crater floor is about two km below the surrounding terrain. A different source presents a topographical map of the northern portion of Venus developed from radar altimeter data. The map is shown on the next page (Fig. 5.9). The high altitude area, colored red, is the Maxwell Montes range.

Infrared Spectrometer

The Infrared Fourier Spectrometers on the Venera 15 and Venera 16 spacecraft were used to measure the thermal emission spectra of the middle atmosphere and clouds of Venus. Venera 15 was configured to measure radiation in the wavelength range of

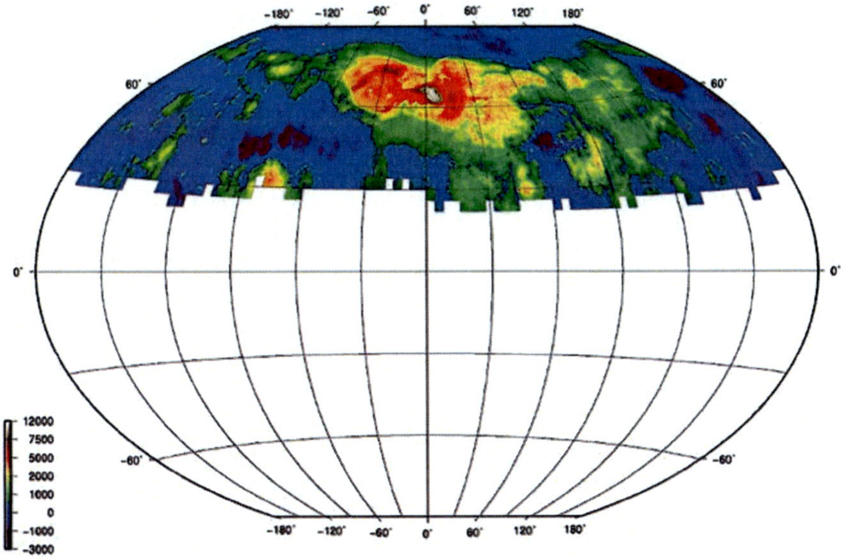

Fig. 5.9 Color coded topographical map of northern area of Venus. (Wikimedia Commons posting by Zamonin)

6.2 μm to 38 μm. The instrument on Venera 16 measured in the wavelength range 6.2 μm to 25 μm.

The field of view of the instrument was a rectangle four degrees by four degrees centered on a line parallel to the Z-axis of the spacecraft. The spacecraft was stabilized in three axes with the Z-axis of the spacecraft pointed in the direction of the south celestial pole. This configuration allowed the spectrometer to view the area in the vicinity of the north pole close to the local vertical and the area in the vicinity of periapsis (62 degrees north latitude) at an angle of 28 degrees.

The optical arrangement of the instrument was that of a Michelson interferometer. A three-position mirror at the input to the instrument selected a view of Venus, deep space, or a calibration heat source. The selected radiation was directed at a beam splitter that directed a portion of the radiation to a fixed flat mirror and a portion of radiation to a flat moveable mirror. The moveable mirror was moved at a constant rate during the measurement cycle and then moved back to the starting position. Reflected radiation from the two flat mirrors were combined and focused onto a bolometer type detector. The interference pattern developed from combination of the two radiation signals was detected by the bolometer and converted to an electrical signal. The frequency (f) of the interference structure was a function of the velocity of the moveable mirror (V) and the wavenumber of a particular emission (υ) as $f = 2\upsilon V$.

The signal at the output of the bolometer was amplified and applied to an analog-to-digital converter. The digitized signal was applied to a computer where it was operated on by a Fourier Transform. The output of the Fourier Transform was a spectrum of amplitudes of the various sources as a function of wavelength.

Some results from the infrared spectrometer experiment are given in a paper by D. Oertel, et al., 1984, *Venera 15 and Venera 16 Infrared Spectrometry: First Results*. They analyzed a series of measurements made on 14 October 1983. The longitude of the orbit was 35.5 degrees at the time and measurements included both day and night conditions. Measurements of the 667-cm^{-1} CO_2 band allowed developing temperature profiles for the altitude range 60 to 90 km. Temperatures ranged from about 170 K at 90 km to about 250 K at 60 km. Lines of water vapor rotational band were observed and emission from Sulphur dioxide (SO_2) and sulfuric acid (H_2SO_4) were identified.

Cosmic Ray Detector

The cosmic ray detector experiment was designed to measure the flux of charged particles in interplanetary space during the mission of Venera 16 to Venus. The experiment, which was managed by the Theoretical and Applied Space Physics Division of the Moscow State University (MSU-TASPD), incorporated six sensors. Two sensors were gas-discharge counters, two were SSD (Si) detectors with 70-micron thickness, and two were SSD detectors with 400-micron thickness.

One of the two gas discharge counters was arranged to count protons with energy greater than 30 million electron volts (MeV) and the other, with additional shielding, counted protons with energy greater than 60 MeV. The solidstate detectors were arranged to count protons with energies of 1–65 MeV, 1–11 MeV, and 5–8 MeV as well as alpha particles with energy of 7–24 MeV. All of these values were in a direction to the sun. In an anti-sun direction, energies for count of protons were: 1–70 MeV, 1–9 MeV, and 5–10 MeV. For alpha particles traveling in the anti-sun direction, the energy for count was 7–22 MeV.

Data published by MSU-TASPD included counts from each detector every four hours from 8 June 1983 to 28 May 1985. There were times when data was not available from some detectors. Data from the gas discharge detectors for energies greater than 30 MeV was present and appeared stable throughout the record. There were a few data dropouts on the >60 MeV detector. The background count varied from about 16 to 18 counts per second for those gas discharge detectors. The highest count noted was 1161 counts per second in the >30 MeV detector on 10 March 1984. The count four hours earlier had been 857 per second on the >30 MeV detector and 141 per second on the >60 MeV detector.

5VK Spacecraft (Vega 1 and Vega 2)

Vega 1 and Vega 2 were impressive spacecraft and the last to be sent to Venus by the Soviet Union. Both spacecraft dispatched a landing capsule to Venus and then the bus went on to gather in-situ measurements and images of the comet Halley. Data and photographs were successfully obtained during flyby of the comet. The landing capsules deployed an instrumented balloon in the cloud deck. Pushed by strong winds, the balloons traveled about a third of the way around the planet. Both landers touched down successfully and provided good data from a number of experiments during the descent and while on the surface.

A photograph of a model of the Vega spacecraft that is displayed in the Smithsonian Udvar Hazy Center in Washington, DC is shown below (Fig. 5.10). The lander and deflated balloons were carried in the spherical heat shield structure on top of the bus.

The Vega mission was managed and operated by the Soviet Union and the spacecraft and launcher were Soviet designs. However, there was significant international presence in planning science for the encounter with Halley and in developing the scientific experiments and in interpreting measurements. A large international presence joined Soviet Union scientists at the Space Research Institute (IKI) in Moscow at eventful times during the mission.

The Vega spacecraft was made up of a bus and landing capsule. The bus structure was essentially the same as the buses used for the Venera 9 through 14 spacecraft. The experiment suite on the bus consisted mainly of instruments to study the comet.

Fig. 5.10 Model of Vega spacecraft in Udvar-Hazy Center. (Wikimedia posting by Daderot)

The landing capsule was essentially the same as the landing capsule used on Venera 9 through 14 except that instead of cameras it carried a balloon and additional experiments. The lander and deflated balloons were carried in the spherical heat shield structure on top of the bus.

Each landing capsule released a balloon at an altitude of 54 km in the thick of the clouds. The balloon, which carried an instrument package, floated in the clouds at about 53 km altitude and it was swept along by the high winds at that altitude. Each landing capsule landed successfully and sent data to Earth during the descent and for nearly an hour after landing.

Mechanical Configuration of Vega 1 and Vega 2

The general construction of the Vega spacecraft was similar to that of Venera 9 through Venera 14. The bus was 2.8 meters long. The center cylindrical section, which contained fuel tanks for the rocket engine, was 1.1 meters in diameter and one meter long. A conical structure at the bottom of the spacecraft in the photograph was 2.35 meters in diameter at its base. The structure necked down below the base of the conical structure to interface with the launch vehicle. The conical structure enclosed the rocket engine and it contained electronics for spacecraft systems and for experiments.

The landing capsule and deflated balloon were contained within the spherical entry enclosure mounted at the top of the bus. The entry enclosure, which was 2.4 meters in diameter, functioned as a heat shield during the high velocity entry into the upper atmosphere of Venus. It was insulated on the inside to protect the lander from excessive heat during entry.

Solar arrays extended from both sides of the bus. The solar arrays were the same as used on the Venera 15 and 16 spacecraft. Each side contained two panels. The arrays were 2.1 meters high. The inner panel was 1.2 meters wide and the outer panel was one meter wide. The solar array panels folded to fit within the shroud of the launch vehicle. A high-gain parabolic antenna that was used to communicate with Earth is shown on the right side of the photograph (Fig. 5.10). The antenna was 1.6 meters in diameter. The narrow beamwidth of the high-gain antenna was pointed towards Earth by maneuvering the vehicle. There were two conical spiral omnidirectional antennas mounted on the facing side of the bus and one similar antenna on the opposite side of the bus.

The telemetry system on the bus contained two independent channels. One channel, referred to as the high-data-rate channel, had a capacity of 65,536 bits per second. The other channel, referred to as the low-data-rate channel, had capacity of 3072 bits per second. The low-data-rate channel could transmit data to Earth over the omnidirectional antennas. The high-data-rate channel required use of the high-gain antenna.

A scan platform holding instruments that investigated the comet Halley was attached to the conical structure at the bottom of the bus. The platform and some of the instruments are at the lower left in the photograph (Fig. 5.10). The scan platform could be steered over a range of −147 degrees to +126 degrees in the ecliptic plane and −60 degrees to +20 degrees in a plane perpendicular to the ecliptic. The steering accuracy was about 0.13 degrees. The platform supported a weight of instruments of 80 kg.

There were several mast-mounted instruments on the bus portion of the spacecraft. Sensors for the wave and plasma analyzer (APV-N) were mounted at the ends of a V shaped structure on top of a five-meter-long mast located on the upper portion of the spacecraft and out of sight in the photograph. Sensors for the magnetometer were mounted on a two-meter long mast extending to the right on the outer solar array at the right side of the photograph.

A dust shield was mounted at the bottom of the bus for protection as the spacecraft passed through dust associated with the coma of the comet Halley. The shield is not shown on the model. Two separate shields were used. One shield, consisting of an aluminum sheet one mm thick, was mounted 5–10 cm from the spacecraft. The other shield, a multilayer sheet 100 microns thick, was mounted 20–30 cm from the spacecraft. The area of the shields was about five square meters.

Flight of the Vegas

The Vega-1 and Vega-2 spacecraft were launched from the Baikonur Cosmodrome in December 1984 by Proton K launch vehicles with Block D upper stages. Flight operations were conducted from the flight control center at Evpatoria in Crimea. The Evpatoria site contained a 70-meter diameter parabolic antenna for deep space communications. A 64-meter diameter antenna at Medvezh'I Ozera near Moscow also received the telemetry signals. Telemetry data from the Vegas were relayed to the Space Research Institute (IKI) in Moscow. Many of the Soviet Union scientists involved with the mission, along with a large number of international scientists, gathered at IKI during important times of the mission.

Flight of Vega 1

Vega 1 was launched successfully on 15 December 1984. A midcourse correction was made on 20 December 1984. The spacecraft approached Venus on 9 June 1985 and released the spherical heat shield containing the lander and balloon system. The spherical heat shield reached the atmosphere of Venus on 11 June on the dark side of the planet. Its velocity at entry into the atmosphere was about 11 km per second. It was slowed by the upper atmosphere to about 280 meters per second by the time it had descended to an altitude of 65 km. The top cap of the heat shield was detached from the lower portion at that altitude and a drogue parachute deployed to pull off the top cap. The drogue and top cap were then jettisoned. A parachute was deployed to pull out the balloon package from the top of the lander at an altitude of 61 km. The lander continued descent towards the surface.

The drogue on the balloon package pulled out a parachute to further slow descent of the balloon package. After extraction, the balloon was inflated by 2.1 kg of helium stored in a tank in the balloon package. The altitude of the balloon was about 54 km at that time. The parachute and inflating system were jettisoned after inflation. The balloon, which was 3.6 meters in diameter, was made of a Teflon cloth matrix and coated with Teflon material to withstand the sulfuric acid laden clouds. A battery powered instrument package hung down below the balloon on a 13-meter long tether.

Ballast was released from the balloon when it had descended to 50 km and the balloon rose to its cruising altitude of about 53 km. The pressure and temperature were benign at that altitude at about 0.5 bars and 310 Kelvin respectively. The Vega 1 balloon was released at about midnight on the dark side of the planet at latitude of 8.1 degrees north and longitude of 176.9 degrees east. The balloon was carried westward at about 69 meters per second at nearly constant altitude by the zonal winds. It traveled 11,600 km and crossed from night to day before transmission from its instruments ceased on 13 June. The instrument package had operated for 46 hours. Transmission of data to Earth was made every two hours while battery power lasted.

The lander deployed a parachute at an altitude of 64.5 km and jettisoned the lower portion of the heat shield. The parachute was jettisoned at an altitude of about 47 km and the lander descended to the surface while being further slowed by an aerobrake disk on the lander acting on the dense atmosphere. It landed safely on 11 June 1985 at 7.5 degrees North, 177.7 degrees East in the Venusian coordinate system. The lander transmitted data from its sensors for 56 minutes after landing. Telemetry signals from the lander were relayed to Earth through the bus during its fly-by of Venus.

The Vega 1 bus swung around Venus with closest approach of 39,000 km on 11 June 1985. It took advantage of gravitational assist from the planet to set a trajectory to intersect the Comet Halley. Its trajectory looped out beyond the orbit of earth and returned to intercept the trajectory of Halley as the comet swung by the sun and headed back out on its long journey in space.

Halley, which has been observed since ancient times, orbits the sun with a period of 75.3 years. It returned for periapsis in March 1986 at a distance of 0.586 astronomical units (AU) from the sun. AU is the average distance of Earth from the sun, which is about 1.496×10^8 km. Halley came within 0.421 AU of the Earth as the comet was heading back out after periapsis. At apoapsis in the year 2024, Halley will be 35.082 AU from the sun and well beyond the orbit of Neptune.

The Vega 1 bus made midcourse corrections on 25 June 1985 and 10 February 1996 to refine its trajectory for intercept of Halley. Observations of Halley began at a distance of 14 million km from the comet on 4 March 1996. Most detailed observations of gas and particles around the comet and photographs of the nucleus were made in span of three hours around the time of closest approach on 6 March 1986.

The distance of closest approach to the comet was 8889 km. The relative velocity between the fast-moving comet and Vega was 79.2 km per second at the time of encounter. Over 500 pictures were taken through different filters during the three-hour period around encounter. Additional photographs were taken on 7 and 8 March as the distance between the comet and spacecraft rapidly opened. Extensive measurements of the properties of gas and particles in the coma were made during the encounter.

Flight of Vega 2

Vega 2 was launched from the Baikonur Cosmodrome on 21 December 1984. The launch was six days after the launch of Vega-1. Vega-2 followed nearly identical procedures at Venus and comet Halley as Vega-1. The spacecraft approached Venus on 13 June 1985 and released the spherical heat shield containing the lander and balloon system. The spherical heat shield reached the atmosphere of Venus on 15 June on the dark side of the planet. After being slowed by descending through the atmosphere, the heat shield was split apart and the lander and balloon were deployed.

The balloon was deployed at latitude of 7.5 degrees South and longitude of 179.8 degrees East. The balloon floated at an altitude of about 53 km and it was carried westward by zonal winds. The balloon traveled about 11,100 km and passed from the night side of Venus to the day side before radio transmissions from its instrument package ceased on 17 June 1986. The instrument package had operated for 46 hours. Transmission of data to Earth was made every two hours while battery power lasted.

The lander descended to the surface and made a successful landing at Venus coordinates of 8.5 degrees South and 164.5 degrees East on 15 June 1986. The lander transmitted data from its sensors to Earth for 57 minutes after landing.

The Vega 2 bus swung around Venus with closest approach to the planet of 24,500 km on 15 June 1985. It took advantage of gravitational assist from Venus to set a trajectory to intersect the Comet Halley. Observations of Halley began at a distance of 14 million km from the comet on 7 March 1996. Most detailed observations of gas and particles around the comet and photographs of the nucleus were made in span of three hours around closest approach on 9 March 1986. The nearest distance to the nucleus was 8030 km during the passage. Over 700 photographs were taken through different filters during the encounter. Additional photographs were taken on 10 and 11 March as the distance between the comet and spacecraft rapidly opened.

Experiments on Vega Bus for Exploration of Comet Halley

The Vega bus carried 14 scientific experiments designed to conduct in-situ measurements of the coma of Halley and image its nucleus. A listing of experiments on the bus and key objectives for each is given in Table 5.6 below. The total weight of the scientific instruments was 120 kg.

Table 5.6 Experiments that investigated Comet Halley

Experiment	Purpose
Television system (TVS)	Image the coma and nucleus of Halley by a narrow-angle camera. Track the comet by a wide-angle camera and use it to point a platform containing the narrow-angle camera and other instruments towards the comet.
Infrared spectrometer (IKS)	Determine the size and thermal emission of the nucleus and determine chemical composition of gas and particles in the coma.
Three-channel spectrometer (TKS)	Examine coma of Halley at ultraviolet, visible, and near-infrared wavelengths to determine chemical properties and density of gas and dust.
Dust mass spectrometer (PUMA)	Measure chemical composition, size, and density of particles in coma.
Dust particle counter (SP-1)	Measure mass of particles by detecting electric charge on plasma generated by particles impacting a gold target.
Dust particle counter (SP-2)	Measure mass of particles by acoustic sensors on a plate and by a plasma sensor that detected electrons and ions in a plasma caused by particles striking a chamber.
Dust particle detector (DUCMA)	Measure mass of dust particles by determining volume of polarized material released by impact of particles.
Energetic particle analyzer (TUNDE-M)	Measure energy and flux of cometary ions accelerated by magnetic fields.
Neutral gas mass spectrometer (ING)	Measure composition of neutral gasses in the coma.
Plasma energy analyzer (PLASMAG)	Measure energy levels of ions and electrons in the solar wind and from the comet.
Wave and plasma analyzer (APV-N)	Measure waves in the frequency range 0.1 to 1000 Hz excited by the comet.
Wave and plasma analyzer (APV-V)	Measure waves in the frequency range 0 to 300 kHz to determine plasma density and frequency spectrum.
Energetic particles (MDSU-TASPD)	Measure particles with energies in the order of a million electron volts.
Magnetometer (MISCHA)	Measure magnetic field in the interaction zone of the coma and solar wind.

Three of the experiments were mounted on the steerable platform and had their fields of view pointed at the comet during approach and flyby. Those instruments were: television camera system (TVS), infrared spectrometer (IKS), and three-channel spectrometer (TKS). Particulars of those instruments are given in the following subsections. A sketch of instruments mounted on the scan platform is shown on the next page (Fig. 5.11).

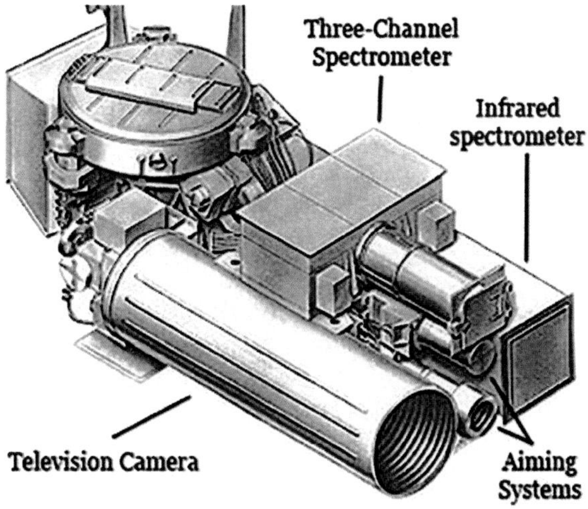

Fig. 5.11 Instruments mounted on Vega scan platform. (NASA PDS collection)

Television System (TVS)

Development of the television system was an international effort by scientists from the Hungarian Central Research Institute for Physics (KFKI), French Laboratoire d'Astronomie Spatiale (LAS), and the Soviet Union Space Research Institute (IKI). The television system contained two television cameras. One, with a narrow field of view and high angular resolution, was used to image the nucleus. The other, with a wide field of view, was used to detect and track the comet. It steered the scan platform to point the narrow field of view camera towards the nucleus.

The narrow field of view camera used Ritchey-Chretien reflective optics. It had an aperture diameter of 240 mm and focal length of 1200 mm. The field of view of the camera was about 0.5 degrees. The average angular resolution at the output of the detector of the camera was 3.6 arcseconds. That resolution corresponds to an image resolution of about 155 meters at the flyby distance of 8889 km from the nucleus for Vega 1. The detector was a charge-coupled device (CCD) with a field of 512 × 512 pixels. Exposure time could be set between 0.01 and 163 seconds. The data rate of the narrow field of view camera could be reduced by employing a floating-point window of 128 × 128 pixels centered on the brightest area in the camera field of view. Most of the photographs taken used this form of image data reduction.

The wide field of view camera used refractive optics with a focal length of 150 mm. The diameter of its aperture was 50 mm. The field of view was about two degrees. The angular resolution at the output of the camera was 0.5 arcminutes. The detector was a CCD with a field of 512 by 256 pixels. Exposure time could be set between 6.0 and 800 milliseconds.

The cameras contained a beam splitter that divided the optical output of the telescopes into two paths. One path was applied to a filter with bandpass of 0.63 to 0.76 microns (red light). The other path was applied to a filter wheel with eight different filters that covered a total spectral range of 0.4 to 1.1 microns. Each filter had a band pass of about 80 nanometers. The wavelength range of 0.4 to 1.1 microns corresponds to a range of green visible light to the near-infrared.

Both cameras were mounted on a two degree of freedom scan platform on the spacecraft. A far-infrared spectrometer and a three-channel spectrometer were mounted to the same platform. Information from the wide-angle camera was the primary means of steering the platform toward the nucleus of Halley. As a backup, an independent eight-segment light sensor could be used to provide steering signals. The platform could also be steered from Earth.

Analysis of photographs of Halley taken by Vega 1 and Vega 2 indicate that the nucleus was a potato-shaped body about 14.5 × 7.5 × 7.5 km in size. Comparing images taken by Vega 1 and later by Vega 2, it was determined that the nucleus rotates with a period of about 53 hours. A raw image of the nucleus of Halley, which was taken by Vega 2 on 9 March 1986, is given below (Fig. 5.12). The rays coming off the nucleus are jets of gas and particles.

Infrared Spectrometer (IKS)

The infrared spectrometer was designed to measure radiation from the coma of Halley in the wavelength band of 2.5–12 microns. It contained two spectral channels for chemical analysis of gas and particles and an imaging channel to determine the size of the nucleus and thermal emission from it.

The instrument used a Cassegrain telescope with a focal length of 538 mm and objective lens diameter of 140 nm. The field of view was one degree. Radiation gathered by the telescope was split into three paths. Each path was focused on a different annular segment of a rotating disk. The three annular segments were concentric on the disk. One path from the telescope was focused on a segment having a

Fig. 5.12 Photograph of nucleus of comet Halley taken by Vega-2. (NASA image from International Halley Watch IHW CD-ROMS, cropped by author)

circular variable filter covering the band of 2.5–5 microns. The filter progressively changed from minimum wavelength to maximum wavelength and started over at minimum wavelength over a full 360-degree rotation. The second path was focused on a segment with a circular variable filter covering the band of 6–12 microns. The third path, referred to as the imaging channel, was focused on a segment of the disk that encoded (chopped) the signal as a function of rotation angle of the disk. That channel responded to radiation in the wavelength range of 7 microns to 14 microns.

The disk was mounted to a shaft that rotated at about seven revolutions per second. The signal from each circular variable filter was sampled at 180 points each half track and digitized. The digitized spectra were averaged over 128 rotations of the disk, corresponding to time duration of 18 seconds. The averaged spectra were telemetered to Earth.

Detectors used for the long wavelength and imaging channels were Mercury Cadmium Telluride (HgCdTe) type. The detector for the short wavelength channel was an Indium Antimonide (InSb) type. The detectors needed to be cooled to about 77 Kelvin to have sufficient sensitivity for the mission. They were mounted in a cryogenic enclosure and cooling was accomplished by expanding nitrogen gas. Nitrogen was stored in four tanks at high pressure. There was sufficient nitrogen gas to cool the detectors for three hours.

The cooling system for the detectors failed on Vega-2 and no useful data was obtained from the instrument. Useful data became available from Vega-1 about 53 minutes before closest approach to Halley at a range of 276,000 km. Cooling of the detectors became a problem at about eight minutes before closest approach at a range of 36,000 km and the data became unusable for a period from that point on. Then at five minutes before closest approach the instrument received an erroneous shut-down command. Commands from Earth restarted the instrument at about 33 minutes after closest approach at a range of 162,000 km going away. Useable data was received from that range to a range of 222,000 km.

Results from the experiment were reported in a paper by Combs, et al., 1988, *The 2.5–12 μm Spectrum of Comet Halley from the IKs-Vega Experiment.* Data from the 2.5 to 5.0 micron channel of Vega-1 showed strong spectral peaks at wavelengths of 2.7 microns and 4.25 microns. These peaks are characteristic of water vapor (H_2O) and carbon dioxide (CO_2) respectively. Weaker responses at 3.6 microns and 4.7 microns correspond to formaldehyde (H_2CO) and carbon monoxide (CO) respectively. The production rate of water was computed to be 10^{30} molecules per second and the production rate of carbon dioxide was computed to be 2.7×10^{28} molecules per second. Spectral data from the 6.0 to 12-micron channel showed a strong response between 8.0 and 12.0 microns due to silicates. There was no response between 6 and 8 microns.

The imaging channel was not intended to generate an image of the nucleus but rather to measure the size and temperature of the emitting region of the nucleus. Data from the imaging channel revealed the temperature of the nucleus to be greater than 360 Kelvin. This temperature is substantially higher than the sublimating temperature of ice. Researchers concluded that there was a dark crust covering a significant portion of the nucleus. Data from the imaging channel coupled with modeling of the encounter indicated that the diameter of the nucleus was in the order of five to ten km.

Three Channel Spectrometer (TKS)

The three-channel spectrometer was designed to investigate the chemical properties and density of gas and dust in the coma of the comet Halley. The experiment was a result of cooperative efforts of researchers in Bulgaria, France, and the Soviet Union.

The instrument used a Cassegrain telescope with a focal length of 500 mm and objective lens diameter of 140 mm. The secondary mirror of the telescope was scanned by steps in two axes to map out an angular region 2.0 degrees by 1.5 degrees in extent for spectral analysis. The scan program consisted of seven lines with 15 angular positions in each line. The result was 105 discrete angular points. The seven scan lines were stacked in the 1.5-degree scan dimension. The dwell time at each angular position was five seconds. Three spectra were formed during each dwell period. Data from angular scanning of the field of view of the spectrometer allowed generating an image of the comet at wavelengths corresponding to various emission lines.

The three-channel spectrometer incorporated two scan modes referred to A and B. The A mode was a mapping mode that included all 105 discrete points. Mode B was a single line mode where line four, which included the center of the overall field of view, was scanned in a back-and-forth manner and covered the 15 angular positions. Mode A was used until 30 minutes before closest approach whereupon a switch to mode B was made. Mode B was activated for 35 minutes and included the time of closest approach to the comet.

The platform that held the three-channel spectrometer, cameras, and infrared spectrometer was steered towards the comet nucleus by the image signal from the wide-angle camera. Apparently, there was a 1.4-degree difference between the boresight of the wide-angle camera and three-channel spectrometer. As a result, the nucleus was slightly outside of the scanned field of view of the spectrometer.

The optical output of the telescope was separated into three channels by three narrow slits located in the focal plane of the telescope. The wavelengths covered by the three channels were: ultraviolet light (120–350 nm), visible light (275–715 nm) and infrared (950–1900 nm). Spectral analysis was performed for each band. The dimensions of the slit for the visible light channel were 2.6 mm by 0.05 mm. The equivalent angular dimensions were 26 minutes of arc by 0.5 minutes. The effective viewing area, including detector effects, was 46 km by 2 km at a closest approach distance of 8000 km. Angular dimensions of the slit for the infrared channel was 100 minutes of arc by 10 minutes. The effective viewing area of the infrared channel was 230 km by 23 km at a closest approach distance of 8000 km.

The ultraviolet and visible spectrometers used concave gratings to disperse the light as a function of wavelength onto a linear array of 1024 charge-coupled devices (CCD). The visible light band of 275–715 nm was covered by 737 detectors. The output amplitudes of individual detectors in the array formed the spectrum. The resulting spectrum included molecular responses superimposed on a continuous spectrum of sunlight scattered from dust particles. The infrared channel used a circular variable filter spectrometer similar to that for the infrared spectrometer described in the previous section.

Unfortunately, no data was obtained from the three-channel spectrometer on Vega 1. Conjecture was that a power lead failed on that instrument. Good data was obtained from the visible light and infrared channels of Vega 2. The ultraviolet channel did not function.

Some results of the three-channel spectrometer experiment on Vega 2 are given in a paper by Krasnopol'skii, et al., 1986, *The Vega 2 TKS experiment: some spectroscopic results for Comet Halley.* With the field of view pointed 300 km from the nucleus, spectra near 1.38 microns yielded data for production rates of H_2O and OH of 4×10^{29} and 1.7×10^{30} molecules per second respectively. Water vapor was determined to be ejected in a cone of width 40 ± 10 degrees towards the sun. The particle density of dust was determined to be 0.35 grams per cubic centimeters. The total dust production rate was estimated to be 10 tons per second. Emissions in the visible band with the field of view about 4000 km from the nucleus indicated molecules with the production rates given in Table 5.7 below.

Balloon Borne Exploration of Venus

A balloon carrying an instrument package was released into clouds in the upper atmosphere of Venus by the landing capsules of both Vega 1 and Vega 2. The balloon and instrument package were designed to float at an altitude of about 54 km and gather information about horizontal and vertical winds and measure properties of the clouds. The balloons were pushed by the high winds aloft of Venus and data was received from both balloons during travel of about 11,000 km around the planet. The deployment and flight of the balloons was described in the *Flight of the Vegas* section of this chapter. A photograph of a model of the Vega Balloon system that hangs in the Smithsonian Udvar-Hazy Museum in Washington, DC is shown on the next page (Fig. 5.13).

The Vega 1 balloon was released at about midnight on the dark side of Venus at latitude of 8.1 degrees north and longitude of 176.9 degrees east. The balloon was

Table 5.7 Production rates of molecules from Comet Haley

Molecule	Common name	Production rate, molecules/sec
OH	Hydroxide	2×10^{30}
CH	Carbon-hydrogen	1×10^{29}
C_2	Diatomic carbon	6×10^{27}
C_3	Cyclopropatrine	3×10^{27}
CN	Cyanicde	1×10^{27}
NH	Imidogen	2×10^{26}

Fig. 5.13 Model of Vega balloon and instrument package in Smithsonian Udvar-Hazy Museum. (Wikimedia posting by Geoffrey Landis)

carried westward at about 69 meters per second at nearly constant altitude of 54 km by the zonal winds. It traveled 11,600 km and crossed from night to day before transmission from its instruments ceased on 13 June 1986.

The Vega 2 balloon was deployed at latitude of 7.5 degrees south and longitude of 179.8 degrees east. The balloon floated at an altitude of about 50 km and it was carried westward by zonal winds. The balloon traveled about 11,100 km and passed from the night side of Venus to the day side before radio transmissions from its instrument package ceased on 17 June 1986.

The balloon was made from a Teflon cloth matrix with an outer Teflon film to resist the corrosive effects of the acid laden clouds of Venus. It was a super-pressure type balloon with an inelastic envelope. It was filled with helium to a pressure higher than the expected atmospheric pressure at the cruising altitude of the balloon. The balloon was 3.4 meters in diameter when inflated to an overpressure of 30 millibars. The total weight of the inflated balloon and suspended instrument package was 21 kg. The balloon was sized to float in the upper atmosphere of Venus at an altitude of 53–54 km.

An instrument package was suspended below the balloon on a tether 13 meters long. The tether in the photograph was shortened for display purposes. The uppermost item of the instrument package was a conical spiral antenna 37 cm long and base diameter of 14 cm. The antenna pattern at the transmitted frequency of 1667.9 MHz was such that, including all transmission losses, the gain factor was 1.1 in the direction of the longitudinal axis and 0.5 at 80 degrees off of the longitudinal axis. The instrument package transmitted data directly to Earth. The balloon was transparent to the transmitted signal.

A package containing the radio transmitter and electronics along with meteoroidal instruments and associated telemetry was suspended below the antenna. It was 40.8 × 14.5 × 13.0 cm in size. A package containing batteries and a backscatter nephelometer was suspended below the middle package. It was 9.0 × 14.5 × 13 cm in size. The batteries, which were lithium type, had capacity of 250 watthours. They weighed one kg. The expected life of the battery when powering the balloon systems was 46–52 hours.

The transmitter output was 4.5 watts at a frequency of 1667.9 MHz. The transmitter had two modes of operation. In the telemetry mode (TM) it transmitted telemetry and periods of pure carrier signal. In the coordinate measurement mode (KI) it transmitted two sideband tones separated by 6.5 MHz. The telemetry mode began with transmission of pure carrier for 30 seconds followed by 270 seconds of telemetry data gathered during the last 30 minutes. The telemetry data was followed by a second transmission of pure carrier. The total TM mode transmission time was 332 seconds.

The purpose of transmitting a pure (unmodulated) carrier was to enable tracking the position of the balloon by interferometry on Earth and tracking the Doppler shift on the signal. The signal transmitted in the KI mode was used for very long baseline tracking interferometry (VLBI) of balloon position and velocity.

During the first 10 hours of flight of the balloon, three consecutive TM mode transmissions were made in a two-hour period. Each transmission represented 30 minutes of stored data from the instruments. Once each two -hour period a KI mode transmission was made. The KI transmission was made about 30 minutes after transmission of the last of the three TM mode transmissions. During the time interval 12–24 hours since start of the balloon flight and from 36 hours to the end of the battery life, one TM transmission and one KI transmission were made each two hours to prolong the life of the battery. The original sequence was transmitted in the interim.

Experiments on the Balloon

Meteorology package – The meteorology package included instruments to measure temperature, pressure, ambient illumination, and vertical wind during flight of the balloon.

The temperature sensor was mounted on a boom that was extended from the instrument package after the package was deployed. The sensor was a metal-film resistance type whose resistance changed with temperature. It had a resistance of 35 ohms under standard conditions. The sensor was incorporated into a bridge circuit with a stable standard resistor. The dc analog signal out of the bridge resulting from temperature change was digitized by a 12-bit analog-to-digital converter. The accuracy of the temperature measurement was within 0.5 degrees Kelvin over a range of 263 to 353 degrees Kelvin.

Vertical wind was measured by a propeller type anemometer mounted to the extended boom to space it 24 cm from the instrument package. It measured flow of the atmosphere in a direction parallel to the longitudinal axis of the instrument package. The number of revolutions was counted and direction of rotation was recorded each 75 seconds. The count, which was proportional to flow velocity, was encoded by six-bit words with a maximum of ±1.5 meters per second (±5.4 km per hour). The ambiguity of measurements outside of this range was usually resolved by atmospheric pressure measurements. Data from the vertical anemometer indicated vertical winds as high as ±2 meters per second as the balloon traveled over Venus.

The variable element of the pressure sensor was a quartz beam that had a resonant frequency proportional to atmospheric pressure on the beam. A motion sensor detected the vibration frequency and its output was applied as positive feedback to an oscillator. The frequency of the oscillator was counted and the count was related to the atmospheric pressure on the beam. The accuracy of the pressure sensor was within 0.5 millibars over a pressure range of zero to two bars.

The light sensor was designed to measure both ambient light and lightning flashes. It received light within a ± 60-degree cone looking downward from the balloon. It was sensitive to light in the range of 4000 Å to 11,000 Å. The output from the detector was amplified and digitized by 12-bit analog-to-digital converter. Only the last six bits of the digital word were recorded. The illumination value was transmitted twice each 30-minute period. Data from the light sensor showed small variations as the balloon traveled on the night side of Venus. The light levels rose very rapidly as the balloons crossed the terminator into the day side of the planet.

Data from the temperature and pressure sensors indicated strong correlation between pressure and temperature as the balloons bobbed in the atmosphere. After the Vega 1 balloon settled out at the cruise altitude of about 53 km, the pressure was about 535 millibars (mb) and the temperature was about 312 Kelvin. Excursions of about 70 mb were noted in the pressure readings of Vega 1 with corresponding temperature excursion of about 0.7 Kelvin as the balloon traveled. On average, the average temperature measured by the Vega 2 balloon was about 6.5 Kelvin lower than that measured by the Vega 1 balloon.

Backscatter nephelometer - The nephelometer was designed to measure density of the cloud layer. The instrument, which was mounted on the inside of the battery box, contained a pulsed light source operating at 9300 Å. Optics transferred the pulsed light to a radiating aperture on the outside of the box. Another optical aperture on the box received reflected light from cloud particles. The reflected light was transferred through a filter to a detector. The filter rejected sunlight. The detector output was amplified and digitized by a 12-bit analog-to-digital converter. The accuracy of the nephelometer was 10^{-5} per meter steradian.

Data from the backscatter nephelometer indicated variations in backscatter up to 20% of the mean value of cloud density. There were no clear regions in the clouds.

Data Transmission from Experiments

Data from the experiments were sampled and stored during a 30-minute interval between transmissions to Earth. Temperature, pressure, vertical wind, and nephelometer readings were sampled at 75-second intervals. Ambient light level was sampled every 10 minutes.

Each telemetry transmission interval included 852 bits of data from the experiments that had been stored in memory during the previous 30 minutes and a 48-bit synchronizing word. Data was transmitted at a rate of four bits per second for the first 840 bits and one bit per second for the last 60 bits.

Determination of Venus Winds by Earth-Based Tracking of Balloons

An array of 20 tracking stations around the Earth was employed a part of a very long baseline radio interferometry (VLBI) arrangement to measure the trajectory of the balloon in the atmosphere of Venus. The motion of the balloon gave a measure of the zonal wind velocity. The tracking stations took advantage of several large radio telescopes around the world used by radio astronomers. The wavelength chosen for the experiment of 18 cm was commonly used by those astronomers.

The LBI array included three large radio telescopes and three medium size radio telescopes in the Soviet Union, three large antennas of the U.S. Deep Space Network (Goldstone, Madrid, and Canberra), and three medium size antennas in the United States, the very large Arecibo radio telescope in Puerto Rico, and seven radio telescopes in other countries throughout the world. The Centre National d'Etudes Spatlales (CNES) in Paris coordinated the international VLBI network except for radio telescopes in the Soviet Union.

The transmitter in the balloon instrument package generated 4.5 watts of power at a frequency of 1667.9 MHz (18 cm wavelength). The transmitter alternately operated in a telemetry mode (TM) and a coordinate measurement mode (KI). The KI mode was used by the VLBI receiving stations on Earth. The transmitter signal in the KI mode was suppressed carrier with double sideband tones 6.5 MHz apart. The KI mode was transmitted for 30 minutes every two hours during the first 10 hours of balloon flight and from 24 hours to 36 hours of flight. From 12 hours to 24 hours from the beginning of flight and after 36 hours, the KI mode was transmitted for 60 minutes each two hours.

Processing of the VLBI data indicated that the average wind velocity pushing the Vega 1 balloon was 69 ± 1 meters per second. The average wind velocity experienced by the Vega 2 balloon was 66 ± 1 meters per second.

Vega Lander

The Vega landers were essentially the same as the landers used for the Venera 9 through 14 missions except for experiments carried. The Vega landers did not carry cameras but they had a more extensive suite of scientific experiments than the previous landers. The general appearance of the lander was the same as for the Venera 13 lander shown in in Fig. 5.1 previously. A photograph of a model of a Vega lander inside a cutaway of the entry enclosure is shown below (Fig. 5.14) the photograph was taken by Russian blogger Denis in an unnamed museum in Russia. The conical spiral antennas in the right side of the photograph are part of the Venera bus in the background.

The instrument capsule portion of the lander was a strong spherical enclosure 0.8 meters in diameter made from titanium. The spherical pressure chamber was covered with thermal insulation several centimeters thick and then by an outer titanium shell. The spherical capsule contained electronics and a sequencer to manage capsule operation along with signal processing equipment for the experiments. It also contained the communications system. The capsule was supported by a series of legs that were attached to a rim-like landing structure. The rim contained crushable material that absorbed some of the landing shock.

A disk 2.1 meters in diameter that served as an aerobrake during descent was mounted just above the spherical capsule. A large helical transmit antenna was mounted above the aerobrake. The hollow inner portion of the cylindrical helix antenna contained the parachutes and some of the scientific instruments.

The lander and deflated balloon were protected during entry into the atmosphere of Venus by the spherical entry enclosure that served as a heat shield. The heat

Fig. 5.14 Vega lander in entry enclosure. (Photograph by blogger Denis)

shield slowed the velocity of the capsule from about 11 km per second before entry into the atmosphere to a velocity of about 280 meters per second by the time an altitude of 65 km above the surface was reached. The top cap of the entry enclosure was detached from the lower portion at that altitude and a drogue parachute was deployed to pull off the top cap. A parachute was deployed to pull out the balloon package from the top of the lander. The lander continued descent towards the surface.

The Vega 1 lander deployed a parachute at an altitude of 64.5 km and jettisoned the lower portion of the heat shield. The parachute was jettisoned at an altitude of about 47 km and the lander descended to the surface while being slowed by the aerobrake disk encountering the dense atmosphere. A successful landing was made at 7.5 degrees north, 177.7 degrees east in the Venusian coordinate system. The lander transmitted data from its sensors to Earth for 56 minutes after landing.

The Vega 2 lander performed a landing sequence similar to Vega 1 and touched down successfully at 8.5 degrees south and 164.5 degrees east in the Venusian coordinate system. The lander transmitted data from its sensors to Earth for 57 minutes after landing.

Experiments on Vega Landers

The scientific instrument group on the landers included:

- Ultraviolet absorption spectrometer (ISAV-S)
- Optical aerosol analyzer, Nephelometer (ISAV-A)
- Aerosol particle-size counter (LSA)
- Aerosol mass spectrometer (MALAKHIT-M)
- Hygrometer (VM-4)
- X-ray fluorescence spectrometer (IFP)
- Gas chromatograph (SIGMA-3)
- Gama ray spectrometer (GS-15-SCV))
- Soil X-ray fluorescence spectrometer (BDRP-AM25)
- Penetrometer/soil ohmmeter (Prop-V)
- Temperature sensor (four sensors)
- Pressure sensor (three sensors)

Data from the experiments were collected by the telemetry system and transmitted to the bus for relay to Earth. The bus, which did a flyby of Venus, had a view of Earth as well as the lander. The VHF spiral antenna on top of the lander radiated the telemetry signals to the bus at a frequency of 122.8 for one lander and at 138.6 MHz for the other. The bus picked up the signals from the lander by a conical spiral antenna and relayed the data to Earth by its high-gain antenna. The transmission frequency to Earth was 992.76 MHz and the data rate was 3072 bits per second. Lander data was relayed to Earth in real time as well as being recorded on a tape recorder in the orbiter for later playback.

A summary of operation of the instruments and results of the experiments conducted by the Vega landers are provided below

Ultraviolet Absorption Spectrometer (ISAV-S)

The ultraviolet spectrometer measured absorption of ultraviolet light by gaseous constituents in the atmosphere of Venus. The wavelengths and amount of absorption of major absorption regions identified particular gasses and their relative abundances.

The instrument consisted of an interaction tube 85 cm long with a gas inlet on one end and gas outlet on the other end. Atmosphere entered the tube due to vertical velocity as the lander descended. Scaling from drawings, the tube was about 5 cm in diameter. A xenon strobe light with energy in the ultraviolet band of 2300 Å to 4000 Å was mounted inside the pressure vessel along with a spectrometer. A strobe of light was applied to the interaction tube through a lens. The light traveled to the other end of the tube and bounced back and was applied to the spectrometer after passing through about 170 cm of gas.

The spectrometer consisted of a concave diffraction grating in conjunction of an array of 512 photodiodes. The 512 channels covered the spectral range of 2300 Å to 4000 Å. The outputs of the 512 channels could be scanned and stored in on-board memory in 60 seconds. At altitudes below 25 km the scan rate was increased and the spectrum was scanned and stored in 36 seconds.

A paper by Bertaux, et al., 1986, *Active spectrometry of the ultraviolet absorption within the Venus atmosphere*, gives some results from the experiment. The data indicated a broad absorption band whose spectral shape and signal level was depended on altitude of the lander. At altitudes below 10 km the main absorbing constituent completely absorbed the ultraviolet light. The paper by Bertaux, et al., suggests that the principal absorbent may be an allotrope of sulfur in the gas phase.

Optical Aerosol Analyzer (ISAV-A)

The optical aerosol analyzer was used to investigate aerosol content of the clouds of Venus. The instrument contained two measurement types. One type was an aerosol spectrometer that measured light scattered by particles at different azimuth angles. That portion of the instrument determined the number of aerosols in a given volume, a spectrum of their sizes, and estimates of their shapes. A separate section of the instrument measured the backscattering cross section of particles. A 5-watt incandescent lamp furnished light for both types of sensors.

A portion of the light from the incandescent lamp was focused by mirrors into an aerosol sampling region. Motion through the atmosphere as the lander descended caused a sample of the atmosphere to flow through an intake nozzle and into the sampling region. The volume of the sampling region was one cubic millimeter. The flow continued out through an exit tube. The spectrometer contained four optical detectors to measure the amplitude of scattered light from aerosols at azimuth angels of 7–17 degrees, 25–65 degrees, 165–175 degrees, and 295–335 degrees.

When an aerosol particle passed through the sampling region, the light scattered from it was sensed by the four optical detectors. The size of the particles was related to the amplitude of the scattered signals. The outputs of the detectors were amplified by logarithmic amplifiers, digitized, and recorded in memory. The readings were held in memory until the next particle passed through the sampling region. The amplitudes of the four pulses from each particle detected were telemetered to Earth. The four-channel spectrometer output was interrogated by the telemetry system twice each telemetry frame. Data telemetered to earth in conjunction with calibration data allowed determining the number density of particles as a function of particle size, refractive index of the particles, and their shape.

The shape of the particles was determined by comparing the scattered signal strength in the lateral channels (25–65 deg. and 295–335 deg). If the particles were spherical, the scattered signal in both lateral channels would be the same. If the particles were aspherical, the ratio of signal levels in the lateral channels would depend on orientation of the particle. It was found that particles with size greater than one micron were spherical. The amplitude of the signal in the backscatter channel (165–175 degrees azimuth) was used to determine refractive index of the particles.

The backscatter sensor focused a portion of the light from the incandescent lamp out through a window in the capsule to the atmosphere. The light was focused about three centimeters beyond the window. The backscattered sight at angles between 160 and 175 degrees was captured by an optical detector. The effective working volume of the instrument in the atmosphere was about five cubic centimeters.

Results of the experiment were given in a paper by Moshkin, et al., 1986, *Vega 1, 2 optical spectrometry of Venus atmospheric aerosols at the 60–30 km levels: preliminary results*. The experiment indicated peaks in aerosol density that corresponded to a three-layer cloud deck with a haze layer below the clouds. The number densities measured by Vega 2 peaked at lander altitudes of 57 km, 50 km, and 46 km. At 57 km, the density of particles larger than 0.4 microns was about 250 particles per cubic centimeter. The density of particles with size greater than 0.7 microns was about 90 particles per cubic centimeter at that altitude.

From comparison of signal levels in the lateral azimuth channels it was determined that particles with size greater than one micron were spherical while smaller particles were somewhat aspherical. Data from the backscattered azimuth channel yielded refractive index of particles larger than one micron of 1.4 ± 0.1.

Data from the backscatter sensor on Vega 1 indicated a change of about a factor of 1.8 in backscatter coefficient as the altitude decreased from 60 km to 30 km. The peak was at about 40 km altitude. Data from Vega 2 indicated a peak in backscatter coefficient at an altitude of 51 km that was about 20 times greater than the value at 57 km and at 33 km.

Aerosol Particle-Size Counter (LSA)

The purpose of the aerosol particle-size counter was to determine the number density and the spectrum of size of aerosols as a function of altitude above Venus. The measurements complemented measurements made by the optical aerosol analyzer.

Light for the instrument was generated by a halogen incandescent lamp. The light was focused onto a diaphragm with a rectangular slit 1.2 mm by 0.22 mm in size. The slit image was focused by a mirror into a sensing chamber. Light scattered by an aerosol was gathered by optics having a 55-degree solid angle field of view of the sensing chamber. The combination of the slit of light entering the chamber intersected by the field of view of the scattered light gathering optics defined the operating volume for the instrument.

When an aerosol passed through the sensing chamber, a pulse of scattered light was gathered by the optics and applied to a photomultiplier tube. The photomultiplier tube generated a current proportional to the intensity of the scattered light and hence proportional to the size of the aerosol. The current pulse was converted to a voltage pulse, the amplitude of which was proportional to the size of the aerosol.

The voltage pulses were applied to a four-channel size discriminator. The first channel counted all pulses corresponding to aerosols with size greater than 0.35 microns in a time period of 0.43 seconds. The second channel counted pulses corresponding to aerosol sizes greater than 0.4 microns in a time period of 0.43 seconds. The counts from both discriminators were applied to the telemetry system. The third channel performed a spectral analysis of amplitudes of the pulses using 10 different levels. The counts in each of the ten discrete sub channels were gathered at 4.3 second intervals and applied to the telemetry system.

The fourth channel was the most elaborate. It contained circuitry that measured individual pulse amplitude within 256 different levels. It also measured the width of each pulse and the time until the next pulse. A complete pulse height spectrum was performed each 100 seconds during the descent through the clouds. In addition, the fourth channel recorded the calibration pulses that were inserted in the optical signal at the beginning of each new height spectral analysis.

A brief description of the instrument and results of the experiment were given in a paper by Zhulanov, et al., 1986, *Aerosol counts in Venus clouds: preliminary Vega 1,2 density profiles, H = 63–47 km*. The experiment indicated peaks in aerosol density that corresponded to an upper cloud layer above 57 km altitude, an intermediate layer from 57 to 50 km, and a lower cloud layer at 50 to 47 km. The number densities measured by Vega 1 peaked at lander altitudes of 58 km and 47 km. At 58 km, the density of aerosol particles larger than 0.4 microns was about 140 particles per cubic centimeter and at 47 km the density was 190 aerosol particles per cubic centimeter. Vega 2, which passed through a different portion of the atmosphere and four days after Vega 1, measured about 140 particles per cubic centimeter at 47 km altitude and about 30 particles per cubic centimeter in the upper cloud layer at 57 m, which was substantially lower than that measured by Vega 1.

Aerosol Mass Spectrometer (MALAKHIT-M)

The mass spectrometer on the Vega landers was designed to determine the chemical composition of aerosols in the cloud layers of Venus. An aerosol collecting chamber was sealed before launch and unsealed by a pyrotechnic operation while the lander was in the clouds of Venus at an altitude of about 62 km. The instrument collected a sample of the clouds and separated the aerosols by weight into two groups: light and heavy. The dividing line between light and heavy was equivalent to aerosol particles of radius of 1.5 microns. The selection was performed inertially.

The two groups of aerosols were collected by separate filters made of carbon fibers. Aerosol droplets were collected by the fibers as the atmospheric material was pumped through the chambers. Once the lander had descended to about 56 km, the chambers were sealed and analysis of the first sample began. The aerosols clinging to the fibers were decomposed by pyrolysis. Pyrolysis involves heating a material in the absence of oxygen to high temperature causing the material to decompose into smaller volatile molecules. The pyrolysis process in the Vega instrument heated the material to about 400 °C.

The gasses resulting from the pyrolytic process were fed to a mass spectrometer. The mass spectrometer was a radio frequency (RF) type with basic operating principle similar to the RF mass analyzer successfully used on the Venera 11 and 12 landers and described in Chap. 3 of this book. The gas molecules resulting from pyrolysis were ionized by a pulsed electron beam and accelerated into a chamber with electrodes with electric potentials including a RF component. The arrangement caused only ions of a given mass-to-charge ratio to pass through the device and strike a detector. Ions with different mass-to-charge ratios would strike the surfaces of the chamber and would not be counted. A mass-to-charge spectrum of the ionized gas molecules were generated by varying the frequency of the RF field and counting the number of ions striking the detector at each frequency setting. The instrument generated a mass spectrum from 10 to 150 atomic mass units (amu).

After the first sample was taken and analyzed, the collecting chamber was evacuated by a vacuum pump and a second sample was taken and analyzed. The second sample was gathered as the lander descended between 56 km and 48 km altitude.

A paper by Surkov, et al., 1986, *Vega 1 mass spectrometry of Venus cloud aerosols: preliminary results*, presents some results from the experiment on Vega 1. A peak in the mass spectrum at 64 amu was thought to be due to sulfur anhydride. The researchers concluded that a lower limit of abundance of sulfuric acid (H_2SO_4) in the aerosols in the cloud layer between altitudes of 62–54 km was 2.0 milligrams per cubic meter. The researchers also concluded that chlorine abundance was about 0.3 milligrams per cubic meter. The mass spectrometer on Vega 2 failed and no data was obtained from it.

Hygrometer (VM-4)

The VM-4 hygrometer installed on the Vega landers used two different methods of measuring humidity: a coulombmetric humidity sensor and a thermoelectrolytic humidity sensor. The measurement ranges of the two sensors largely overlapped and provided independent measurements of humidity as the spacecraft descended. The two sensors were mounted on the same base and were fed atmospheric samples through an inlet port opened by pyrotechnics. The timer-programmer in the lander turned on both sensors and fired the squib to open the inlet port at an altitude of 62 km above the surface of Venus. Humidity measurements were made from that altitude down to about 25 km.

The principle of operation of the coulombmetric humidity sensor was based on electrolysis of water into hydrogen and oxygen. The amount of hydrogen liberated related to the amount of water vapor in the atmosphere. The humidity sensor ported the Venusian atmosphere into two channels. One channel contained a chamber with a very sensitive thermometer that measured the temperature as atmospheric gas flowed by a temperature sensitive element. The other channel contained a coulomb tube that had a pair of helical platinum electrodes separated by a film of phosphorus pentoxide (P_2O_5). Moisture in the atmosphere was absorbed by the P_2O_5 and the moisture was electrolyzed by a voltage potential between the two helical elements.

The output of the coulombmetric tube, which contained atmosphere gas plus hydrogen plus oxygen liberated by electrolysis, was applied to a chamber containing a very sensitive thermometer. The temperature of the gas, including hydrogen, flowing by in that chamber was at a different temperature due to the high thermal conductivity of hydrogen. The temperature measurements of the two channels were telemetered to Earth whereupon the humidity in the atmosphere was determined. The coulomb sensor was capable of measuring humidity over a range of 0.01% to 0.20% by volume.

The thermoelectrolytic humidity sensor included a heated volume of saturated lithium chloride (LiCl) solution and relied on the temperature dependence of water vapor pressure above the solution. Moisture evaporated from the heated saturated solution but moisture was absorbed from the atmosphere by the hydroscopic nature of the LiCl salt solution. Equilibrium was reached at a given temperature of the solution and that temperature was accurately measured and telemetered to Earth where the relative humidity was determined.

A paper by Surkov, et al., 1986, *Vega 1, 2 humidity profiles for the Venus atmosphere*, gives results of the humidity measurements. Those results are given in Table 5.8 on the next page.

X-Ray Fluorescence Spectrometer (IFP)

The X-ray fluorescence spectrometer was designed to determine the chemical elements in the aerosol droplets in the clouds of Venus. The instrument used radioisotopes as X-ray sources to induce fluorescence in the aerosols. The aerosol particles

Table 5.8 Measurement of water vapor in the atmosphere of Venus

	Water vapor, percent by volume	
Altitude, km	Vega 1	Vega 2
55–60	0.2 ± 0.05	0.05 ± 0.03
45–50	0.1 ± 0.03	0.1 ± 0.03
25–30	0.02 ± 0.005	0.012 ± 0.005

absorbed energy from the incident X-ray radiation and that energy caused an electron to be ejected from a shell near the nucleus of an atom. Electrons from outer shells progressively dropped down to fill the holes giving up energy in the form of a photon of X-ray radiation. The frequency of the X-ray identified the element from which the photon emanated.

The instrument made measurements continuously as the lander descended from 63 km altitude to 47 km. The analysis compartment was unsealed at an altitude of 63 km and the atmosphere was allowed to pass through. Aerosol particles in the clouds were captured by a filter having a size of 10 square centimeters. The filter was effective in capturing all aerosols with diameters of one micron and larger. The filter allowed a metered amount of atmosphere to pass through it as the lander descended.

The aerosols were exposed to X-ray radiation from three different radioisotope sources. A ^{55}Fe radioisotope of iron was used to excite radiation characteristic of chlorine, sulfur, and phosphorus atoms. A ^{109}Cd radioisotope of cadmium was used to induce radiation from iron atoms. A ^{3}H/Zr radioisotope was used to measure filter transmission. X-ray energy generated by fluorescence of the atoms was detected by radiation detectors. The detectors were proportional counter types that produced an output pulse with amplitude proportional to the energy of the X-ray striking it. The abundance of the element was determined by noting the number of pulses of the characteristic energy for that element.

A paper by Andreichikov, et al., 1986, *Element abundances in Venus aerosols by x-ray radiometry*, presents some results of the experiment. The researchers determined that sulfur occurs at altitudes from 63 km to 47 km with average concentration of 5.8 milligrams per cubic meter. Chlorine was found to be present in the altitude range of 61 km to 52 km at average concentration of 4.1 milligrams per cubic meter. Phosphorus was found at altitudes from 52 km to 47 km at average concentration of 7.7 milligrams per cubic meter. A trace of iron was found in the altitude range of 54 km to 47 km.

Gas Chromatograph (SIGMA-3)

The SIGNA-3 gas chromatograph was included in the experiments suite of the Vega landers to determine the chemistry of the aerosol particles in the clouds of Venus. The intent was to verify that the aerosols were droplets of sulfuric acid (H_2SO_4) as had been conjectured. The SIGMA-3 instrument was similar to the SIGMA-2 gas chromatograph that flew in the Venera 13 and Venera 14 landers but it was arranged to analyze aerosols rather than atmospheric gasses.

The SIGMA-3 instrument included three sets of reaction cells and two gas analysis channels. One of the analysis channels used nitrogen as the carrier gas and the other used helium. The reaction cells were fitted with carbon coated glass fibers. Identification of H_2SO_4 employed the reaction of sulfuric acid with carbon on the filter at elevated temperature. The reaction was:

$$\mathrm{C} + 2\mathrm{H_2SO_4} \rightarrow CO_2 + 2\mathrm{H_2O} + 2\mathrm{SO_2}.$$

The atmosphere of Venus also contains these same reaction gasses and a method of discriminating against atmospheric SO_2 was implemented. This involved preheating two of the reaction chambers to 80 °C before the atmosphere was allowed to enter while the third chamber was gradually brought up to 50 °C as the spacecraft descended. After filling with atmosphere, including aerosols, the chambers were sealed and the second chamber was heated to 300 °C. These processes allowed isolating the reaction gasses from atmospheric gasses.

A paper by Gel'man et al., 1986, *Reaction gas chromatography of Venus cloud aerosol*, estimated that the concentration of sulfuric acid (H_2SO_4) in the altitude range 63 km to 48 km was about one milligram per cubic meter.

Gamma Ray Spectrometer

The gamma ray spectrometer measured gamma rays naturally emitted by radioactive constituents in the surface of Venus. A basic description of the instrument was given in the section of experiments carried by Venera 8 in Chap. 3 of this book. The percentages of naturally radioactive uranium, thorium, and potassium were determined. The spectrometer was mounted within the spherical pressure enclosure but gamma rays were able to penetrate the walls of the enclosure. The spectrometer was mounted near the bottom of the enclosure to be close to the surface of Venus. Background gamma ray impingement from the enclosure was measured and reported back to Earth during the travel to Venus.

Amounts of potassium (K), uranium (U) and thorium (Th) measured by the Vega landers and reported in a paper, *Landing on Venus: Past and Future*, by Basilevsky et al., 1986, given in Table 5.9 on the next page.

Soil X-Ray Fluorescence Spectrometer (BDRP-AM25)

The X-Ray fluorescence spectrometer carried by the Vega 2 lander determined properties of the soil and rocks in the highland area of Venus near Aphrodite Terra. Previous landers had found basalt rocks in the low- lying regions of Venus.

Principles of operation of X-ray fluorescence spectrometers were described in the account of the X-ray fluorescence spectrometer (IFP) for aerosols given above.

Table 5.9 Measurements of abundances of elements in the surface of Venus

Element	Vega 1	Vega 2
K (mass %)	0.454 ± 0.22	0.40 ± 0.20
U (ppm)	0.64 ± 0.47	0.68 ± 0.38
Th (ppm)	1.5 ± 1.2	2.0 ± 1.0

The spectrometer was mounted on the inside of the lander pressure capsule and a series of airlocks was used to transfer a sample of rock chips or soil generated by a drill to the instrument. The drill was a telescoping assembly that lowered a drill bit onto the surface. The apparatus was capable of drilling into rock and creating a sample of fine rock chips. Particles and dust from drilling were carried through a tube to airlocks. The tube was initially at a much lower pressure than the atmosphere of Venus and when seals on the tube were breached by pyrotechnics, the sample of rock was drawn through the tube to airlocks and into the X-ray fluorescence chamber.

The sample of rock chips was irradiated by X-ray radiation from two ^{55}Fe (iron-55) radioisotope sources and one ^{238}Pu (plutonium-238) source. The plutonium-238 source had good efficiency of exciting the lighter elements in the rock sample and the iron-55 sources were effective in exciting heavier elements. The impinging X-ray radiation from the radioactive sources caused the elements in the rock samples to fluoresce and emit X-ray radiation. The energy of the secondary X-rays was an indication of the chemistry of the sample.

The secondary X-rays were detected by four gas-discharge proportional counters. The analyzer was arranged with several channels, each with a narrow region of pulse height. The number of pulses per unit time was counted for each channel of the analyzer. The result was a spectrum of number of counts as a function of energy of the X-ray radiation.

The rock chip samples were irradiated simultaneously by the three radioisotope X-ray sources and the resulting fluorescence radiation was detected by the four proportional counters at the same time. The combination of X-ray sources and detectors allowed detecting elements from magnesium ($Z = 12$) to bromine ($Z = 35$) and from gold ($Z = 79$) to bismuth ($Z = 83$) where Z is atomic number.

The spectrometer in the Vega 2 lander was turned on at an altitude of about 25 km above the surface. During the 35-minute descent from that altitude to the surface, the instrument was calibrated and a background measurement was made with the soil sample chamber empty. After reaching the surface, the drill was deployed and it operated successfully. Rock chips were transferred to the spectrometer for analysis. A total of ten molecules were identified and measured. The results were reported in a paper by Surkov, et al., 1986, *Venus Rock composition at the Vega 2 landing site.* The five largest amounts of material in the sample are given in Table 5.10 on the next page.

A transient event caused the drilling device to deploy early during the descent of the Vega 1 lander and it had conducted its programmed events before the lander

Table 5.10 Composition of rocks on Venus

Material	Percent by weight
SiO_2	45.6 ± 3.2
Al_2O_2	16.0 ± 1.8
MgO	11.5 ± 3.7
FeO	7.7 ± 1.1
CaO	7.5 ± 0.7

reached the surface. As a result, no data was obtained from the Vega 1 soil X-ray spectrometer.

Temperature and Pressure Sensors (METRO)

There were two platinum resistance temperature sensors on the lander. The resistance changed predictably with temperature and this change was measured to determine temperature. One sensor consisted of a platinum wire wound on a ceramic form. It was exposed to the atmosphere and it had a short time constant of 0.1 seconds. The winding of the other temperature sensor was covered by a thin ceramic shield. The time constant of that protected temperature sensor was about three seconds. Both temperature sensors had a measurement range of 200 K to 800 K and accuracy of 0.5 K. The temperature sensors were mounted in the space between the sealed capsule and the landing structure. The atmosphere flowed by the sensors as the lander descended. The temperature measured by the two sensors agreed within 2 Kelvin during the mission and much of this difference was thought to be due to the different time constants.

There were three pressure sensors on the lander. Two were designed to provide accurate data in the altitude range of the clouds and down to the middle atmosphere. The measuring element in those sensors was a quartz resonator with an associated electrical oscillator. Pressure changes on the quartz material caused the resonant frequency to change with accompanying change in the output frequency of the oscillator. The oscillator frequency gave a measure of pressure. The third pressure sensor was designed to provide accurate data at lower altitudes and on the surface of Venus. It used a membrane that deflected due to ambient pressure. The position of the membrane was sensed by a capacitor that in turn determined the frequency of an oscillator. Once again, the oscillator frequency gave a measure of pressure.

Data from the temperature and pressure sensors as the spacecraft descended were reported in a paper by Linkin et al., 1986, *Vertical thermal structure in the Venus atmosphere from provisional Vega 2 temperature and pressure data*. The paper presented results in the form of plots of temperature and pressure as a function of altitude. Information was scaled from those plots by the author and presented in the form of a table on the next page (Table 5.11). Temperature was converted to degrees Centigrade.

Table 5.11 Temperature and pressure during descent of Vega 2 lander

Altitude, km	Temperature, °C	Pressure, bars
60	−23	0.28
50	67	1.0
40	132	3.5
30	217	10
20	292	22
10	362	50
0	460	89

A pronounced temperature inversion was present between altitudes of 64 and 62 km. At 63.6 km the temperature was 269 K (−4 °C) and at 62 km altitude the temperature decreased to 254 K (−19 °C). The normal trend was an increase in temperature with decreasing altitude.

Conclusion

The ambitious Vega missions were very successful. The buses deployed landers and then made flybys of the comet Halley to photograph the nucleus and measure constituents of the coma. Each lander released an instrumented balloon into the atmosphere before successfully landing on the surface. The balloons were driven about a third of the way around the planet by high zonal winds. The landers gathered detailed measurements of properties of the clouds and of the surface of Venus.

Bibliography

Andreichikov, et al., *Element abundances in Venus aerosols by x-ray radiometry*, Pis'ma Astron. Zh. 12 Feb. 1986.

Andreev, R. A., et al., Mean zonal winds on Venus from Doppler tracking of the Vega balloons, Soviet Astronomy Letters, vol. 12, Jan–Feb 1986.

Basilevsky, A. T., et al., Landing on Venus: Past and Future, Planetary and Space Science, Vol 55, Issue 14, November 2007.

Bertaux, J. L., et al., Active spectrometry of the ultraviolet absorption within the Venus atmosphere Soviet Astronomy Letters, vol. 12, Jan–Feb 1986.

Combes, M., et al., The 2.5-12 μm Spectrum of Comet Halley from the IKS-VEGA Experiment Icarus, Vol 76, 404–436, 1988.

Gel'man, B. G., et al., Reaction gas chromatography of Venus cloud aerosols Soviet Astronomy Letters, vol. 12, Jan–Feb 1986.

Istomin, V. G., et al., Mass Spectrometry on the Venera-13 and Venera-14 Landers-Preliminary Results, Soviet Astronomy Letters, Vol 8, July–Aug 1982.

Kotelnikov, V. A., Bogomolov, A. F., Rzhiga, O. N., Radar Study of Venus Surface by Benera-15 and Venera-17 Spacecraft, Adv. Space Res., Vol. 5 No. 5 1985.

Krasnopol'skii, V. A., et al, The Vega 2 TKS experiment: some spectroscopic results for Comet Halley, Soviet Astronomy Letters, vol. 12, July–Aug 1986.

Kremnev, R. S. et al., The Vega balloons: a tool for studying atmosphere dynamics on Venus, Soviet Astronomy Letters, vol. 12, Jan–Feb 1986.

Linkin, V. M., et al., Thermal structure in the Venus middle cloud layer, Soviet Astronomy Letters, vol. 12, Jan–Feb 1986a.

Linkin, V. M., et al., Vertical thermal structure in the Venus atmosphere from provisional Vega 2 temperature and pressure data, Soviet Astronomy Letters, vol. 12, Jan–Feb 1986b.

Moroz, V. I., et al., The Venera 13 and Venera 14 Spectrophotometry Experiments, Soviet Astronomy Letters, Vol 8, July–Aug 1982.

Moshkin, B. E., et al., Vega 1, 2 Optical spectrometry of Venus atmospheric aerosols at the 60–30 km levels: preliminary results Soviet Astronomy Letters, vol. 12, Jan–Feb 1986.

Mukhin, L. M., et al., Venera 13 and Venera 14 Gas-Chromatography Analysis of the Venus Atmosphere Composition, Soviet Astronomy Letters, Vol 8, July–Aug 1982.

NASA Technical Memorandum, NASA TM-77712, *Contemporary Achievements in Astronautics: Salyut-7, the Vega Project and Spacelab*, March 1986.

Oertel, D., et al., Deutsce Akademia der Wissenschaften, East Berlin and Moroz, I, et al at Institute for Space Research, USSR, *Venera 15 and Venera 16 infrared spectrometry: first results*, NASA Astrophysics Data System.

Oertel, D., et al., *Venera 15 and Venera 16 Infrared Spectrometry–First Results*, Soviet Astronomy Letters Vol. 10, Mar–Apr 1984.

Sagdeev, R. Z, et al., Meteorological data along the Vega 1,2 float paths, Soviet Astronomy Letters, vol. 12, Jan–Feb 1986a.

Sagdeev, R. Z., et al., The Vega balloon global ground network, Soviet Astronomy Letters, vol. 12, Jan–Feb 1986b.

Siddiqi, Asif A., *Beyond Earth A Chronicle of Deep Space Exploration, 1958–2016*, NASA Report NASA SP-2018-4041.

Surkov, Y. A., et al., Element Composition of Venus Rocks-Preliminary Results from Venera 13 and Venera 14, Soviet Astronomy Letters, Vol 8, July–August 1982.

Surkov, YHUu. A., et al., Venus Rock Composition at the Vega 2 Landing Site, Journal of Geophysical Research, Vol 91, No 813, November 30, 1986a.

Surkov, Yu. A., et al., Mass spectrometry of Venus cloud aerosols: preliminary results, Soviet Astronomy Letters, Vol. 12, Jan–Feb. 1986b.

Surkov, Yu. A., et al., Vega 1, 2 humidity profiles for the Venus atmosphere, Soviet Astronomy Letters, vol. 12, Jan–Feb 1986c.

Taylor, F W, et al., Infrared Remote Sounding of the Middle Atmosphere of Venus from the Pioneer Orbiter, Science, Vol 203, 23 Feb. 1979.

Zhulanov, Y. V. et al., Aerosol counts in Venus clouds – Preliminary VEGA 1,2 density profiles, H = 63–47 km Soviet Astronomy Letters, vol. 12, Jan–Feb 1986.

Chapter 6
U.S. Magellan Spacecraft that Explored Venus 1980–2020

Planetary exploration missions conducted by the United States concentrated on Mars and only a few spacecraft were sent to Venus. The only U.S. spacecraft mission to explore Venus in the 1980 to 2020 time frame was Magellan. Magellan was a very successful polar orbiting spacecraft that reached Venus in August 1990. It mapped 98% of the surface of Venus with high-resolution synthetic aperture radar.

Other U.S. spacecraft made flybys of Venus in the time period to obtain gravity assist from the planet. Those spacecraft were Galileo enroute to Jupiter in 1989, Messenger with flybys in 2006 and 2007 to set up subsequent orbits of Mercury, and the Parker Solar Probe that made multiple flybys of Venus for gravity assist in the process of adjusting the probe's orbit around the sun.

Background of Magellan Program

NASA studies conducted in the 1960s and 1970s revealed benefits of radar mapping of the surface of Venus from orbit. Dense clouds over the planet precludes optical mapping as was done extensively for Mars and Earth's moon. Ambitious plans for a radar mapping spacecraft named Venus Orbiting Imaging Radar (VOIR) were published in 1978. However, the spacecraft design that emerged was deemed too costly and the program was cancelled in 1982.

A lower cost alternative program named Venus Radar Mapper was planned by NASA and funding for that program was approved by Congress in 1983. Significant cost savings were achieved by using existing hardware or the design of existing hardware for much of the basic spacecraft. For example, the hexagon main structure of the spacecraft was a spare from the *Voyager* program. The high-gain antenna, low-gain antenna, and thrusters were also taken from *Voyager*. The *Voyager* program included two spacecraft that were launched in 1977 on missions to Jupiter and Saturn with extended missions to Uranus and Neptune.

T. Lund, *Spacecraft that Explored the Inner Planets Venus and Mercury*,
Springer Praxis Books, https://doi.org/10.1007/978-3-031-29838-7_6

Further use of existing equipment included the command and data subsystem, attitude control computer, tape recorder, electric power distribution unit, and pyrotechnic control subsystem from the *Galileo* program. *Galileo* was a successful spacecraft that orbited Jupiter for 13 years. It was launched by the Space Shuttle in October 1989. As a further cost savings for Venus Radar Mapper, the large high-gain antenna on the spacecraft was used for both the radar mapper and for communications with Earth. Again to reduce cost, a suite of experiments to study the atmosphere and ionosphere of Venus that had been planned were eliminated. The imaging radar was the only experiment carried by the spacecraft.

NASA assigned overall system responsibility for the Venus Radar Mapper program to the Jet Propulsion Laboratory (JPL). JPL contracted out for the two main components of the program: the spacecraft and the radar imager. Martin Marietta Astronautics was selected to build the spacecraft and Hughes Aircraft Company was selected to develop and build the radar. Contracts for those items were awarded in the fall of 1983.

NASA Headquarters changed the name of the spacecraft to *Magellan* in December 1985. The name change continued the NASA practice of naming spacecraft after noted scientists and explorers. Development and test of the radar and spacecraft continued into 1988. The completed spacecraft was shipped to the Kennedy Space Center in September 1988. It was the first planetary spacecraft to be taken up to Earth orbit by the Space Shuttle. It was launched from the cargo bay of Space Shuttle *Atlantis* on a long trajectory to Venus in May 1989. Magellan began a productive polar orbit of Venus in August 1990.

Mechanical Configuration of the Magellan Spacecraft

A photograph of the Magellan spacecraft being readied for test at the Martin Marietta facility near Denver is shown on the next page (Fig. 6.1).

Much of the spacecraft was wrapped in white thermal blankets to assist in temperature control. The diameter of the high-gain antenna at the top of the spacecraft was 3.7 meters. The total length of the spacecraft was 6.4 meters. The solid rocket motor shown at the bottom in the photograph was used to slow the spacecraft for insertion into orbit around Venus. The rocket motor was jettisoned after orbit was achieved.

A labeled and dimensioned drawing of the spacecraft is given on the following page (Fig. 6.2). The drawing includes the Inertial Upper Stage that launched the spacecraft from the Space Shuttle on a trajectory to Venus.

A different perspective of the spacecraft with the solar arrays deployed is shown in the sketch on the next following page (Fig. 6.3).

The coordinate system of the spacecraft was such that the +Z axis was in the direction of the boresight of the high-gain antenna. The X-axis was parallel to the rotation axis of the solar arrays. The +Y-axis was in the nominal direction of the star

Fig. 6.1 Magellan spacecraft being readied for test. (NASA photograph)

scanner. The spacecraft velocity vector was in the –Y direction when the spacecraft was mapping to the left of the direction of travel.

The forward equipment module contained electronics for the imaging radar and radar altimeter, telecommunications electronics, star scanner electronics, two batteries, gyroscopes, attitude reference unit, and three reaction wheels for maneuvering the spacecraft. The 10-sided equipment bus, which was originally built as a spare for the Voyager program, was 42.4 centimeters high and about two meters across. Each of the 10 sides contained a 42 × 47 × 18-centimeter enclosure for electronics. The enclosures contained the flight computers, interface between the computers and other subsystems, bulk memory, tape recorders, solar array controls, power distribution, and pyrotechnic controls.

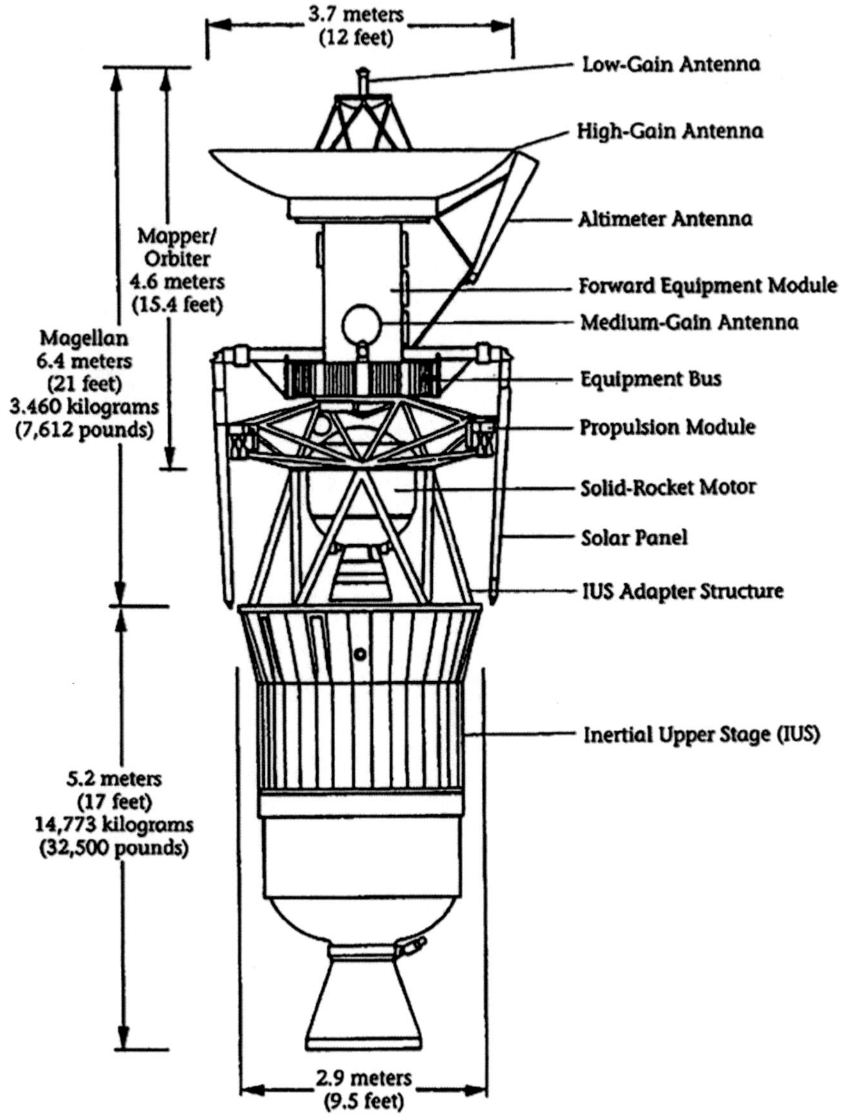

Fig. 6.2 Drawing of Magellan spacecraft mated to inertial upper stage. (NASA graphic)

The solar arrays protruding from the sides of the spacecraft were 2.5 by 2.5 meters in size and generated up to 1200 watts of power in the vicinity of Venus. The arrays were steerable in one axis to provide the best orientation towards the sun as the spacecraft orbited Venus.

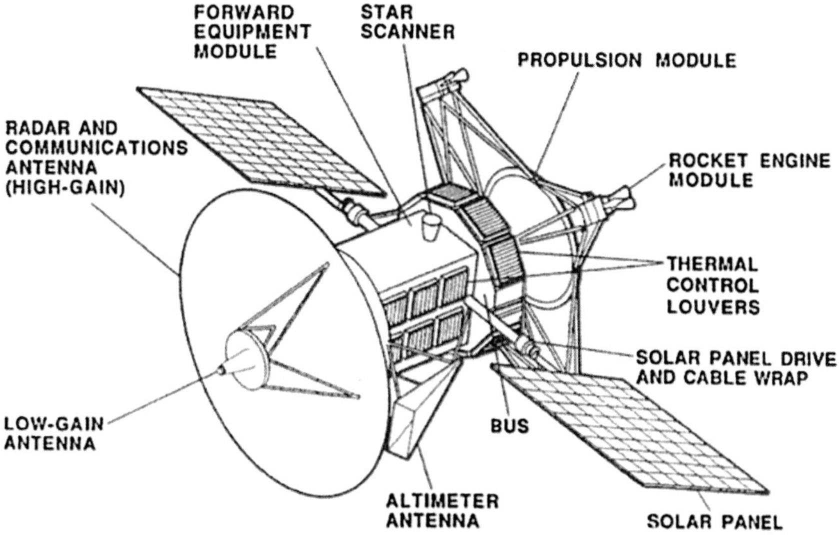

Fig. 6.3 Sketch of spacecraft with solar panels deployed. (NASA graphic)

Magellan Systems

Propulsion System

Solid rocket motor – As Magellan approached Venus, the solid rocket motor was ignited at the proper time to reduce velocity and allow the spacecraft to be captured by the gravitational field of Venus and enter a near polar orbit. The rocket motor was a Star 48 type build by Thiokol Corp. The number 48 denoted the diameter of the motor in inches and "star" denoted the shape of the open cavity at the center of the motor that served as the combustion chamber. The arms of the star afforded a large surface area of propellant at the beginning of the burn. The rocket generated 67,758 newtons (15,232 pounds) of thrust. The thrust was nearly constant from ignition to burnout.

The rocket motor was 1.25 meters in diameter and 2.03 meters long to the end of the expansion nozzle. It weighed 2146 kg of which 2014 kg was propellant. Burn time of the motor was 87 seconds and specific impulse was 286 seconds. The propellant was contained in a titanium case. The rocket was ignited by a squib that fired upon receiving an electrical signal from the Pyrotechnic Control subsystem.

Liquid fuel thrusters – A total of 24 hydrazine-powered thrusters were used to control attitude and dynamics of the spacecraft. Hydrazine is a monopropellant that decomposes when it passes over a bed of catalyst. Thrust was produced by gases resulting from decomposition of the hydrazine exiting through the nozzles of the thruster units. Valves activated by the attitude control system controlled turn-on and turn-off of propellant to individual thrusters. The hydrazine was stored in a tank that held 132.5 kg of the monopropellant.

Thrusters were mounted to the four corners of the propulsion module. The propulsion module is the truss-like structure just above the heads of the workers in the photograph of Fig. 6.1. Two 445 newtons thrusters can be seen pointed downward from each corner of the propulsion module in the photograph.

Each of the four corners contained six thrusters. Two of the six thrusters generated 445 newtons (100 pounds) of thrust, one had 22 newtons (five pounds) of thrust, and three had 0.9 newtons (0.2 pounds) of thrust. The two 445-newton thrusters and two of the three 0.9-newton thrusters were pointed parallel to the –Z axis (opposite to the high gain antenna boresight). The single 22-newton thruster at each corner was pointed perpendicular to the Z axis and in a direction to impart roll to the spacecraft. These thrusters were operated in pairs at diagonal corners to generate roll torque. A 0.9-newton thruster was similarly aligned at each corner to provide fine roll control. The 445-newton thrusters were sized to counteract maximum expected torque due to slight misalignment of the thrust axis of the solid rocket motor with the center of gravity of the spacecraft.

Attitude Control System

Magellan was a three-axis-stabilized spacecraft. The high-gain antenna was fixed mounted with its antenna beam aligned with the longitudinal axis of the spacecraft. As a result, the spacecraft needed to be maneuvered to point the antenna in the desired direction. This maneuvering was performed by the attitude control system. The attitude control system set the side looking angle of the antenna beam during radar mapping, pointed the beam towards earth to transfer data after a mapping pass, and maneuvered to a preset attitude to allow the star scanner view selected stars.

The near polar elliptical orbit of Magellan was 85.5 degrees from the equator of Venus. The orbit was set slightly off of 90 degrees from the equator so that the poles could be mapped when scanning to the side. Periapsis was at latitude of 9.5 degrees north at an altitude of 294 km. Apoapsis of the orbit occurred below the planet at an altitude of 8540 km.

Radar mapping for the primary mapping cycle commenced for each orbit as the spacecraft descended over the north pole at an altitude of about 2000 km. Mapping continued past periapsis to at about 75 degrees south latitude. Best radar mapping performance was obtained with the side looking angle as large as possible consistent with achieving adequate signal to noise ratio. The side looking angles selected for the primary mapping cycle varied from 16 degrees at high latitude, where the altitude was high, to 45.7 degrees at minimum altitude at periapsis. The spacecraft was maneuvered by the attitude control system to present the best look angle to the radar throughout the mapping passes.

The several maneuvers of the spacecraft required each orbit were commanded and managed by the attitude control computer. The main controlling functions were orchestrated by two redundant ATC-16MS computers in the attitude control system. The ATC-16MS computers, built by ITEK Corporation for space use, were also used in the Galileo spacecraft. The computers were 16-bit, floating point machines with total address space of 65,536 words. The basic cycle time was 250 nanoseconds. The throughput of the computer was about 500 instructions per second.

A star scanner, activated near apoapsis of each orbit provided an absolute attitude reference. The spacecraft was oriented so that the star scanner could view selected stars. The attitude control computer kept track of spacecraft attitude between updates from the star scanner. The attitude control system contained a rate gyro in each axis that sensed rotation rate of the spacecraft. The rates were integrated by the attitude control computer to provide continuous attitude information. Slight drifts in the gyros were corrected each orbit by data from the star scanner.

The star scanner contained a conical baffle that viewed a section of space and a mirror and lens arrangement that focused the scene onto a surface with two slits. One slit was vertical (aligned with the Z-axis of the spacecraft) and the other was inclined 30 degrees from it. A photodiode was located behind each slit. The spacecraft was rotated about the Z-axis to scan the star field in azimuth. An electrical pulse was generated by the photodiode when the rotating Vertical slit passed by a star. A second pulse was generated from the inclined slit as it passed by the star. The elevation angle of the star was determined by measuring the time difference between the two pulses. Azimuth was determined by noting the rotational orientation of the spacecraft when the star appeared in the vertical slit.

Two widely spaced, bright stars about 90 degrees apart were selected for reference for the star scanner. The expected brightness and clock position of each star were preset into the computer memory. Clock position was the rotation angle of the spacecraft in inertial space. The position of each star as computed from gyro-based attitude information was compared with the measured positions of the stars. The difference was used by the computer to update its attitude computation.

Attitude of the spacecraft was changed by changing the speed of reaction wheels. There were three 36-cm diameter reaction wheels, one aligned with each axis of the spacecraft. Rotation of the spacecraft about a particular axis was induced by rotating the reaction wheel for that axis in the opposite direction. The reaction to changing speed of the wheel was an equal and opposite reaction.

If a net bias in restoring angle was encountered that required speeding up the reaction wheel in the same direction, the speed of the wheel could eventually approach its maximum and no further correction would be possible. To prevent this, at some point near apoapsis of the orbit, the thrusters were engaged to hold that particular attitude and the reaction wheels were run down to stop. The reaction wheels were then able to control attitude for the next orbit. The reason that the thrusters were not used all the time for attitude control is that the total amount of propellant needed for the mission would be prohibitive.

Command and Data Subsystem

The command and data subsystem was the brains of the Magellan spacecraft. It received commands from Earth and processed the commands to control major activities of the spacecraft. It also assembled the massive amount of data from the radar as well as engineering data for transmission to Earth. Radar data was routed to tape recorders during the radar mapping pass and then transmitted to Earth later in the orbit.

The computing function for the command and data subsystem consisted of four RCA 1802 microprocessors arranged in two redundant pairs. The RCA 1802 was a single chip central processing unit (CPU) in a rectangular 40-pin package. It was a special radiation hardened device developed by RCA and Sandia National Laboratories. The microprocessor was built using silicon on sapphire technology, which has good resistance to radiation.

The RCA 1802 microprocessor had 8-bit parallel organization with an 8-bit bidirectional data bus. It interfaced with standard RAM and ROM memory with direct memory addressing up to 65,536 bytes. It performed the functions of fetching, interpreting, and executing instructions stored in the memory units. It interfaced with standard input-output (I/O) devices through the bidirectional data bus and control lines.

Spacecraft operation was controlled for up to three days at a time by commands sent from Earth and stored in memory in the command and data subsystem. Data from Earth included accurate navigation data for the spacecraft as developed by tracking stations on Earth. Spacecraft position information was used to reference location of radar mapping data and to set the side looking angle of the antenna beam during the mapping phase.

Telecommunications

The telecommunications system received commands and navigation data from Earth and transmitted a vast amount of data from the radar mapping operation and engineering data to Earth. The telecommunications system used both X- band and S-band data links. The X-band link operated at frequencies near 8.4 GHz and the S-band link operated at frequencies near 2.3 GHz. The X-band link was used to downlink radar mapping data to Earth. S-band was normally used to uplink commands and downlink engineering data.

The high-gain antenna was normally used both for uplink of commands and downlink of data. That antenna was pointed towards Earth during Magellan's orbit around Venus after the radar mapping process was completed. The antenna had a beamwidth of 0.6 degrees at X-band and 2.2 degrees at S-band. The antenna gain for downlink was 48.1 dB at X-band and 34.7 dB at S-band. The high gain at X-band with attendant narrow beamwidth was needed to transmit the large amount of radar data in the time available. The spacecraft also contained a medium gain antenna that was used for uplink commands and downlink engineering data during cruise to Venus and during orbit insertion. The beamwidth of the medium gain antenna was about 18 degrees.

Data and commands were normally transmitted from Earth at S-band. However, when the spacecraft was within about 10 degrees of the sun, the S-band link was not reliable and uplink and downlink transmissions were made at X-band. The uplink data rate was nominally 62.5 bits per second. A lower data rate of 7.8 bits per second could be used during emergencies.

The real-time data rate from the radar mapper was much higher than could be transmitted directly to Earth. To accommodate the high data rate, radar data was recorded by two tape recorders during a mapping pass and played back at a lower rate and transmitted to Earth after the mapping pass. Radar data was transmitted at X-band at a normal data rate of 268.8 kilobits per second. A lower data rate of 115.2 kilobits per second was available for emergencies. Engineering data was transmitted to Earth at S-band at a normal data rate of 1200 bits per second. A lower data rate of 40 bits per second could be used for emergencies.

Receiver and Transmitter – The telecommunications system used a redundant pair of NASA S-band Standard Transponders (NST). The NST contained an S-band receiver, command detector, exciter, and low power amplifier for both S-band and X-band transmission. The S-band transmitter output of either of the two redundant transponders could be applied to a traveling wave tube amplifier that provided output power of 5 watts. Likewise, the X-band transmitter output of either of the two redundant transponders could be applied to a traveling wave tube amplifier with output power of 20 watts.

The telecommunications system received S-band and X-band transmissions from Earth simultaneously. A downconverter was used to convert the X-band received signal to S-band so that it could be processed by the NST. The telecommunications system transmitted simultaneously on S-band and X-band. The frequency of the transmissions could be derived from an on-board crystal oscillator or the frequency could be derived coherently by phase locking with an offset from transmissions received from Earth. The coherent return signal allowed measurement of two-way Doppler shift on the return signal. Processing the Doppler shift allowed very accurate tracking of the spacecraft.

The S-Band uplink frequency was 2116 MHz and the S-Band downlink frequency was 2296 MHz. The X-Band uplink frequency was 7171 MHz and the X-Band downlink frequency was 8425 MHz. The transmitted power after losses was about 18 watts at X-band and 5 Watts at S-band.

Spacecraft Power

Solar arrays were the primary source of power for the spacecraft. Back-up and supplemental power was available from two batteries that were charged by the solar arrays. The active area of each solar array was about 6.3 square meters. The solar arrays could provide about 1200 watts of power at Venus at the start of the mission.

The available power was expected to degrade with time to about 1029 watts by the end of the nominal mission. The arrays could be rotated about an axis parallel to the X-axis of the spacecraft to achieve the best orientation to the sun as the spacecraft maneuvered. Outputs from a sun sensor controlled rotation of the arrays to the best orientation to the sun as the attitude of the spacecraft changed during the mission.

The output of the solar arrays was conditioned by a power control unit (PCU). The PCU used shunt regulators to dissipate excess power from the solar arrays in resistive networks to maintain voltage from the solar arrays in the range of 28 to 35 volts DC. That bus was referred to as the raw power bus. Further regulation developed the 28-volt power bus that was regulated to 28 ±0.6 volts DC. An inverter provided alternating current power at a frequency of 2.4 kHz at 50 volts AC to spacecraft systems. Each spacecraft system stepped down and rectified the 2.4 kHz power form to voltage levels required by their circuits.

Two batteries augmented power from the solar arrays when required and they provided all the spacecraft power during intervals when the sun was occluded by Venus. The batteries were nickel cadmium type with capacity of 30 ampere hours. The nominal voltage during discharge was 31 volts. One battery alone could support the spacecraft with some loss in mapping performance. The batteries were charged by power from the solar arrays.

Radar Sensor

The radar sensor was the only scientific instrument carried by the Magellan spacecraft. It was a very capable instrument that performed three significant tasks while in orbit around Venus: Synthetic Aperture Radar (SAR) imaging, radar altimetry for topography, and radiometry for mapping thermal emission from the surface. The three functions were performed sequentially within a repetitive data acquisition time of 250 to 1200 milliseconds, depending on altitude, as Magellan sped above the surface. The radar, which operated in S-band, was a substantial instrument that weighed 154 kg. It included 17 electronics units contained in a rack 1.5 × 0.9 × 0.3 meters in size.

The synthetic aperture radar processed Doppler information in the received echo signal to achieve an effective (synthetic) aperture of the antenna much larger than the physical aperture of the antenna in the along-track direction. The large synthetic aperture resulted much better along-track resolution than would be normally result from the beamwidth of the antenna.

The radar used the same high-gain antenna for the SAR mode as was used by the telecommunication system to send data to Earth. The antenna was a parabolic reflector type with diameter of 3.7 meters. It had a gain of 36 dB at the radar frequency and the beamwidth was 2.1 by 2.6 degrees. The boresight of the antenna was along the Z-axis of the spacecraft. The spacecraft was maneuvered to point the antenna beam at the desired elevation angle relative to the local vertical during mapping and in the direction of Earth when transmitting data to Earth.

The radar altimeter used a rectangular horn antenna that had gain of 19 degrees and beamwidth of 10 by 30 degrees. The 30-degree beamwidth was in the cross-track direction while mapping. The boresight of the horn was in the X-Z plane and displaced 25 degrees from the Z-axis in the cross-track direction.

The radiometer also used the high gain antenna but with the radar in a receive-only mode. It received natural thermal emissions from the surface in the frequency range of the radar receiver. The emission data assisted in determining the type of surface that the spacecraft was passing over.

The radar was time shared among the three sensors. The data burst protocol was made up of the SAR mode followed by the radar altimeter mode followed by the radiometer mode. The cycle then repeated. A NASA diagram of events during the burst period is shown in the graphic below (Fig. 6.4).

The radar operated at a frequency of 2.385 GHz (wavelength of 12.58 cm). It was a pulsed radar with peak power of 325 watts and a pulsewidth of 26.5 μs. The pulse was biphase modulated by a 60-element code with a chipping rate of 2.26 MHz. The waveform presented the amount of energy per pulse corresponding to the 26.5 μs pulsewidth but, after processing of the echo signal, the effective pulsewidth for resolution purposes was 1/60 of that or about 0.44 μs.

Radar Modes – The SAR radar mode transmitted a burst of 150 to 800 pulses during a time interval from 25 μs to 250 μs. The received pulses became interleaved with the transmit pulses but the spacecraft had knowledge of altitude above the terrain and elevation angle of the antenna beam so that range ambiguities could be resolved. The pulse repetition frequency and echo pulse receive gate timing were adjusted continuously as a function of altitude and antenna angle.

In the radar altimeter mode, a burst of 17 pulses were transmitted in a one millisecond time period. The receiver was not switched on until the burst of 17 pulses had been sent. In the radiometer mode, no pulses were transmitted and the receiver was gated on for 50 milliseconds.

The synthetic aperture radar mode used the high-gain antenna pointed 90 degrees to the side of the flight path and at varying angles from the local vertical. The cross-track range resolution improved as the angle of incidence of the antenna beam on

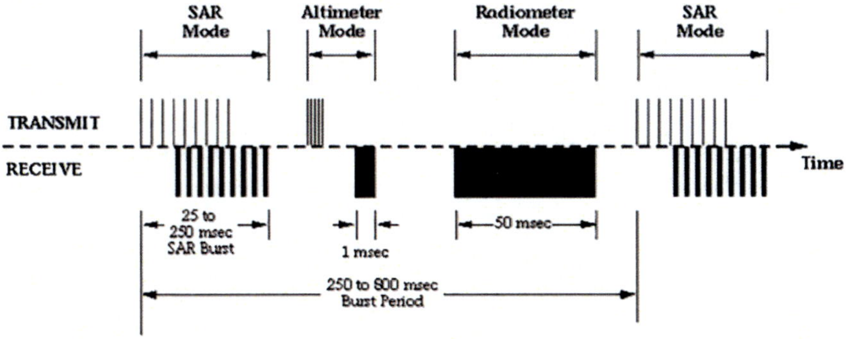

Fig. 6.4 Radar activity during burst interval. (NASA graphic)

the surface increased. However, the signal-to-noise ratio decreased with increasing incidence angle. The signal-to-noise ratio also decreased with increasing altitude of the spacecraft. Consequently, to maintain adequate signal-to-noise ratio, the elevation angle of the antenna beam was programmed to be about 16 degrees from the vertical near the poles increasing to about 45 degrees from vertical near periapsis.

The elevation angle of the SAR antenna beam had priority and so the angle between the boresight of the radar altimeter antenna and the local vertical varied during the orbit. The boresight of the altimeter antenna was 25 degrees from the Z-axis and the 30-degree beamwidth of the antenna was in the cross-track direction. When mapping near the poles, the high gain antenna was pointed 16 degrees from vertical and the boresight of the altimeter horn was 9 degrees from the local vertical. This put the view of the nadir within the beamwidth of the horn. Near periapsis with the high-gain antenna pointed 45 degrees from the vertical, the boresight of the horn was 20 degrees from the local vertical but the lower altitude resulted in signal-to-noise ratio adequate for altimeter operation.

Radar equipment – The transmit signal path of the radar began with a stable oscillator (STALO) that provided a reference frequency to a PRF/timing unit. The PRF/timing unit generated clocks and timing for the radar and other systems in the spacecraft. The STALO also provided a precise reference frequency to a frequency synthesizer that generated the S-band carrier frequency for the radar. The S-band carrier was modulated by a 60-bit biphase code at a chipping rate of 2.26 MHz. The modulated carrier was gated to form a modulated pulse 26.5 μs wide. The modulated pulse was applied to a power amplifier transmitter that amplified it to a peak power level of 325 watts.

The transmitter output was switched to the high-gain antenna through a duplexer for the SAR mode, to the horn antenna through a duplexer for the altimeter mode, and no pulses were sent in the radiometer mode. The duplexers protected the receiver from the high power transmit pulse and allowed the receiver to be connected to the antenna during the receiver interval.

The echo signal from the duplexers was applied to the receiver. The receiver was a double conversion type where the first conversion translated the received signal to an intermediate frequency (IF). The IF signal was amplified and applied to a second downconverter that translated it to baseband. The baseband conversion was done in phase quadrature so that the sign of the Doppler information in the signal could be identified.

The signals in the two quadrature channels were applied to analog-to-digital converters and the digital representations were applied to data formatters and then on to the Command and Data subsystem. The radar generated data at a rate of about 806.4 kilobits/second during SAR mapping. The data stream was fed to two redundant tape recorders.

Transmitted data from the spacecraft received on Earth included the two quadrature channels. Signal processing was performed on the ground using the quadrature data to develop radar images from the synthetic aperture radar and altitude data from the radar altimeter. The data was processed by the Radar Data Processing Subsystem that ran on a collection of computers at JPL.

The side-looking synthetic aperture radar gathered a strip of radar data during each orbit. The area covered on the surface on each orbit was about 20 km wide and 17,000 km long. The resolution of each pixel of the image after processing on the ground was about 110 meters in the along-track direction. Resolution varied in the cross-track direction as the incidence angle of the antenna beam changed. Cross-track resolution was 101 meters at latitude of 10 degrees north and 250 meters at latitude of 90 degrees north or 70 degrees south.

Redundancy was provided for main elements of the transmitter and receiver chains. For example, there were two STALOs, two PRF/timing units, two transmitters, two receivers, two baseband processors, and two data formatters. The various elements were cross strapped to allow operation in case of failure of a particular element. One or the other of the two transmitters was selected to provide transmitter pulses to the antennas. Likewise, one or the other of two receiver channels could be selected.

Flight of Magellan

Magellan was carried into Earth orbit by Space Shuttle STS-30 (Atlantis) on 4 May 1989. It was launched on a trajectory to Venus from the cargo bay of Atlantis on 5 May 1989. The tragic loss of Space Shuttle Challenger and its crew in December 1986 resulted in a 32-month delay in flights of the Space Shuttle. As a result, launches of planetary spacecraft Magellan to Venus and Galileo to Jupiter by the Space Shuttle had to be rescheduled. Magellan was rescheduled for a May 1989 launch. That launch date required Magellan to fly a Type-IV trajectory that took it around the sun 1.5 times before arriving at Venus 15 months later. Had the spacecraft been able to launch in May 1988, a Type-I trajectory would have been used with a flight time to Venus of only four months.

The Magellan spacecraft, with solar arrays folded and attached to the inertial upper stage, fit within the cargo bay of the Space Shuttle. Equipment in the shuttle, referred to as the Airborne Support Equipment, supported the spacecraft and raised it on a tilt table to 52 degrees for deployment. Launch from the Shuttle involved releasing springs that thrust the spacecraft away from the cargo bay. A photograph taken by the crew of Atlantis just as the spacecraft departed the shuttle is shown on the next page (Fig. 6.5). About an hour after being released from the space shuttle, the inertial upper stage was ignited to send Magellan on a trajectory to Venus.

Tests and calibrations were performed during the long cruise to Venus. Three trajectory correction maneuvers were performed. The first occurred 17 days after launch and involved a velocity increment of 2.94 meters per second. This was accomplished by a 5.35 second burn of four 445-Newton rocket engines. The second trajectory correction occurred on 13 March 1990 with a burn of 0.967 seconds of four 445-Newton engines. A velocity increment of 0.469 meters per second was obtained. The third trajectory correction took place on 25 July 1990 by a 321-second

Fig. 6.5 Magellan spacecraft attached to inertial upper stage in launch position from Space Shuttle. (NASA photograph)

burn of all eight aft facing 0.9-Newton engines. A velocity increment of 0.692 meters per second was obtained.

The solid rocket motor was fired for 85 seconds on 10 August 1990 to slow the spacecraft by 2660 meters per second and allow it to be captured by the gravitational field of Venus. The resulting elliptical orbit, which had a period of 3.26 hours, was inclined 81.7 degrees from the equator. The eccentricity of the orbit was 0.292. Periapsis occurred at latitude of 9.5 degrees north and altitude above the surface of 294 km. Apoapsis occurred at an altitude of 8543 km from the surface.

An extensive in-orbit checkout of spacecraft systems and the radar was performed and the spacecraft was prepared for the mapping operation. Venus rotates with a period of 243 Earth days (5832.6 hours). The near polar orbit of the spacecraft allowed mapping 360 degrees of longitude by the radar as Venus rotated under the orbiting spacecraft. Considering the orbit period of 3.26 hours, a total of 1789 orbits were executed while Venus rotated once on its axis. Mapping was performed on the side of Venus lying under periapsis of the orbit. Mapping data was transmitted to earth during the long portion of the orbit around apoapsis.

The first mapping cycle, referred to as Cycle 1, began on 15 September 1990 and lasted until 14 May 1991. Cycle 1 was planned to map from the north pole to about 57 degrees south latitude. The side-looking radar was pointed to the left of the flight path during Cycle 1. Details of the mapping operation and data transmission to Earth are given in the next section of this chapter.

The Cycle 2 mapping operation began on 15 May 1991 and lasted until 14 January 1992. Cycle 2 mapped from about 75 degrees north latitude to the south pole. The side-looking radar was pointed to the right of the flight path during Cycle 2.

The Cycle 3 mapping operation began on 15 January 1992 and lasted until 13 September 1992. Cycle 3 mapped from 75 degrees north latitude to 45 degrees south latitude and its images were used to create stereo pairs with images from Cycle 1. The side-looking radar was pointed to the left of the flight path during Cycle 3.

Cycle 4 was designed to obtain data on the gravitation field of Venus. The radar was shut down and the high-gain antenna was pointed towards Earth. The Doppler shift on the X-band signal transmitted to Earth was used to derive extremely precise spacecraft velocity. Small variations in velocity were related to mass changes under the spacecraft. The accuracy of gravity measurements was better at lower altitudes. The altitude at periapsis was lowered to 184 km by a 54-minute burn of the thrusters on 14 September 1992. Cycle 4 gravity data acquisition began on 14 September 1992 and lasted until 25 May 1993.

The altitude of the orbit was further lowered by causing the spacecraft to dip into the atmosphere of Venus, a technique referred to as "aerobraking." The resulting orbit had periapsis of 180 km and apoapsis of 540 km. The period of the orbit decreased to 95 minutes. The lower and more circular orbit benefited the accuracy of gravity measurements. Cycle 5, which lasted from May 1993 to August 1994, provided enhanced gravitational data at the lower altitudes of the orbit. The enhanced gravimetric mapping covered over 95% of the surface of the planet.

Cycle 6, which lasted from April 1994 to October 1994, collected additional gravimetric data and radio science experiments were conducted. At the end of Cycle 6, the orbit of the spacecraft was lowered further and the solar arrays were oriented such that they would cause drag in the tenuous upper atmosphere of Venus. The torque from the thrusters necessary to counteract the torque on the spacecraft due to drag of the atmosphere on the solar arrays was measured and gave a measure of the structure of the upper atmosphere of Venus. Following those experiments, the orbit of the spacecraft was lowed further on 11 October 1994 and the spacecraft was allowed to burn up in the atmosphere.

Mapping Mission

A sketch of the Magellan spacecraft showing its appearance during the mapping interval is given below (Fig. 6.6). The NASA sketch, which had the antenna pointed upwards, was rotated by the author to better visualize the attitude of the spacecraft

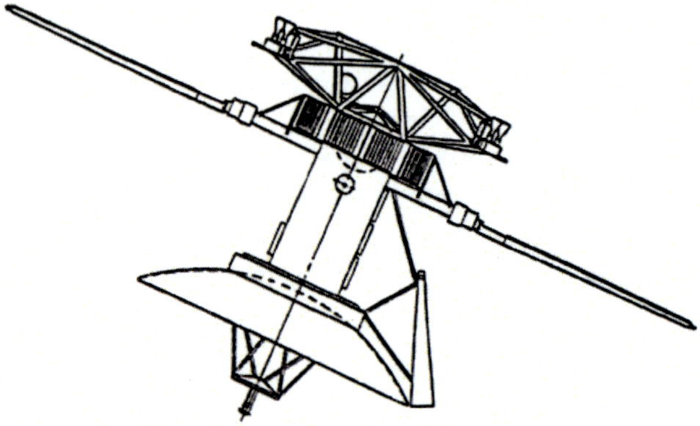

Fig. 6.6 Sketch of Magellan spacecraft in mapping mode (NASA graphic rotated by author)

in the mapping mode. The horn antenna for the radar altimeter is located to the right in the sketch with the aperture end of the horn near the edge of the high-gain antenna. The spacecraft traveled parallel to the Y coordinate axis, which is perpendicular to the plane of the sketch. The +Z-axis is along the longitudinal axis and along the high-gain antenna boresight.

Cycle 1, Cycle 2, and Cycle 3 of the mission mapping operations employed the synthetic aperture radar (SAR), radar altimeter, and radiometer functions of the radar to develop data for mapping terrain and topographical features of Venus. Measurements of the three modes were multiplexed in time with SAR, radar altimeter, and radiometer in sequence in a burst. The burst durations ranged from 250 to 1200 milliseconds in length.

A NASA graphic showing the path of the orbit around Venus during Cycle 1 mapping is given on the next page (Fig. 6.7). Events associated with the mapping operation during the orbit are identified in the graphic. The eccentric orbit brought the spacecraft close to the surface of Venus during the 37-minute mapping interval and then swung it far out towards apoapsis. The remaining 158 minutes of the 195-minute orbit period was occupied by positioning the spacecraft to point the high-gain antenna towards Earth, transmitting mapping data to Earth, performing a star scan, updating memory, and finally returning the antenna beam towards the surface for the next mapping pass.

Synthetic Aperture Radar Mapping

Cycle 1 of the mapping operation was referred to as the prime mission. The synthetic aperture radar beam was directed to the left of the flight path during Cycle 1. Mapping commenced after the spacecraft passed near the north pole at an altitude of about 2000 km. The orbit took the spacecraft down to periapsis at 9.5 degrees north latitude and altitude of 294 km, and then the altitude increased as the

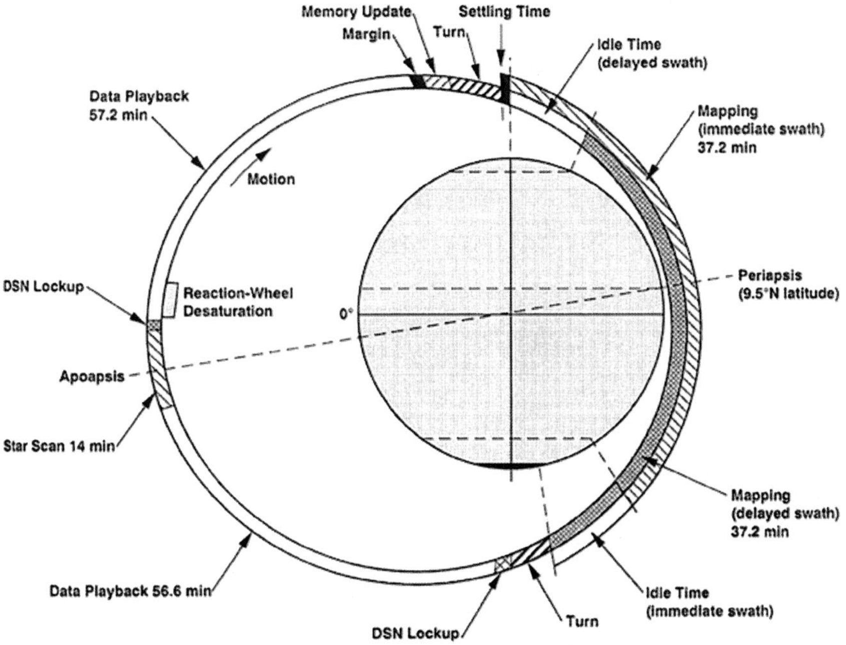

Fig. 6.7 Magellan mapping orbit around Venus. (NASA graphic)

spacecraft flew towards the south pole. The first mapping orbit in the cycle began at the north pole and continued for 37.2 minutes, stopping at 57 degrees south latitude. The start of mapping was purposely delayed by seven minutes during the second orbit, starting at latitude of 70 degrees north and continuing to 74 degrees south latitude. This alternating pattern, which was continued for the 243- day cycle, took advantage of adequate overlap achieved near the poles by mapping every other orbit and it allowed mapping further to south latitudes.

The cross-track range resolution improved as the angle of incidence of the antenna beam on the surface increased. However, the signal-to-noise ratio decreased with increasing incidence angle. The signal-to-noise ratio also decreased with increasing altitude of the spacecraft. Consequently, to maintain adequate signal-to-noise ratio, the side looking angle of the antenna beam was programmed to be about 16 degrees from the vertical near the poles increasing to about 46 degrees near periapsis. Since the antenna was fixed to the spacecraft, the attitude of the spacecraft was continuously adjusted to follow the programmed look angle profile. The synthetic aperture radar imaged a swath on the surface typically 20 km wide and 17,000 km long. Data gathered by the synthetic aperture radar, radar altimeter, and radiometer was stored in two tape recorders along with engineering data during the mapping interval.

At the end of the mapping interval, the spacecraft attitude was changed to point the high-gain antenna towards Earth. After a brief period to allow equipment in the

Deep Space Network (DSN) on Earth to acquire the signal, the first playback of radar data from the tape recorders commenced and lasted for 56.6 minutes. The spacecraft was approaching apoapsis after the first playback interval. A star scan to update attitude information was then performed that lasted for 14 minutes. The last portion of radar data playback then followed that lasted 57.2 minutes. The spacecraft was approaching the north pole at the end of the second data playback. An update of the memory was performed and then the spacecraft attitude was changed to direct the antenna beam to the mapping orientation for the next mapping pass. Instructions for several days of mapping were uplinked from Earth to the spacecraft about once a week and stored in memory.

The radar generated data at a rate of 806.4 kbps during the mapping mode. Since the high-gain antenna was needed for radar mapping, the data was sent in parallel to two 4-track tape recorders, DMS-A and DMS-B. DMS stands for Data Management Subsystem. Each tape recorder was capable of storing the 1.8×10^9 bits of data generated in one 37.2-minute mapping pass.

Tracks 1 and 2 were used on each tape recorder. To minimize the amount of tape required, track 1 was recorded with the recorder running in the forward direction for a portion of the mapping interval and then the direction of tape travel was reversed and track 2 was recorded during the reverse direction of tape travel for the remainder of the mapping interval. The time interval for stopping and then reversing the direction of tape travel was staggered between the two tape recorders to prevent losing data in the dead time during reversal. A significant error rate was detected in DMS-A in December 1990 so that tape recorder was turned off. The data gap during reversal of the tape in the one tape recorder amounted to a loss of about 40 km of mapping out of the 17,000 km total mapping swath.

The spacecraft was maneuvered to point the high-gain antenna towards Earth after the mapping interval. Recorded data was transmitted to Earth in two segments at a data rate of 268.8 kbps. Track 1 from both tape recorders were played back sequentially for a total of 56.6 minutes. After the star scan operation, which lasted for 14 minutes, Track 2 from both recorders was played back sequentially for 57.2 minutes. A total of 1.835×10^9 bits of data could be transferred during the total 113.8-minute playback time.

At the end of the 243-day mapping interval of Cycle 1, about 84% of the surface of Venus had been mapped. A mosaic of SAR images gathered during Cycle 1 is shown on the next page (Fig. 6.8). The mosaic appears in the NASA report *Guide to Magellan Image Interpretation* by John Ford et al.

The numbers along the horizontal portion of the image denote longitude and the numbers on the vertical position denote latitude in the Venus coordinate system. The gap between longitude of 30 degrees and 55 degrees was caused by loss of data when Venus passed behind the sun (superior conjunction).

Cycle 2 of the mapping mission began at the end of Cycle 1. The goals of Cycle 2 were to fill in the gap left by superior conjunction during Cycle 1 and map the south pole area. Most of the mapping operation was conducted with the radar antenna beam directed to the right of the direction of travel. Mapping was begun at latitude of 65 degrees north and it continued to the south pole. The angle of

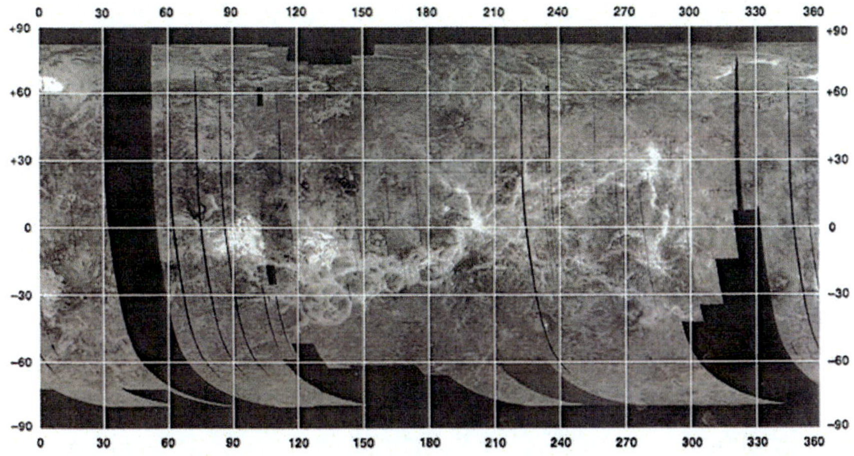

Fig. 6.8 Mosaic of SAR imagery from Cycle 1 of the Magellan mapping mission. (NASA graphic)

incidence of the antenna beam was maintained at 25 degrees from latitude of 65 degrees north to 45 degrees south and then began decreasing to 13 degrees when approaching the south pole. A test to determine the feasibility of stereo mapping was conducted for eight orbits with the antenna beam directed to the left of the direction of travel and viewing the same area as during Cycle 1. Channel 2 mapping by the SAR covered about 54% of the surface. The surface coverage of Cycle 1 and Cycle 2 combined was 96%.

Cycle 3 mapping began at the end of Cycle 2. The main goal of Cycle 3 was to obtain images that could be used as stereo pairs when combined with images from Cycle 1. Much of the mapping process overflew areas mapped by Cycle 1 but with a smaller look angle. Cycle 3 also filled in some areas missed by Cycle 1. The total coverage obtained by combining data from all three cycles was 98% of the surface of Venus.

A synthetic aperture radar image of interesting terrain is shown on the next page (Fig. 6.9). The image shows a system of lava flows where lava broke through a ridge belt. The image is about 550 km wide in the east-west direction and 630 miles in the north-south direction.

Radar Altimeter Mapping

The radar altimeter measured time delay from the leading edge of the transmitted pulse to the first return of the waveform from the surface. The first return came from the terrain directly under the spacecraft and the time delay was proportional to altitude. The Magellan radar altimeter mode used the same radar waveform as the synthetic aperture radar mode with a coded pulsewidth of 26.5 microseconds and peak

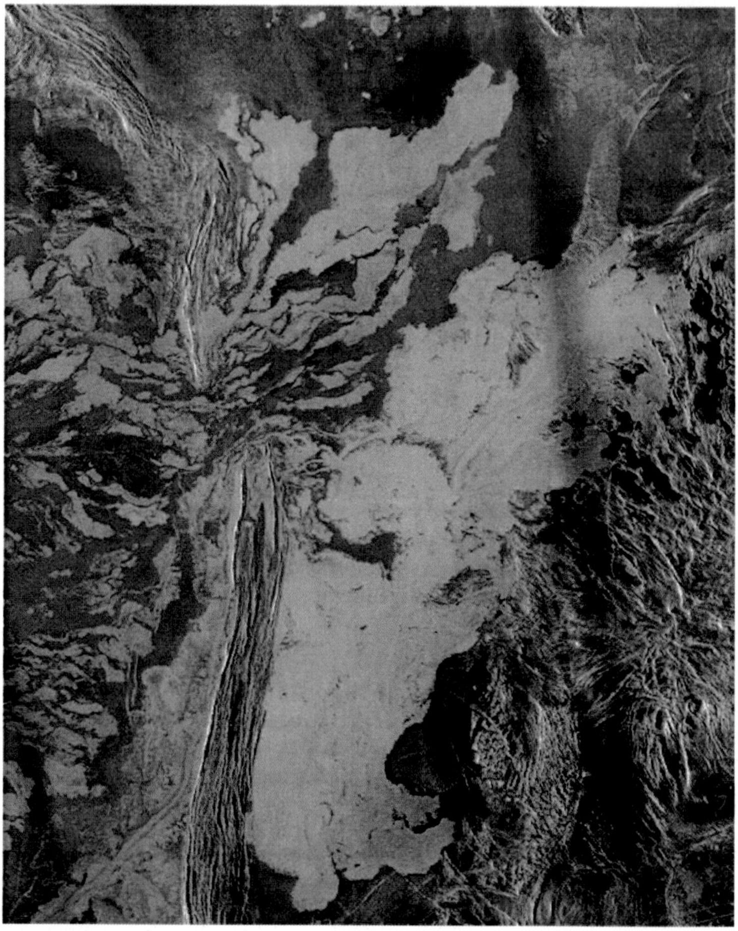

Fig. 6.9 SAR image of lava flows on Venus. (NASA graphic)

power of 325 watts. The radar was switched to the radar altimeter horn antenna during altitude measurements. The radar sequentially switched from synthetic aperture radar mapping, to radar altimetry, and to radiometric measurements in one data gathering burst period that lasted from 250 to 1200 milliseconds depending on altitude.

The accuracy of the radar altimeter measurement was about five meters over relatively smooth terrain. However, the uncertainty of the orbit of the spacecraft resulted in an uncertainty of about 50 meters in absolute altitude determination. The resolution cell on the surface for the altitude measurement was about 10 by 12 km near periapsis and 20 by 29 km near the poles.

Radar altimeter measurements were made from latitudes of 85 degrees north to 80 degrees south. Data from the radar altimeter combined with data from the side-looking synthetic aperture radar allowed three- dimensional representations of the terrain of Venus to be developed.

The radar altimeter horn antenna had beamwidth of about 10 degrees in the along-track direction and 30 degrees in the cross-track direction. The gain of the antenna was 19 dB. The boresight of the antenna was positioned 25 degrees from the +Z axis in the X-Z plane and parallel to the X-axis. The high-gain antenna look angle varied from 16 degrees from the local vertical axis near the poles to 45 degrees near periapsis. With the look angle to the left of the flight path, a look angle of 16 degrees near the poles would result in a line to the local vertical (nadir) lying 9 degrees off boresight of the horn. This is well within the 30 degrees half power beamwidth of the horn. With the spacecraft near periapsis, and a look angle of 45 degrees, the nadir would lie 20 degrees from boresight of the horn. Although beyond the half power beamwidth, the altitude was relatively low and there was adequate signal-to-noise ratio for radar measurements along the nadir.

Altimeter data developed by the radar was assembled by the command and data subsystem and recorded on the tape recorders for later playback during transmission to Earth. A color-coded topographical map of the terrain of Venus developed from radar altimeter data is shown on the next page (Fig. 6.10). The map is shown as a Mercator projection covering from 69 degrees north latitude to 69 degrees south latitude. A view of the topography of Venus looking from above the north pole is shown on the next page (Fig. 6.11). Color coding of altitude is the same as indicated for Fig. 6.10. The high terrain region in the northern portion of Venus is called Ishtar Terra. It is dominated by the highest mountain on Venus, Maxwell Montes, which rises 11 km above the mean elevation of the planet.

Radar altimeter data and synthetic side-looking radar data could be combined to develop 3-dimentional views of the terrain. One such view is of Maat Mons shown on the following page (Fig. 6.12). The vertical scale of the image has been increased 10 times. Maat Mons rises about 5 km above the surrounding terrain and 8 km above the mean elevation of the surface of Venus. It is located at 0.9 degrees north latitude and 194.5 degrees east longitude. The simulated color hues were based on color photographs of the surface taken by Soviet Union Venera 13 and Venera 14 spacecraft that landed on Venus in 1981.

Radiometric Mapping

The radiometer measured the natural electromagnetic radiation from the surface. Any object or surface above a temperature above absolute zero (zero Kelvin or −273 deg C) radiates with intensity that is a function of the temperature of the body, wavelength of the radiation, and emissivity of the surface. A special state of an object called a black body, which is a perfect radiator and perfect absorber of energy, is the basis for radiometric work. The radiation from an ordinary surface is lower than from a black body by a factor known as the emissivity. The emissivity of bare soil on earth at the 12.6 cm wavelength used by the Magellan is greater than 0.95.

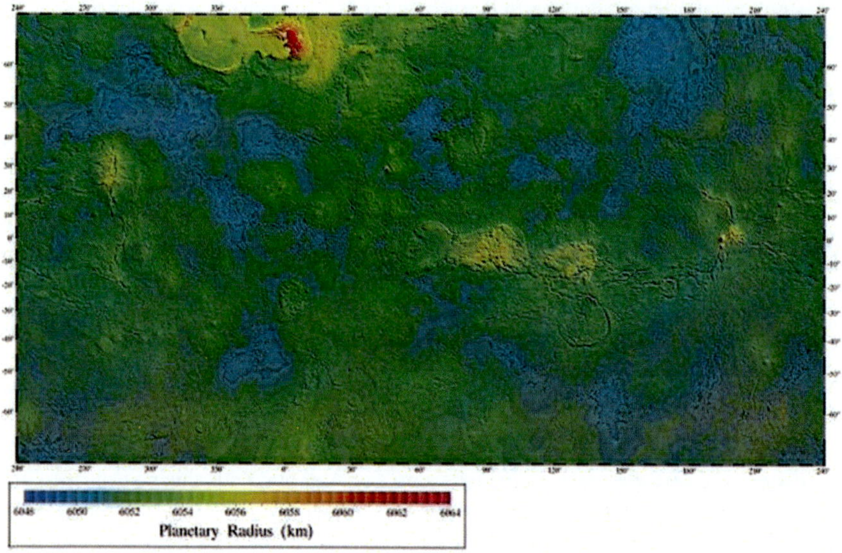

Fig. 6.10 Topographical map of Venus from radar altimeter measurements. (NASA graphic)

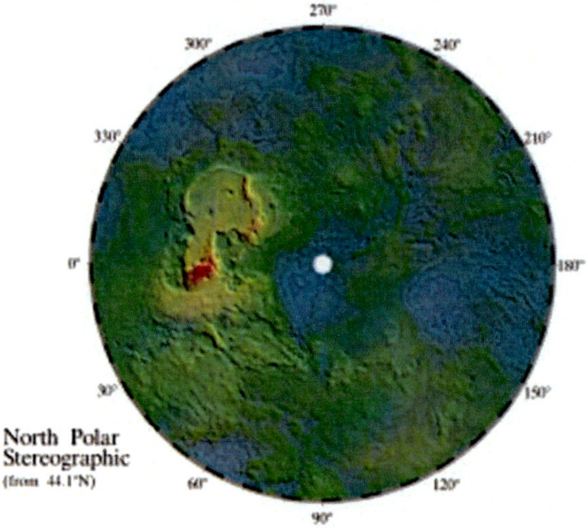

Fig. 6.11 Topographical view of Venus from above north pole. (NASA graphic)

Radiometer data during the mapping orbits were used by researchers on Earth to develop maps of emissivity of the surface.

The temperature of the surface of Venus and its dependence on elevation of the terrain were known from measurements by previous spacecraft. Also known were characteristics of the high-gain antenna and wavelength used for the measurements.

Fig. 6.12 3-D view of Maat Mons developed from Magellan radar data. (NASA graphic)

This information, along with radiometer signal strength, allowed computing the emissivity of the surface during data processing on Earth.

Radiometer measurements were interleaved with SAR and radar altimeter measurements. The radiometer input to the radar receiver was switched from the high-gain antenna to a dummy load at a known temperature every other data gathering burst cycle. The signal from the load was used as a reference to compare with the radiometer signal from the surface. The accuracy of the temperature measurement was about 2 Kelvin. The noise bandwidth of the receiver was 10 MHz and the integration time per measurement was 50 milliseconds. The emissivity measurements were accurate to about 0.02 and local variations as small as 0.005 could be detected.

The antenna beamwidth and antenna look angles resulted in viewing areas on the surface for Cycle 1 mapping that varied from 23 km cross-track and 15 km along-track at 10 degrees north latitude to 91 km cross-track and 77 km along-track at 90 degrees north latitude.

The mean value of emissivity for the surface of Venus was determined from Magellan data to be about 0.85 over the three mapping cycles of the mission. The principal investigator for the radar, Gordon Pettengill, stated that the value of 0.85 was consistent with dry basaltic minerals. Basaltic rocks are typical of lava flows. Variation in emissivity over the surface reached a low value of 0.3 in isolated areas.

The output data from the radar in the radiometer mode was assembled by the command and data subsystem and recorded on the tape recorder for later playback and transmission to Earth. A plot of emissivity of the surface derived from Magellan

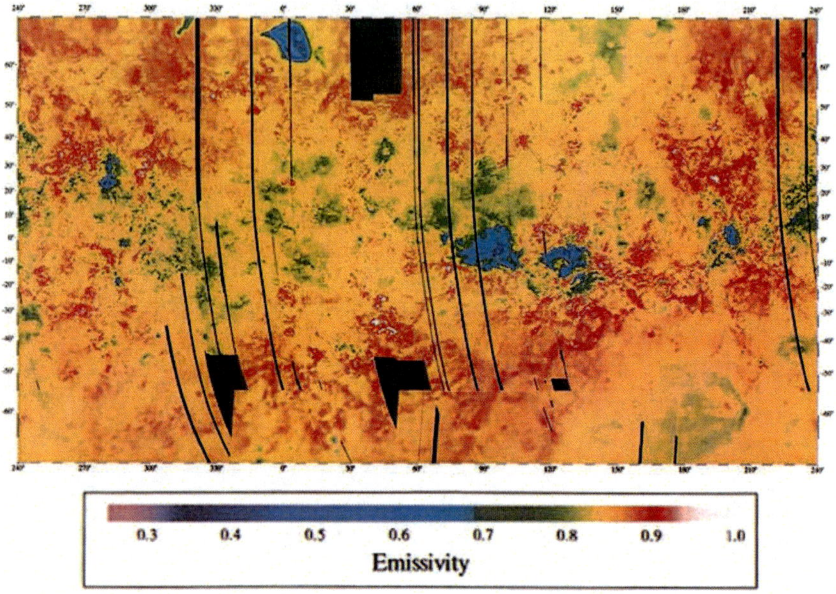

Fig. 6.13 Emissivity of the surface of Venus. (NASA graphic)

data is given as a Mercator projection above (Fig. 6.13). The scale of the graphic for longitude and latitude are the same as in Fig. 6.10 where the numbers are easier to read.

Gravimetric Mapping

Gravimetric data was developed during Cycle 4 and Cycle 5 of the mapping process. The radar was shut down and the high gain antenna was pointed towards Earth during those cycles. The two-way Doppler shift on the X-band transmission from the spacecraft was used to develop precise velocity of the spacecraft relative to Earth. Variations in velocity in the orbit were related to mass changes under the spacecraft. The spacecraft sped up slightly when passing over a denser region.

Accuracy of the gravimetric measurements was better at lower altitudes. Before the Cycle 4 orbits began, the altitude at periapsis was lowered to 184 km. Later in the mission, before Cycle 5 orbits began, the orbit was further lowered to periapsis of 180 km and apoapsis of 540 km. The period of the orbit decreased to 95 minutes. Cycle 4 gravity data acquisition began on 14 September 1992 and lasted until 25 May 1993. Cycle 5 began in August 1993 and ended in August 1994. Gravitational mapping data was obtained over 95% of the surface of Venus by the two gravimetric cycles.

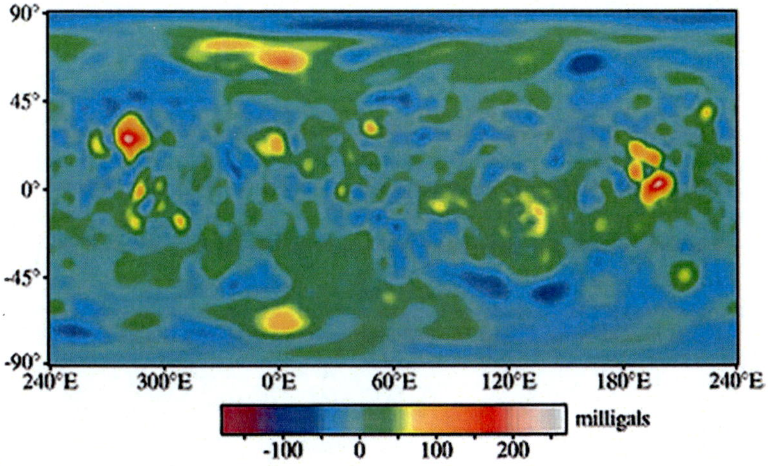

Fig. 6.14 Gravimetric map of Venus. (NASA graphic)

A Mercator projection chart of the variations in gravity on the surface of Venus is shown above (Fig. 6.14). The term milligal is used to measure local variations in the gravity of the planet. A milligal is equal to about 10^{-6} g, where g is the standard gravitational attraction of Earth. The mean gravity of Earth is about 9.81 m/s^2 (981,000 milligal). The mean gravity of Venus is 8.87 m/s^2. The map shows higher than average gravity in the vicinity of volcanoes, possibly due to upwelling of lava.

It is remarkable that such a fine-grained map of gravimetric variations over the surface of Venus was obtained from the distant Earth. The sensitivity comes about from the extremely accurate velocity measurement of the spacecraft by two-way Doppler processing.

Conclusion

The Magellan mission was an outstanding success. The spacecraft mapped over 98% of the surface of Venus using its synthetic aperture radar (SAR) and radiometer instruments. Image resolution for the SAR maps was about 110 meters in the along-track direction and varied from 100 to 250 meters in the cross-track direction depending on latitude. Topographical measurements were made over 95% of the planet by the radar altimeter. The resolution of terrain measurements was about five meters over smooth terrain. Magellan data taken during the time period 1990 to 1994 remains today the best available data on the topography of Venus.

An attractive logo was developed for the Magellan mission. The logo appears on the cover page of *Magellan Spacecraft Final Report*. A copy of the logo is shown on the next page (Fig. 6.15).

Fig. 6.15 Logo for Magellan mission. (NASA graphic)

Bibliography

Butrica, Andrew J., *To See the Unseen: A History of Planetary Radar Astronomy*, Chapter 7 Magellan, NASA SP-4218, 1996

Ford, J. P., et al., *Spaceborne Radar Observations: A Guide for Magellan Radar-Image Analysis.*

Ford, John P., et al., *Guide to Magellan Image Interpretation* JPL Publication 93–24, November 1993.

Jet Propulsion Laboratory, *Magellan: The Unveiling of Venus*, JPL Report JPL 400-345 3/89, 1989

Jet Propulsion Laboratory, *Magellan Mission to Venus,* https://nssdc.gsfc.nasa.gov/planetary/magellan.html, February 1995

JPL Website, *Magellan Mission to Venus,* https://www2.jpl.nasa.gov/magellan/fact.html

Martin Marietta Technologies, *Magellan Spacecraft Final Report*, MGN-MA-011, January 1995

NASA, *Mission Operation Report, Magellan,* Report No. E844-89-30-01, 1989

NASA-NSSDCA, *Magellan*, NSSDCA/COSPAR ID: 1989-033B

Tanaka, Kenneth L., et al., *The Venus Geologic Mappers' Handbook,* U.S. Geological Survey Report 93-516, July 1993

V-GRAM, *Magellan Bulletin About Venus and the Radar Mapping Mission*, JPL Publication, NASA-CR-192923, April 1993

Wall, Stephen D., et al., *User Guide to the Magellan Synthetic Aperture Radar Images,* NASA Publication 1356, March 1995

Williams, David R., *Magellan Fact Sheet,* NASA-NSSDCA https://nssdc.gsfc.nasa.gov/planetary/factsheet.html, February 2019

Williams, David R., *Magellan Mission Plan,* NASA-NSSDCA https://nssdc.gsfc.nasa.gov/planetary/mgncycles, January 1997

Young, Carolynn, (Editor), *The Magellan Venus Explorer's Guide*, JPL Publication 90-24, August 1990

Chapter 7
U.S. Spacecraft that Explored Mercury: Mariner 10 and Messenger

The United States spacecraft Mariner 10 and Messenger are the only two spacecraft that have explored the planet Mercury as of 2021. Mariner 10 explored Mercury in the time period 1974–1975 and Messenger explored from 2008 to 2015. The BepiColombo spacecraft is enroute to Mercury and is expected to enter orbit around the planet in December 2025. The BepiColombo mission is an international effort between the European Space Agency and the Japan Aerospace Exploration Agency. BepiColombo is not covered here since its main exploration of Mercury is still a few years off.

Mariner 10 Spacecraft

Mariner 10 was sent on an ambitious mission to gather data and photograph Venus during a flyby and then travel on to the unexplored planet Mercury. The mission was successful. Mariner 10 took the first photographs of clouds of Venus in ultraviolet light during a flyby. The ultraviolet images showed good detail of the clouds. Mariner 10 then made three flybys of Mercury. Thousands of high quality photographs of the surface of Mercury were transmitted back to Earth during the flybys along with data on the magnetic field and other properties of the planet.

An artist's conception of Mariner 10 in deep space is shown on the next page (Fig. 7.1).

Background of Mariner 10

The Mariner 10 mission took advantage of studies of gravity assisted spacecraft trajectories in the 1960's. Those studies showed that a trajectory could be designed to take advantage of gravity assist from Venus to send a spacecraft on a path to

T. Lund, *Spacecraft that Explored the Inner Planets Venus and Mercury*, Springer Praxis Books, https://doi.org/10.1007/978-3-031-29838-7_7

Fig. 7.1 Mariner 10 spacecraft. (NASA image collection)

intercept Mercury in its orbit. Flybys of both Venus and Mercury could be achieved and the spacecraft would orbit the sun and return for repeated flybys of Mercury.

The mission to Mercury was approved by NASA in 1969 and work went forward to take advantage of a launch opportunity in November 1973. The Jet Propulsion Laboratory (JPL) was assigned responsibility for the mission and a Mariner-Venus-Mercury project office was set up at JPL in January 1970. A contract was awarded to the Boeing Company in July 1971 to build a test model and a flight model of the spacecraft. The test model was eventually updated to a full flight model and served as backup to Mariner 10.

The spacecraft design borrowed heavily from flight-proven hardware from previous Mariner spacecraft. The most recent of those was Mariner 9, which was the first spacecraft to orbit Mars. Its capable cameras made a nearly complete photography map of the surface of Mars in 1972.

Mariner 10 was launched from Cape Canaveral on 3 November 1973. A flyby of Venus occurred on 5 February 1974 and the first flyby of Mercury occurred on 29 March 1974. A second flyby of Mercury occurred on 21 September 1974 and a third flyby occurred on 16 March 1975. Photographs of the surface of Mercury and scientific sensor measurements were made during each flyby.

Mechanical Configuration of Mariner 10

A drawing of the spacecraft, labeled to show locations of major elements, is shown on the next page (Fig. 7.2). The spacecraft coordinate system sited the Z-axis along the longitudinal axis of the spacecraft with +Z in the direction of the rocket motor exhaust. The +X-axis was along the centerline of the right-hand solar array in the figure. The –Y-axis was outward along the boom that held the magnetometers.

The spacecraft was built around a hexagonal magnesium frame similar to that used for previous Mariner spacecraft. The frame was 139 cm across the diagonals measured from the corners and 127 cm across the flat surfaces. It was 45.7 cm high. Each of the eight sides of the frame contained an electronics bay about 11.7 cm deep. Five of the outer surfaces of the octagonal sides of the spacecraft contained adjustable louvers for temperature control.

The spacecraft normally cruised to the planets with the sun screen, labeled in the drawing (Fig. 7.2), facing the sun. The sunscreen shaded the spacecraft to assist in thermal control. A fuel tank for the propulsion rocket motor was mounted in the center of the hexagonal frame and the rocket motor was mounted below the tank with its exit protruding through the sunscreen. The spacecraft was oriented in three axes by a series of orthogonal nitrogen gas jets.

The prominent boom on the right side of the drawing in Fig. 7.2 was 6.1 meters long. The boom held two magnetometers, one near the end and the other about mid-span. A low-gain omni-directional antenna was mounted at the end of another boom 2.85 meters long on the upper side of the spacecraft in the drawing. A steerable high-gain parabolic antenna 1.37 meters in diameter was mounted to an articulated boom mounted on the lower side of the spacecraft in the drawing. A scan platform that could be steered in the vertical and horizontal directions was mounted on the

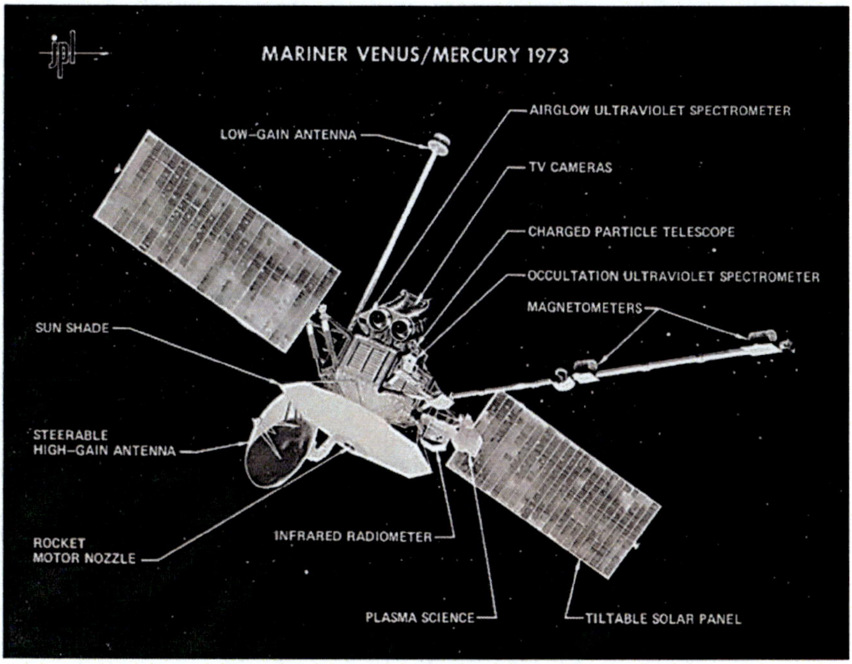

Fig. 7.2 Labeled drawing of Mariner 10. (NASA/JPL graphic)

upper side of the frame. Two television cameras and an ultraviolet airglow sensor were mounted to the scan platform.

Two solar arrays provided electrical power for the spacecraft. The arrays could be independently rotated around their long axis. Each array was 2.69 meters long and 0.97 meters wide. The total span of the spacecraft across the solar arrays was 7.6 meters.

The weight of the spacecraft at launch was 504 kg including 29 kg of hydrazine propellant for the rocket motor. In addition, the adapter to mate the spacecraft to the launch vehicle weighed 30 kg. Experiments made up 78 kg of the total spacecraft weight.

Mariner 10 Systems

Trajectory Correction Rocket Motor

The rocket motor, which fired through a hole in the center of the sunscreen, developed 222 newtons (50 pounds) of thrust. The specific impulse was 233 seconds. The motor was powered by hydrazine which is a monopropellant liquid that spontaneously combusts when in contact with an oxidizer such as nitrogen tetroxide. A small amount of oxidizer was sprayed into the combustion chamber along with the hydrazine to cause ignition. Once ignited, the hydrazine continued to burn in the presence of a catalyst. The catalyst consisted of a bed of iridium in the combustion chamber.

Hydrazine fuel was contained within a bladder in a spherical tank that held 29 kg of fuel. Nitrogen from a spherical tank was applied through a shutoff valve and then ported between the inner walls of the tank and bladder to force the fuel into the combustion chamber. Valves in the fuel line and in the oxidizer line were opened simultaneously for ignition by signals from the central computer and sequencer (CC&S).

When the rocket motor was firing, spacecraft attitude was maintained by deflecting four small vanes in the exhaust of the motor. These vanes had the ability to deflect the thrust vector by ±5 degrees. The vanes were deflected by signals from the autopilot to maintain a constant spacecraft attitude.

The motor could be shut down by stopping the flow of hydrazine and then restarted several times. The propulsion system was capable of changing vehicle velocity by a total of 119 meters per second. The motor was fired two times for trajectory correction enroute to Venus. The first firing, for 19.9 seconds, resulted in a speed change of 7.8 meters per second. The second firing for 3.8 seconds resulted in a speed change of 1.3 meters per second. It was fired a third time before reaching Mercury to adjust the Mercury encounter distance. That firing was for 51 seconds and resulted in a speed change of 17.8 meters per second. After the first encounter with Mercury, the motor was fired six additional times to adjust trajectory as the spacecraft orbited the sun and made two more close approaches to Mercury.

Attitude Control Subsystem

The attitude of the spacecraft was adjusted by means of nitrogen gas jets managed by the attitude control subsystem. A set of jets was located on one tip of the solar panels to provide roll and yaw control. Another set of jets was mounted on the spacecraft frame near the Y axis to provide pitch control. Pairs of jets were used to induce rotation about a given axis of the spacecraft. The nitrogen tank held a total of 2.7 kg of gas.

The spacecraft was oriented during cruise so that the +Z axis of the spacecraft was directed towards the sun and the Canopus sensor was pointed towards the star Canopus. Pointing the +Z axis towards the sun oriented the solar arrays towards the sun.

Steering signals used to point the +Z axis of the spacecraft towards the sun were provided by sun sensors. A pair of sun sensors referred to as acquisition sun sensors were located on the tips of the solar arrays and a sun sensor referred to as the cruise sun sensor was located on the sun-facing side of the spacecraft outside of the sunshade. The acquisition sun sensors in total provided complete angular coverage about the spacecraft and they were used during the initial sun acquisition process to orient the +Z axis of the spacecraft towards the sun. Steering error signals from the Canopus sensor were used to rotate the spacecraft about the longitudinal or Z axis to maintain track of Canopus in the roll plane.

Articulation and Pointing Subsystem

The articulation and pointing subsystem controlled the pointing angles of the high-gain antenna, the tilt of individual solar arrays, and the cone and clock angles of the scan platform. Two cameras and an ultraviolet spectrometer were mounted to the scan platform. A drawing showing the configuration of the moveable elements and location of actuators is shown on the next page (Fig. 7.3).

The scan platform holding the cameras and ultraviolet spectrometer could be rotated around the cone axis (Z-axis of the spacecraft) from 25 to 280 degrees and around the clock axis (perpendicular to the cone axis) from 58 to 180 degrees. Cone angle was measured from the +Z axis and clock angle was measured from the line of sight of the Canopus tracker. To protect the camera, the platform was restricted from pointing the camera any closer than 53 degrees to the sun. Each of the solar arrays could be tilted around the long axis up to 76 degrees from Z-axis. A zero-degree tilt would position the solar arrays directly at the sun when the Z-axis was aligned with the sun. The solar arrays could be tilted independently. This capability was used later in the mission to orient the spacecraft by means of the slight force of the solar wind.

The high-gain antenna was attached to a deployable boom shown in Fig. 7.3. The antenna could be pointed by exercising two separate actuators referred to as the boom and dish actuators. The boom actuator was located at the end of the boom and it rotated the antenna about the axis of the boom. The dish actuator rotated the

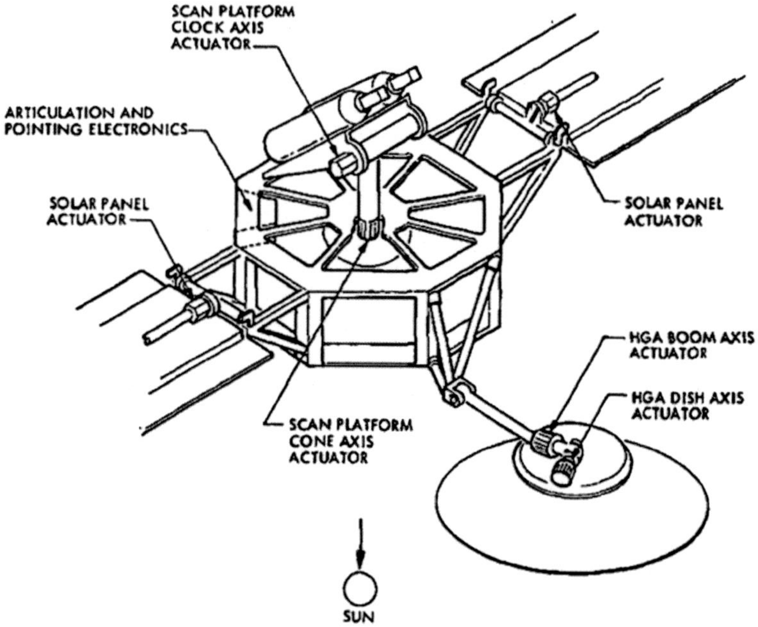

Fig. 7.3 Moveable elements and location of actuators. (NASA graphic)

antenna about an axis perpendicular to the axis of the boom. The boom actuator allowed rotation from zero degrees to 255 degrees. The dish actuator allowed rotation from −4.5 to +184.5 degrees. The actuators could be operated in either a position mode or an incremental mode.

Spacecraft Power

Electrical power for the spacecraft was provided by a battery during launch and early portion of flight and by the solar arrays during the long cruise to the planets and during most of the encounter time at the planets. The battery was a nickel-cadmium type with capacity of 20 ampere-hours at 39 volts (780 watthours).

Mariner 10 carried two solar array panels each 2.7 meters long and 0.97 meters wide. The two panels contained a total of 5.1 square meters of solar cells. The total peak output power from the solar arrays was 500 watts. The spacecraft required between 340 and 490 watts of power depending on the activity at the time. The solar arrays were deployed shortly after the spacecraft had been inserted into a trajectory towards Venus. Acquisition of the sun by the sun sensors was achieved soon after. The attitude of the spacecraft was oriented by the attitude control system to direct the +Z-axis of the spacecraft at the sun and that pointed the solar arrays at the sun.

The solar intensity was a factor of 6.7 higher at Mercury than at Earth and temperature control of the spacecraft was an important consideration. The solar arrays panels were designed to pivot about their long axis to allow them to be tilted to reduce the heat load on the panels as the spacecraft flew nearer to the sun.

The output voltages of the solar arrays were limited to 51 volts by shunt Zener diodes. The electrical system seamlessly switched to the battery or solar array power source having the highest voltage by diodes in series with each source. Normally, the solar array had the highest voltage and it was used to power the spacecraft. The output of the power source selected was applied to a boost regulator that stepped the voltage up to a closely regulated 56 volts dc. A second boost regulator, referred to as the standby boost regulator, provided redundancy.

The output of the boost regulator was applied to three inverters. The largest inverter converted the input voltage to a 2400 Hz square wave at a voltage of 50 volts rms. This was the main source of power for spacecraft systems and experiments. The switching frequency was established by a synchronizing signal from the central computer and sequencer. Each of the main electronic chassis in the spacecraft contained a transformer and rectifier to condition the 2400 Hz square wave power to various dc power forms required by its electronic circuits. Another inverter converted the input voltage to a 400 Hz sinusoidal power form at 28 volts rms. A third inverter converted the input voltage to three-phase 400 Hz sinusoid at 27 volts rms that powered the gyros.

The battery supplemented power from the solar arrays during midcourse maneuvers when the solar arrays were not pointed at the sun and when the spacecraft was in the shadow of a planet. A battery charger, powered by the solar arrays, charged the battery after each usage.

Central Computer and Sequencer

Major spacecraft functions were controlled and sequenced by a central computer and sequencer (CC&S). The CC&S functioned as a timer/sequencer that initiated a series of hard-wired commands to control the spacecraft during the mission. Parameters for the several midcourse correction events were computed on the ground transmitted up to the spacecraft, and loaded into the CC&S memory for execution later by the CC&S. The commands from the ground pertaining to midcourse correction included: pitch turn duration and polarity, roll turn duration and polarity, motor burn duration, and initiate maneuver sequence.

The computer portion of the CC&S included a 512-word memory with 22-bit words. The computer was programmable from the ground to control the various phases of the mission and it controlled the experiments. The sequencer portion of the CC&S contained a fixed set of sequences to autonomously control the spacecraft should commands from the ground not be available. The sequencer carried a detailed program for the next mission phase and a basic program to complete all phases of the mission if communication from Earth were lost. The detailed program was loaded in stages to fit within the available memory. During the initial phase of observing the Earth and Moon, the Venus encounter sequence was uploaded. During the Venus phase the Mercury encounter sequence was uploaded, and so on.

Flight Data Subsystem

The Flight Data System (FDS) was a digital computer-based telemetry system. It received and managed engineering data from various elements of the spacecraft along with outputs from scientific instruments. The FDS conditioned the various inputs and assembled them into serial data streams that were applied to modulators in the telecommunications system for transfer to Earth. The data streams could be sent directly to Earth or to a data storage subsystem for storage on a digital tape recorder for later transmission to Earth. The FDS could generate several different telemetry formats.

The flight data subsystem was usually controlled by commands from the central computer and sequencer. It could also be controlled by commands sent from Earth. Major functions within the FDS were: timing and control, memory, memory input/output, and management of engineering data, imaging data, nonimaging data, and data output.

Telecommunications System

The telecommunications link received commands and data from Earth and it transmitted telemetry data from experiments and spacecraft systems to Earth. Uplink to the spacecraft from Earth operated in S-Band at a frequency of 2113 MHz. The main downlink transmitter in the spacecraft operated in S-band at a frequency of 2295 MHz. The transmitter output power was 20 watts. The downlink signal was modulated by telemetry data and by range modulation extracted by the receiver.

A transmitter operating in X-band at 8415 MHz was carried by the spacecraft to allow dual radio frequency probing of the atmosphere of the planets during times just before and just after of occultation. The power output of the X-band transmitter was 200 milliwatts. The frequencies of the S-band and X-band transmitters were maintained at a precise ratio of 3/11.

The uplink S-Band signal from Earth contained both commands and range modulation. The signal was received by the omnidirectional antenna throughout the mission and that allowed communication during maneuvering of the spacecraft. The S-Band receiver in the spacecraft extracted command data and range modulation. It also generated a phase coherent signal to drive the two transmitters. The range modulation, extracted by the receiver, was applied to both transmitters so that two-way range could be measured on Earth.

S-Band and X-Band signals were transmitted to Earth by means of a dual-feed parabolic antenna. The antenna, which was 1.37 meters in diameter, had a beamwidth of 0.9 degrees at the X-Band transmitter frequency and about 4 degrees at the S-Band frequency. The output of the S-Band transmitter was switched to the low-gain omnidirectional antenna while the spacecraft was maneuvering.

Image data could be transmitted to Earth in real time at a rate of 117.6 kilobits per second. That rate corresponded to one image being sent every frame time of the camera of 42 seconds. Image data could also be played back out of the tape recorder at a rate of 22.05 kilobits per second.

Command Decoder

Command signals from Earth were in the form of biphase modulation of the uplink carrier. The receiver demodulated the biphase modulation and the resulting command signals were applied to a command decoder. The command decoder decoded the command messages and based on their address sorted them into commands that were to be applied to particular subsystems on the spacecraft. Those "direct commands" were prefixed by "DC." There was a total of 86 such commands available although not all were used. A few of these commands are given in the table below (Table 7.1) as examples.

Science Experiments Carried by Mariner 10

The Mariner 10 spacecraft carried eight science experiments: television cameras, infrared radiometer, plasma analyzer, magnetometer, charged particles telescope, ultraviolet airglow spectrometer, ultraviolet occultation spectrometer, and X-Band /S-Band radio occultation.

Table 7.1 Examples of commands from Earth for Mariner 10

Designation	Command
DC-1	Select redundant memory states in flight data subsystem
DC-10	Select low-gain antenna for transmission
DC-11	Select high-gain antenna for transmission
DC-17	Advance cone angle of Canopus tracker by one step
DC-27	Initiate CC&S maneuver sequence
DC-42	Switch TWT power amplifier to high power mode
DC-43	Switch TWT power amplifier to low power mode
DC-48	Turn OFF TV camera subsystem
DC78	Turn ON TV subsystem
DC-54	Turn ON power to S-band TWT power amplifier
DC-55	Turn OFF power to S-band TWT power amplifier
DC-88	Provide capability to change CC&S computer program

Television Cameras

Mariner 10 carried two identical television cameras. Two cameras allowed one camera to gather an image while the image was being read out of the other camera. The two-camera arrangement also provided redundancy in case one camera failed. Electronic images were formed by applying the scene image to a 2.54 cm diameter vidicon tube. The frame time per picture was 42 seconds. A photograph of the two cameras outside of the spacecraft is shown below (Fig. 7.4).

The cameras were built with a folded optics system (cassegrain) that resulted in a shorter physical length. The main telescope of the camera had a focal length of 1500 mm and f number of f/8.4. The field of view was 0.36 by 0.48 degrees. A wide-angle optics assembly was attached outside of the main telescope. It is the appendage shown at top rear of the cameras in the photograph. The wide-angle image was applied to the vidicon tube by a mirror that occupied one position of the filter wheel in each camera. The wide-angle optics had a focal length of 62 mm at f/8.5. The field of view of the wide-angle mode was 11 by 14 degrees.

A filter wheel with eight positions was located just ahead of the vidicon tube. Most of the positions inserted filters into the optical path. The eight positions were: blue bandpass, UV polarizing, minus UV high-pass, clear, UV bandpass, defocusing lens, yellow bandpass, and a mirror to relay the wide-angle image onto the vidicon. The UV filters were used to obtain images of the cloud structure of Venus.

The television image consisted of 700 scan lines with 832 image elements per line giving a format of 700 by 832 pixels. The frame time was 42 seconds. The analog output of the camera was encoded into 8-bit words.

The readout rate of the image was 117,600 bits per second. The output from the cameras was a series of continuous frames of 42 second duration.

Options for readout of image data included: full frame in real-time at 117,600 bits per second, ¼ frame in real-time at 22,050 bits per second, and full frame with 75% of pixels deleted in real-time at 22,050 bits per second. The amplitude of the pixels

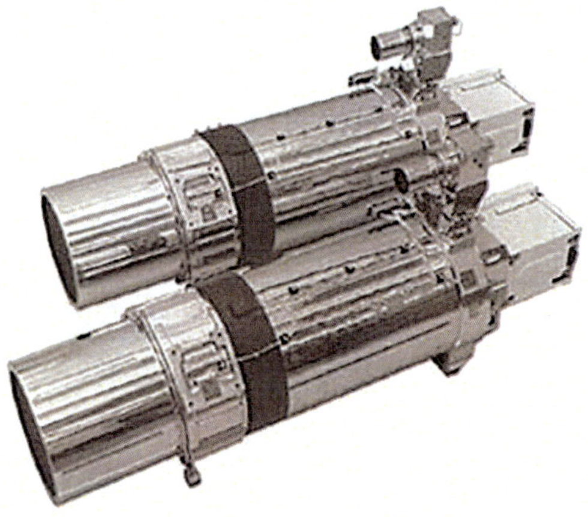

Fig. 7.4 Cameras for Mariner 10. (NASA photograph collection)

was encoded at 6-bits for the lower frame rate format. The full frame could be recorded on a tape recorder at 117,600 bits per second for later playback at 22.05 kilobits per second.

Over 7000 pictures were taken by the camera of Venus and Mercury and transmitted to Earth during the mission. A photograph of the clouds of Venus taken at ultraviolet wavelengths is shown below (Fig. 7.5). The picture was color enhanced by NASA to bring out detail.

Mariner 10 made three photographic passes over Mercury separated by about six months. A picture of the surface of Mercury taken at a range of 12,121 km on the first passage across the planet is shown on the next page (Fig. 7.6). The center of the photograph is at latitude of −8.0 degrees and longitude of 171.8 degrees on the planet. The distances on the surface of the picture are 96 km horizontal and 74 km vertical. The resolution of the photograph is 269 meters. The exposure time was 19.5 milliseconds.

Infrared Radiometer

The infrared radiometer was used to measure temperatures of cloud tops of Venus and temperatures of the surface of Mercury. Principles of operation of radiometers were covered in Chap. 4 in the section "Experiments for Mariner 2".

The infrared radiometer on Mariner 10 operated in two wavelength bands established by filters: 34–54 microns and 7.5–14 microns. The shorter wavelength band

Fig. 7.5 Color enhanced photograph of clouds of Venus taken by Mariner 10. (NASA photograph)

Fig. 7.6 Photograph of surface of Mercury taken on the first flyby of Mariner 10. (NASA photograph)

was sensitive to temperatures in the range of 200–700 K and the longer wavelength band was sensitive to temperatures in the range of 80–300 K.

Separate cassegrain type telescopes provided a field of view of about 0.5 degrees for each wavelength band. A scan mirror alternately directed the field of view to angles 120 degrees apart. When the spacecraft passed over a planet the field of view would alternately view the planet and deep space. The deep space look gave a cold reference for the radiometer.

A temperature profile of the clouds of Venus indicated temperatures of 220–250 K (−53 °C to −23 °C). Measured temperatures of the surface of Mercury were 460 K (187 °C) at mid-afternoon and 100 K (−173 °C) near midnight on the planet. Mercury was nearly the farthest distance from the sun in its orbit at the time of the temperature measurements.

Ultraviolet Airglow Spectrometer

The ultraviolet airglow spectrometer detected and measured emission lines from certain elements in the atmosphere of Venus and Mercury. The emission came from resonance scattering of solar radiation from a particular element. The word "airglow" was coined during many years of investigation of sources of light in the night sky of Earth. The ultraviolet spectrometer was arranged to detect lines from several elements.

The instrument analyzed emissions in the 304–1659 angstrom (Å) wavelength band. A diffraction grating was used to identify emission as a function of wavelength. Detectors were placed in the focal points of 10 selected wavelengths diffracted by the grating. The amplitudes at the outputs of the detectors were proportional to the abundance of particular elements. The effective bandpass of each detector was 20 angstroms. The wavelengths of emission from a few elements include: ionized helium (He+) at 304 Å, helium at 584 Å, hydrogen at 1216 Å, oxygen at 1304 Å and Carbon at 1657 Å.

The instrument was mounted on the same scan platform as the cameras to allow scanning in the clock and cone angles of the spacecraft. Cone angle was measured from the +Z axis and clock angle was measured from the line of sight of the Canopus tracker. The field of view of the instrument was 3.6 degrees in the clock direction and 0.125 degrees in the cone direction.

The comet Kohoutek was visible to Mariner 10 in January 1974 while enroute to Venus. Scans of the comet by the ultraviolet spectrometer were made over a two-week period. Strong emissions at hydrogen Lyman α at 1216 Å were found along with a small amount of carbon. The scans revealed the comet had a large hydrogen corona with diameter of about 20 million km.

Scanning of the clouds of Venus by the ultraviolet spectrometer indicated a substantial amount of hydrogen and oxygen and trace amounts of helium and atomic carbon. Spectrometer measurements of the atmosphere of Mercury indicated a very tenuous atmosphere with a very small amount of helium the most prevalent element.

Ultraviolet Occultation Spectrometer

The ultraviolet occultation spectrometer viewed the ultraviolet flux from the sun at select wavelengths through the atmosphere of Venus and Mercury just before and just after occultation of the sun by the planets. Four spectral bands 150 angstroms wide centered at 485, 740, 810, and 895 angstroms were examined. Atmospheric gasses in that spectral range had relatively large absorption cross-sections and that allowed measurement of their abundances.

The occultation experiment of Mercury was performed as Mariner 10 crossed behind the nightside of the planet and again just before it crossed into the dayside. The results of the experiment indicated that the total gas content in the atmosphere was extremely low. The pressure was estimated to be 2×10^{-9} millibars at the surface. Helium was found to be the primary constituent at a partial pressure of 5×10^{-12} millibars.

Magnetometer

The magnetometer measured magnetic fields in interplanetary space and in the vicinity of Venus and Mercury. The experiment used two triaxial magnetometers mounted on a long boom. One magnetometer was located 2.3 meters from the spacecraft and the other was located at 5.3 meters. The boom, with magnetometers mounted, is the dominant object on the right side of the drawing of the spacecraft given in Fig. 7.2. The use of two spatially separated magnetometers allowed separating out the magnetic field from the spacecraft from the ambient magnetic field.

An operating range of either ±16 nanotesla (nT) or ±128 nT could be selected for the measurements. In addition, bias offsets could be introduced that effectively increased the operating range to ±3188 nT. Tesla is the SI unit for magnetic field strength. For reference, the magnetic field of Earth ranges from about 25 to 65 microtesla (μT).

A maximum magnetic field of about 20 nT was observed during passage of the spacecraft by Venus although investigators did not think that the field was planet based. The magnetometers detected clear evidence of a magnetic field associated with Mercury. A field strength of 100 nT was measured at the time of closest approach on the first encounter with Mercury. The closest approach distance was 703 km and the latitude was near the equator of the planet. During the third encounter with Mercury, a maximum magnetic field of 400 nT was measured at the closest approach distance of 327 km. The latitude of the spacecraft at that time was 70 degrees. A magnetic field model for the planet was generated from magnetic field measurements.

Plasma Science Experiment

The plasma science experiment was planned to gather information on the solar wind and its interaction with Venus and Mercury. The experiment included two different instruments: a Scanning Electrostatic Analyzer and a Scanning Electron Spectrometer. The Scanning Electrostatic Analyzer was designed to measure energies and arrival directions of positively charged particles and negative electrons from the general direction of the sun. The Scanning Electron Spectrometer was designed to count electrons from a direction opposite from the sun. The thermal velocity of electrons was much larger than the bulk velocity of particles flowing from the sun and that allowed measurement of energy of electrons from the anti-sun direction.

The two instruments were mounted on a scanning assembly attached to a short boom affixed to the octagonal frame of the spacecraft. The instrument could be scanned in the ecliptic plane through an angle of 120 degrees centered six degrees from the sunline. The scan rate was one degree per second.

Scanning Electron Analyzer – The Scanning Electrostatic Analyzer had separate channels to measure energies and arrival directions of positively charged particles and of electrons. The positive particle channel could accommodate positively charged particles in the energy range of 0.08–8.0 keV. One keV is 1000 electron volts. The negative channel measured electrons with energies from 4.0 to 400 electron volts. The analyzer used three hemispheric shells nested together. The center shell was maintained at zero potential and the outer shell was connected to a variable positive voltage and the inner shell was connected to a variable negative voltage.

The curved space between the center shell and outer shell acted as a gauntlet for positive particles and allowed measurement of their energies. A specific voltage potential between the shells caused a positive particle with specific energy to curve and follow the curve of the plates. Those particles were captured at the output of the curved plates by a charge collector. Particles with energies less than or greater than the specific amount struck the curved deflection shells and were not counted. The positive voltage applied to the outer shell was changed in steps to cover a range of particle energies from 0.08 keV to 8.0 keV. An angular distribution of charge collectors at the output of the shells allowed determination of direction of arrival.

The inner shell was connected to a varying negative voltage source. Just as in the case of positive particle region, the curved space between the center shell and inner shell acted as a gauntlet for electrons and allowed measurement of their energies. The negative voltage applied to the shell was changed in steps to cover a range of electron energies from 4.0 to 400 electron volts.

Unfortunately, the cover over the entrance to the Scanning Electrostatic Analyzer did not open during the mission and so no information was obtained from that instrument.

Scanning Electron Spectrometer – The Scanning Electron Spectrometer was also a hemispherical type with a stepped electric field to select electrons at different energy levels. The spectrometer had a field of view of 7.0 by 17 degrees. The instrument was mounted back-to-back with the scanning electrostatic analyzer on the scan platform. The platform was scanned in the ecliptic plane through an angle of 120 degrees centered six degrees from the sunline. The scan rate was one degree per second. The narrow field of view was along the scan arc.

The instrument developed a spectrum of electron energies each six seconds. The spectrum of electron flux was generated by 15 logarithmically spaced channels within the energy range 13.4–690 electron volts.

The scanning electron spectrometer was not affected by cover that remained in place over the entrance to the scanning electrostatic analyzer. The instrument worked as intended and it gave a measure of density and energy of electrons in the anti-sun direction.

Charged Particles Telescope

The charged particles telescope experiment was included on Mariner 10 to measure properties of electrons, protons, and alpha particles in interplanetary space and in the vicinity of Venus and Mercury.

The instrument contained two telescopes with associated detectors. The field of view of the telescopes was directed about 45 degrees from the sun line. The telescopes were referred to as the low-energy telescope and the main telescope. The low-energy telescope had a field of view of 38 degrees half angle and it had two detectors. One detector responded to protons with energies from 0.53 to 1.9 MeV. One MeV is a million electron volts. The other detector measured protons in the range of 1.9–8.9 MeV. The low-energy telescope did not respond to electrons.

The main telescope had six co-linear sensors. Five sensors were silicon detectors and the sixth was a Cesium Iodide (CsI) scintillator. The first sensor responded to alpha particles and protons in the range 0.65–10.3 MeV per nucleon and electrons with energies above 170 keV. The field of view for this mode was 47 degrees half angle.

Measurements of electrons and protons in the vicinity of Venus indicated no increase in energy or spectral changes in the vicinity of the planet. This confirmed other observations that Venus does not have a magnetosphere. Proton energies of about 550 keV and electron energies of about 300 keV were measured in the magnetosphere of Mercury.

S-Band and X-Band Radio Occultation

The S-Band and X-Band radio occultation experiment was designed to measure electron density in the atmospheres of Venus and Mercury by measuring propagation properties of their atmospheres. The signal strength and phase and group velocity changes of range code modulated S-Band and X-Band signals were measured as the spacecraft passed into and out of occultation by the planet. The carrier phase was advanced by electron content and the modulation, which travels at group velocity, was retarded by electron content.

An S-Band signal with range modulation was transmitted from Earth and received by the omni directional antenna of the spacecraft. The receiver extracted the range modulation and generated a phase coherent signal to drive the carriers of the S-Band and X-Band transmitters. The S-band and X-Band signals were transmitted by the high-gain antenna of the spacecraft and received on Earth during occultation by the planets. The modulation and Doppler shifts of the two signals received on Earth were processed and the differences were used to develop electron density information.

The experiment indicated that Venus had a distinct layered ionosphere. The night-side of Venus had two sharp ionization layers. One layer with peak electron density of 9000 electrons per cubic centimeter (electrons/cm^3) was at an altitude of 142 km and the other layer with peak electron density of 7000 electrons/cm^3 was at 124 km altitude. The day-side ionosphere was found to have a peak density of 2.9×10^5 electrons/cm^3 at 142 km altitude. The density decreased to about 1000 electrons/cm^3 at 340 km altitude. No evidence of an ionosphere was found around Mercury.

Flight of Mariner 10

Mariner 10 was launched from Cape Kennedy at 05:45 GMT on 3 November 1973 by an Atlas SLV-3D–Centaur D-1A launch vehicle. Atlas SLV-3D used a MA-5 propulsion system that provided thrust of 1900 kilonewtons (427,120 pounds) at sea level. Centaur was an advanced upper stage that burned liquid hydrogen with liquid oxygen oxidizer. It generated thrust of 133 kilonewtons (29,900 pounds).

Atlas performed well and lofted the spacecraft into position for Centaur to place it in orbit around Earth. Atlas was separated from Centaur and Atlas dropped away. Centaur was then fired for 5.1 minutes to place the spacecraft in a parking orbit 188 km above the Earth. After about a third of the way around Earth, Centaur was fired again for 2.25 minutes. That burn accelerated the spacecraft to a speed of about 41,000 km/h and inserted it in a trajectory to intercept Venus. Centaur firing was controlled from the ground using accurate trajectory information derived from tracking stations that were part of the Deep Space Net (DSN). Mariner 10 was separated from Centaur at 06:24 GMT.

Several items that had been folded to fit within the launch shroud including solar panels, antennas, and booms, began to be deployed about four minutes after separation from Centaur. The −X and +X solar panels deployed at 06:29:59 and 06:31:23 respectively. The low-gain antenna deployed at 06:29:59 and the high-gain antenna deployed at 06:32:47. The magnetometer boom deployed at 06:38:43.

Acquisition sun sensors on the tips of the solar arrays were used as signals to drive the attitude control system to orient the spacecraft so that its +Z axis was close to the sunline. The cruise sun sensor then provided signals to point the +Z axis at the sun. Once oriented at the sun, the spacecraft was rolled about the Z axis until the Canopus sensor detected the star Canopus. The star Vega was detected first and lock was obtained. Later, the roll angle was changed to lock to Canopus. Thus stabilized with the sun sensor locked to the sun and the Canopus sensor locked to Canopus, the spacecraft began the long cruise to Venus.

During checkout of systems, it was noted that electrical heaters for the cameras would not turn on. Since the cameras were thermally isolated from the rest of the spacecraft, there was concern that the cameras would become too cold to operate. The operators decided to turn on the cameras and leave them on to maintain the cameras in an operating temperature range. Several pictures taken of Earth and the moon the day after launch confirmed good operation of the cameras. Fortunately, on 17 January 1974 during cycling of other heaters, the heaters for the cameras turned on and continued to operate normally.

A Trajectory Correction Maneuver (TCM) was made ten days after launch to adjust the trajectory so that the spacecraft would pass close to the target point near Venus to be deflected towards Mercury. Maneuvering parameters were uplinked from the ground and stored in the CC&S to orient the spacecraft so thrust of the rocket motor would be in the direction to bring about the necessary correction. The turn rate induced about the pitch and roll axes by the attitude control jets was accurately known. That allowed the jets to be operated for a given time to achieve a given amount of rotation. Parameters sent up from the ground and stored in the CC&S for the TCM included: pitch turn duration and polarity, roll turn duration and polarity, motor burn duration, and initiate maneuver sequence.

The gyros in the spacecraft were turned on and allowed to stabilize to maintain an attitude reference of the spacecraft before maneuvering. The first TCM was then initiated by the CC&S by causing roll of the spacecraft of 49 degrees followed by a pitch of 127 degrees. The rocket motor was commanded to burn for 19.9 seconds to cause a velocity change of 7.84 meters per second. Following the TCM, the spacecraft was maneuvered to its attitude before the TCM began. Lock to the sun and Canopus were regained and the spacecraft continued cruise to Venus. The spacecraft would have passed behind Venus and about 67,000 km from the target point without the midcourse correction. After the correction, Mariner 10 was projected to pass within 1500 km of the target point.

The approach to Venus needed to be quite precise to cause the spacecraft to be deflected on the proper trajectory towards Mercury. The aim point was a 400 km diameter circle about 16,900 km to the right and in front of Venus. Gravity of Venus would then bend the trajectory to that planned. A second trajectory correction maneuver, TCM-2, was made 15 days before encounter with Venus on 21 January 1974. The correction involved a roll angle of 48 degrees, pitch angle of 35.2 degrees and burn time of 3.7 seconds. A velocity increment of 1.38 meters per second was achieved. As a result of TCM-2, the spacecraft passed within 10 km of the target point.

On 8 January 1974 an anomaly occurred when the boost regulator in the spacecraft power system abruptly switched from the main boost regulator to the redundant backup boost regulator. The backup system carried the spacecraft through to the end of the mission.

Mariner 10 approached Venus from the dark side and passed within 5768 km of the planet on 5 February 1974. The spacecraft exercised its experiments and cameras as it approached Venus and as it curved around the planet and exited on the sunlit side. A total of 4165 photographs were taken of Venus in visible and ultraviolet wavelengths. Ultraviolet photographs revealed pronounced structure in the clouds and the fact that the tops of clouds rotate around Venus with a period of four days.

The planned gravity assist at Venus was successful and the spacecraft sped on towards Mercury. A problem was encountered on 18 February when the Canopus tracker apparently locked on to a bright particle of thermal paint that had flaked off of the spacecraft. This problem occurred several times during the mission but each time relock to Canopus was achieved, sometimes with assistance of commands from Earth.

A third midcourse correction was made on 16 March 1974, 13 days before encounter with Mercury. The correction was made with the spacecraft in cruise orientation with the +Z axis pointed at the sun. The reason for not maneuvering the spacecraft prior to the burn was that an oscillation in roll had been observed on a previous midcourse correction and that had used up a considerable amount of nitrogen. The trajectory correction maneuver was delayed until the geometry of the trajectory allowed the required correction to be made with the Z axis of the spacecraft pointed at the sun. The correction involved burn of the rocket engine for 51.2 seconds. The resulting velocity increment was 17.8 meters per second.

The spacecraft approached Mercury from the sunlit side and then curved around into the night side before leaving the planet. A NASA graphic illustrating the approach to Mercury is shown below (Fig. 7.7). Mariner 10 made its closest approach, 703 km above the surface, on 29 March 1974.

Photography of the surface of Mercury began while the spacecraft was 5.3 million km out and continued until Earth was obscured by Mercury near the altitude of closest approach. Photographs continued to be taken as the spacecraft traveled past the planet. A Mosaic of photographs taken as the spacecraft approached Mercury is shown on the next page (Fig. 7.8). The image was composed of 18 photographs taken at 42 second intervals from a distance of about 200,000 km.

Over 2000 pictures were transmitted to Earth during the first encounter with Mercury. Most were transferred in real-time. During a brief period where Mercury blocked the direct view to Earth, image data was recorded on the tape recorder for later playback. The transfer of real-time image data at a rate of 117,600 bits per second from the vicinity of Mercury to Earth was a significant telecommunications achievement.

Measurements of the atmosphere of Mercury by the ultraviolet spectrometer experiment indicated that the atmosphere was very tenuous with surface pressure less than 10^{-10} millibars. The magnetometers detected an intrinsic magnetic field of about 100 nT. The infrared radiometer experiment indicated temperature of the surface to be about 460 K (187 °C) on the sunit side and 150 K (−123 °C) on the dark side of Mercury.

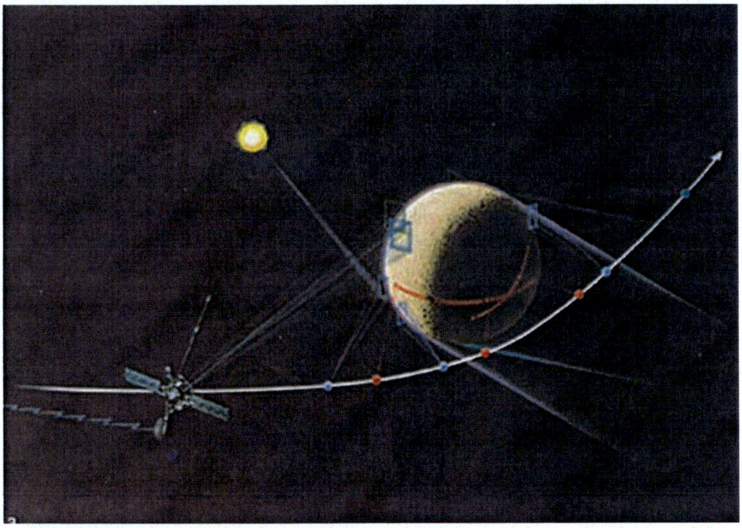

Fig. 7.7 Mariner 10's approach to Mercury. (NASA graphic)

Fig. 7.8 Mosaic of photographs of Mercury as Mariner 10 approached the planet. (NASA graphic)

The trajectory of Mariner 10 was such that after swinging around Mercury it continued in an orbit of the sun. The period of the orbit was a factor of two longer than the 88-day period of the orbit of Mercury around the sun. As a result, a second encounter with Mercury occurred 176 days after the first encounter. A fourth trajectory correction maneuver (TCM-4) was required to reduce the miss distance at the second encounter to the desired value. TCM-4 was broken up into two firings to allow the spacecraft to cool between firings. The first firing for 195 seconds occurred on 9 May 1974 and caused a velocity change of 50 meters per second. The second firing for 199 seconds on 10 May caused a velocity change of 27.6 meters per second. The predicted encounter distance after TCM-4 was 34,00 km over the sunlit side and that would be favorable for photography.

Planning was undertaken for a third encounter. The miss distance during the third encounter was computed to be 400,000 km withoug further trajectory correction. A fifth trajectory correction maneuver (TCM-5) was conducted on 2 July 1974 to reduce that miss distance.

The tape recorder failed in August 1974 and that required camera data to be transmitted to Earth in real time. The Goldstone receiving station was modified to array two 26-meter diameter antennas and the 64-meter antenna together to improve the signal-to-noise ratio of the telemetry data.

The second encounter took place on 21 September 1974. The path of Mariner 10 took it over the sunlit side of Mercury. The altitude of the spacecraft at closest approach was 48,000 km. About 500 pictures were taken during the second flyby. The photographs allowed a mosiac of the southern hemisphere of Mercury to be assembled.

Mariner 10 again departed Mercury and continued its orbit around the sun. Two trajectory maneuvers were conducted during the orbit to move the point of closest approach closer to Mercury during the third encounter. TCM-6, which was performed on 30 October 1974, consisted of an 83 second burn and resulted in a velocity change of 14.8 meters per second. TCM-7, performed on 13 February 1975, consisted of a 22 second burn and resulted in a velocity change of two meters per second. A final trajectory correction maneuver (TCM-8) was made on 7 March 1975 to move the distance of closest aproach about 160 km farther out to lessen the chance of collision with Mercury. TCM-8 consisted of a three second burn that resulted in a velocity change of 0.5 meters per second.

The third encounter took place on 16 March 1975 with the spacecraft passing over the north polar region. Closest approach took place at an altitude of 327 km at 70 degrees north latitude in the night region of Mercury. Photographs of the surface were taken every 42 seconds starting about two hours before closest approach. Photography was halted when the spacecraft passed into the night region about 13 minutes before the point of closest approach. Photography resumed when the cameras could again image sunlit region about five minutes after closest approach. The last series of photographs were taken on 17 March.

The Deep Space Communications site in Canberra, Australia was the only site with line of sight to Mariner 10 during the third flyby of Mercury. The sensitivity of the receiver associated with the 64-meter diameter antenna at Canberra was degraded because of failure of a cooling system for the detector. The failure of the tape recorder in the spacecraft earlier in the mission required that data be sent to Earth in real time. The data rate was reduced to 22,050 bits per second by transmitting only the middle quarter of each frame and that resulted in a tall, narrow picture format. The reduced inage size allowed high quality truncated images to be received.

The flyover near the polar region allowed refining measurements of magnetic field of Mercury. The field strength was found to be 700 nT near the poles and 350 nT near the equator. The charged particle experment disclosed that the a stream of nuclear particles emitted by the sun flow along magnetic field lines that extend outward from the sun. The pasma science experiment indicated interaction between the magnetic field of Mercury and the solar wind similar to that observed with the solar wind and the magnetic field of Earth.

Mainer 10 continued outbound after the encounter and engineering experiments were conducted until the spacecraft ran out of maneuvering gas on 24 March 1975. The transmitter was shut down on 24 March 1995, ending the mission.

Conclusion

The Mariner Venus/Mercury 1973 mission was a great success. Mariner 10 took the first photographs of clouds of Venus in the ultraviolet during a flyby and it made three flybys of Mercury. It transmitted back to Earth thousands of high quality photographs of the surface of Mercury during those flybys along with data on the magnetic field and other properties of the planet. Mariner 10 traveled nearly 1.6 billion kilometers during its incredible 506-day mission. Years after the voyage of Mariner 10, Bruce Murray, who led the Mariner 10 imaging team, stated "The economy class Mariner flyby of Venus and Mercury was one of the most productive space science experiments ever carried out."

Messenger Spacecraft

The United States launched a remarkable spacecraft to explore Mercury in 2004. The spacecraft, called MESSENGER for MEcury Surface, Space ENvironment, GEochemistry, and Ranging, orbited Mercury for four years and mapped the entire surface of the planet with high quality photography. Some images had resolution as fine as 18 meters. Messenger also generated topography of the surface by a laser altimeter and provided a wealth of other scientific data.

Messenger traveled a 6.6-year circuitous path to Mercury. The trajectory involved a flyby of Earth, two flybys of Venus, and three flybys of Mercury before the spacecraft was at the proper location and velocity to be inserted into a near polar orbit around Mercury. The spacecraft had made 15 orbits of the sun before settling into orbit around Mercury in March 2011. Messenger orbited Mercury 4100 times and returned 291,008 photographs to Earth before it ran out of fuel and descended into the planet in April 2015. It performed superbly for over four years in orbit around Mercury.

The general appearance of the spacecraft is shown in an artist's conception of Messenger in orbit around Mercury given below (Fig. 7.9). The prominent white object on the spacecraft is a sunshade to protect the spacecraft from the searing heat from the sun near the orbit of Mercury.

Background of Messenger Program

The first spacecraft to explore Mercury was Mariner 10, which made three flybys of the small planet in 1974 and 1975. Hundreds of good quality photographs covering 45% of the surface of Mercury were transmitted to Earth along with data on the magnetic field and other properties of the planet. Information gathered during the Mariner 10 mission was of great benefit in planning the Messenger mission.

It was apparent that a more productive exploration of Mercury could be carried out by an orbiting spacecraft rather than by flybys. However, orbiting Mercury is more difficult to accomplish than a flyby. The trajectory needed to account for for increase in spacecraft speed resulting from moving in closer to the sun at the orbit of Mercury. The spacecraft had to catch up to Mercury in its orbit and then at the proper time effect a major decrease in velocity to allow the spacecraft to be captured by the gravitational field of the planet.

Fig. 7.9 Artist's portrait of Messenger orbiting Mercury. (NASA graphic)

Trajectories to achieve orbit of Mercury with velocity increments available from existing launch vehicles were developed by Chen-wan Yen of JPL in 1985 and expanded by Yen in 1988. The trajectories obtained gravity assist velocity increments by a flyby of Earth, two flybys of Venus, and three flybys of Mercury. The trajectories were arranged to decrease spacecrqaft velocity and modify direction at each flyby.

Studies and workshops were conducted in the late 1980s to generate options for a Mercury orbiting mission. A basic orbiting mission to Mercury along with science goals for the mission were generated by the Mercury Orbiter Science Working Team. NASA published results of the team's work in February 1991. The possible spacecraft that emerged was named "Messenger."

NASA created a program named Discovery in 1992 to field smaller and less expensive planetary spacecraft than existing craft. Messenger was a candidate for the competition for Discovery spacecraft in 1997 and again in 1998, but other spacecraft had higher priority and Messenger did not make the cut. Messenger was again submitted in the 1999 opportunity for the Discovery program and this time it was accepted.

Sean Solomon of the Carnegie Institution of Washington was chosen to lead the Messenger program as principal investigator. The Johns Hopkins University Applied Physics Laboratory (JHU/APL) was selected to develop and operate the spacecraft. The Carnegie Institution of Washington was founded by famed philantripist, Andrew Carnegie. His goal was "*to encourage, in the broadest and most liberal manner, investigation, research, and discovery and the application of knowledge to the improvement of mankind.*" A portion of the Carnegie Institution's efforts are to seek "*answers to questions about the structure of the universe, the formation of our solar system and other planetary systems.*" The Institute was renamed the Carnegie Institution for Science in 2007.

The Applied Physics Laboratory (APL) of Johns Hopkins University was established in 1942 to mobilize academic research to address challenges of World War II. APL continued to provide technical support to the Department of Defense and to NASA in early years. Presently their work is directed towards nondefense projects. JHU/APL developed and operated several satellites prior to Messenger.

The Applied Phisics Laboratory completed the design of the spacecraft in March 2002 and a Critical Design Review was held with NASA. The review was successful and go-ahead was given to build the spacecraft. Building and testing of the spacecraft was completed in March 2004 and the spacecraft was transferred to Astrotech Space Operations (ASO) near Kennedy Space Center (KSC). ASO is a extensive facility owned by Lockheed Martin that provides facilities to service spacecrtaft before launch.

Thorough testing of the spacecraft and scientific instruments were performed at ASO and the spacecraft was mated to the third stage of the Delta II launch vehicle. The third stage, with the spacecraft attached, was moved KSC where it was assembled to the top of the Delta II launch vehicle. Launch took place on 3 August 2004.

Mechanical Configuration of the Messenger Spacecraft

A photograph of the Messenger spacecraft being worked on by JHU/APL technicians is given below (Fig. 7.10). Much of the spacecraft was wrapped in gold color thermal blankets to assist in temperature control. The large planar objects at the forefront of the picture are solar arrays, folded down to fit within the launch shroud. The arrays were 1.54 meters wide and 1.75 meters long. The sun shade had not been installed when this photograph was taken although the framework holding the shade can be seen on the far side of the spacecraft.

A close-up photograph Messenger being lowered to mate with the third stage of the Delta II launch vehicle is shown on the next page (Fig. 7.11). The inside of the sun shade can be seen at the far side bottom of the spacecraft. The adapter structure to attach the spacecraft to the launch vehicle appears at the bottom of the spacecraft. Several scientific instruments looked through the opening of the adapter once the spacecraft was separated from the third stage.

Messenger was a compact spacecraft with a main body 1.42 meters high, 1.85 meters wide and 1.27 meters deep. The frame was built from a series of vertical and horizontal flat panels of composite material. The frame incorporated three compartments to hold three large propellant tanks. Two of the tanks held hydrazine fuel

Fig. 7.10 Messenger spacecraft in laboratory. (NASA/JHUAPL/Carnegie Institution photograph cropped by author)

Fig. 7.11 Messenger spacecraft being mated to third stage of launch vehicle. (NASA/JHUAPL/Carnegie Institution photograph)

and the other tank held nitrogen tetroxide oxidizer. The tanks were identical in size and construction. They were approximately 0.56 meters in diameter and 1.04 meters long.

A drawing of a rear view of the spacecraft showing placement of major components is given on the next page (Fig. 7.12). The spacecraft coordinate system oriented the +Z axis along the longitudinal axis of the spacecraft and downwards in the drawing. The +Y-axis was parallel to the arm supporting the magnetometer and in the direction of the magnetometer. The +X axis was parallel to the support structure for the solar array and in direction to the right in the drawing.

The three large propellant tanks were mounted in a row along the Y-axis with their longitudinal axes parallel to the Z-axis of the spacecraft. They were mounted

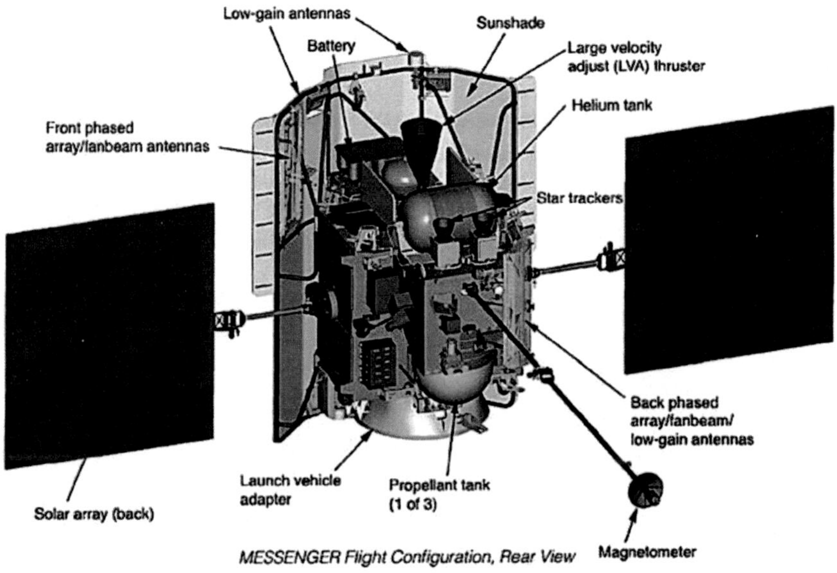

Fig. 7.12 Drawing of rear view of Messenger. (NASA/JHUAPL/Carnegie Institution graphic)

to vertical plates of the frame with the oxidizer tank between the two fuel tanks. The frame had a top cover over the fuel tank compartments referred to as the top deck. The top deck held a helium tank, auxiliary fuel tank, bipropellant rocket motor, monopropellant thrusters, and a battery. Some of those items are labeled on the drawing.

The helium tank was 40.6 centimeters diameter and 76.2 cm long and the spherical auxiliary fuel tank was 30.5 cm in diameter. The bipropellant rocket motor, referred to as the Large Velocity Adjust (LVA) thruster, was mounted with its thrust along the Z-axis.

The sunshade, which was essential for thermal control of the spacecraft, was a segmented section of a cylinder 2.5 meters long and 2.0 meters wide. Temperatures on the sun-facing side of the shade were predicted to reach 370 degrees centigrade when the spacecraft was closest to the sun. The temperature of the shaded main body of the spacecraft behind the sunshade was predicted to be about 20 degrees centigrade. The sunshade had front and back layers of Nextel ceramic cloth with several layers of Kapton insulation in between. The sunshade was supported by a welded titanium tube structure. The tube structure also supported five antennas, four sun sensors, and the sensor for the X-ray spectrometer. Struts supporting the solar arrays and the boom supporting the magnetometer could be folded to fit within the shroud of the launch vehicle.

The Messenger spacecraft weighed 1107 kg at launch. Of that, 599 kg was propellant.

Messenger Systems

Propulsion System

Messenger was fitted with a total of 17 thrusters. The main thruster was a bipropellant rocket engine referred to as Large Velocity Adjust (LVA) thruster. It had a thrust of 667 Newtons (150 pounds). The LAV thruster, which was a type LEROS 1B built by AMPAC-ISP Corp., burned hydrazine (N_2H_4) fuel with nitrogen tetroxide (N_2O_4) oxidizer. It slowed the spacecraft when in the vicinity of Mercury to allow the spacecraft to be captured by the gravitational field of the planet.

The spacecraft also had 12 small hydrazine monopropellant thrusters with 4.4 Newtons (1 pound) of thrust each. Those thrusters were used for fine attitude control and fine velocity control. In addition, there were four larger monopropellant thrusters with 22 Newtons (5 pounds) of thrust. The 22 Newton thrusters were used to control attitude of the spacecraft during burn of the LAV and to impart small velocity increments. The monopropellant thrusters were built by Aerojet-General Corp. The monopropellant thrusters operated by spraying hydrazine fuel over a hot catalyst that caused the hydrazine to decompose into nitrogen and hydrogen gas. The mass of those gases flowing out through the nozzles of the thrusters provided thrust in the opposite direction.

Hydrazine fuel was used for both the bipropellant thruster and the monopropellant thrusters. Two large fuel tanks contained 178 kg of hydrazine each and an auxiliary tank held 9.3 kg for a total of 365.3 kg of fuel at launch. The oxidizer tank contained 232 kg of nitrogen tetroxide for the LAV bipropellant thruster. The helium tank held 2.3 kg of helium at an initial pressure of 26 MPa (3375 psi). Helium was used to pressurize the fuel and oxidizer tanks to promote flow of the propellants. All of the tanks were made from titanium. The total weight of propellants and helium carried by messenger at launch was 599 kg.

The auxiliary fuel tank acted as an accumulator with helium gas on one side of a diaphragm and fuel on the other side. The compressed gas on one side of the diaphragm forced fuel out of the tank when the fuel valve was opened. Fuel was supplied from the auxiliary tank to feed monopropellant thrusters for attitude correction or when a slight velocity change was required. Another important task of the auxiliary tank was to feed hydrazine to monopropellant thrusters on the upper deck of the spacecraft that were activated to settle the contents of the main fuel tanks before fuel was drawn from the tanks. The auxiliary tank was refilled from the main fuel tanks after a period of use.

Electrical Power Subsystem

The main source of electrical power for the spacecraft was two large solar arrays. A 23-ampere-hour battery was charged by the solar arrays and provided power to the spacecraft during periods of low solar illumination as when maneuvering or when orbiting in the shadow of Mercury. The spacecraft required electrical power of between 385 and 485 watts during its long cruise to Mercury and about 640 watts during orbit of Mercury. These requirements were easily handled by the power system. The power subsystem included two solar arrays, solar array junction box, battery, power system electronics, and a power distribution unit.

Solar array -The solar array panels were 1.54 meters wide and 1.75 meters high. Panels supporting the solar cells were made up from 18-mm thick aluminum honeycomb structure with composite face sheets. Each panel was deployed at the end of a 0.76-meter long titanium strut to space it from the spacecraft. The struts were hinged to allow the array to fold against the spacecraft to fit within the launch shroud. Figure 7.11 gives a good view of a solar array and its orientation when stowed. When extended on the struts as in the drawing in Fig. 7.12, the arrays could be rotated about the axis of the struts through angles of −20 degrees to +200 degrees around the X-axis by the solar array drive assembly. Zero-degree angle corresponded to the panels being normal to the sun (-Y direction). Rotation of the arrays allowed ample power to be obtained while limiting solar heating of the array to 150 degrees Centigrade.

The solar cells were triple-junction gallium arsenide type with efficiency of about 28%. Each cell was 0.14 mm thick and 3 cm by 4 cm in size. The cells were mounted in strings of 36 cells in a line across the narrower dimension of the array. There were 18 strings separated by lines of optical reflector mirrors. The size of each mirror appears about the same as the size of the solar cell in photographs. There were two mirrors in the vertical direction in each line of mirrors such that the area covered my mirrors was twice as large as the area covered by solar cells. This arrangement reduced the solar heat load on the panels.

Spacecraft power control – The outputs of the 36 solar cell strings in the two solar arrays were isolated by series diodes and connected together in the solar array junction box. The output voltage from the arrays could vary from about 45–95 volts, depending on the load and solar intensity on the array. The summed power from the arrays was applied to the power system electronics unit. The power from the arrays was divided up among eight peak power tracking modules in the unit. Those modules contained a switching device and filters to perform pulse-width modulation of the incoming power. The duty ratio was automatically adjusted to give an output of about 28 volts at the output of the filter.

The semi-regulated power lines at the output of the eight power tracking modules were summed in the power distribution unit to form the power bus for the spacecraft. The battery was also connected to the power bus. A power distribution unit supplied power to the many different loads of the spacecraft.

Battery – The battery was a nickel hydrogen type that had a capacity of 23 ampere-hours and nominal output voltage of about 28 volts. Nickel hydrogen batteries have relatively high energy density and they can undergo tens of thousands of charge-discharge cycles with little degradation in performance. They are widely used in Earth orbiting satellites that must operate for a time in the shadow of the Earth. They were well suited for the Messenger orbiting mission of Mercury.

The battery was unique in that it used gaseous hydrogen as the anode. The design of the hydrogen electrode was similar to that used in alkaline fuel cells, which also use hydrogen gas. A platinum catalyst was needed to get the hydrogen gas into solution. A substrate made up of platinum particles bonded together with Teflon on a nickel screen formed the negative terminal of the cell. The positive electrode was nickel hydroxide material similar to that used in nickel-cadmium batteries. The electrolyte was a potassium hydroxide solution at about 30% concentration.

The hydrogen gas had to be under high pressure for the cells to function. Pressures of over 5.5 Megapascal (800 psi) are commonly used and require that the cells be contained within a pressure vessel. The battery for Messenger contained 11 pressure vessels with two cells in each pressure vessel. The enclosure for the pressure vessels was a cylinder fitted with hemispherical end caps. Scaling from photographs, the length of each pressure vessel was about 20 cm and the diameter was about 10 cm. The overall size of the battery with 11 pressure vessels was 36.6 cm by 22.83 cm by 49.17 cm. It weighed 24.5 kg.

Nickel hydrogen cells have an open circuit voltage of 1.55 volts and nominal operating voltage under load of 1.25 volts. The two cells within each pressure vessel were connected in series and the leads from the 11 pressure vessels were connected in series. The terminal voltage of the 22 cells in series was about 34 volts at no load and 27.5 volts at nominal load.

Attitude Control

Messenger was a three-axis stabilized spacecraft. One of the important requirements of the attitude control system was to point the sunshade towards the sun when traveling at distances less than 0.85 astronomical units (au) from the sun. An au is the nominal distance of the Earth from the sun (149.6 million km). It was critical that the body of the spacecraft be maintained in the shadow of the sunshade to avoid overheating of spacecraft systems and scientific instruments. The spacecraft's trajectory took it as close as 0.3 au from the sun when near Mercury. Early in the mission, when the spacecraft was greater than 0.85 au from the sun, the spacecraft was maneuvered to place the body of the spacecraft towards the sun to keep it warm. Attitude information was also important for interpretation of data from scientific instruments.

Spacecraft attitude was sensed by a combination of star trackers, sun sensors and an inertial measurement unit. Data from the star tracker and rate data from the gyros in the inertial measurement unit were used by the guidance and control software to derive spacecraft attitude. Data from the sun sensors was used as backup to assure that the Y-axis of the spacecraft was pointed at the sun.

Star trackers – The attitude of the spacecraft was accurately determined by one of two redundant star trackers. One star tracker was operated at a time and the other was kept in unpowered reserve. A complete map of stars was stored in the star tracker memory. A wide-angle camera in the star tracker imaged a section of the sky and that image of star locations was compared with the stored image of the proper pattern based on on-board ephemeris data. The spacecraft was maneuvered to bring the measured star field into coincidence with the stored field for that moment in time. Ephemeris data computed in the spacecraft was updated from Earth regularly.

Inertial measurement unit – The inertial measurement unit (IMU), built by the Northrop-Grumman company, was unique in that it contained four hemispheric resonator gyros. The hemispheric resonator gyro consisted of a high-Q quartz vibrating hemispheric shell resonator that was set into vibration by an electrostatic field. The standing wave vibration pattern on the resonator precessed when the hemisphere was rotated about its symmetry axis. The precession was detected by sensors in the outer shell of the gyro and when scaled was a measure of rotary motion. The hemispheric resonator was housed in an evacuated hermetically sealed case.

The IMU also contained four accelerometers. The accelerometers were used to sense acceleration when trajectory correction maneuvers (TCM) were conducted. Acceleration readings were integrated as a function of time to determine velocity increments. The accelerometer circuitry was only powered when a TCM was being conducted.

Reaction wheels – Attitude of the spacecraft was adjusted by changing the speed of reaction wheels. The response to torque required to change speed of the wheel was an equal and opposite reaction to the spacecraft. Messenger contained four reaction wheels built by Teldix in Germany. Each reaction wheel had a maximum torque of 0.075 newton-meters and each could store 7.5 newton-meter-seconds of angular momentum. The mass of each wheel was 4.2 kg. The wheels were mounted with their spin axis in different directions. Rotation of the spacecraft about a particular axis was induced by selectively changing speeds of a combination of wheels.

If a net bias in restoring angle was encountered that required speeding up the reaction wheel in the same direction, the speed of the wheel would eventually approach its maximum and no further correction would be possible. To counter this, a momentum dump was performed during which the thrusters were engaged to hold a particular attitude and the reaction wheels were run down to stop. The reaction wheels were then able to control attitude again.

Attitude control could also be achieved by selectively operating the 4.4-Newton monopropellant thrusters. This method would eventually use up the propellant stores and so the use of reaction wheels was the preferred method of attitude control.

Sun sensors – Messenger contained two redundant sets of sun sensors that served as attitude control backup to assure that the sunshade was pointed towards the sun. Each sun sensor subsystem consisted of three sun sensor elements and an electronics box. Two of the sun sensor heads of each set protruded through opposite corners of the sunshade and the third was mounted to the back of the spacecraft. The two sets of sun sensors comprised four sensor heads at the corners of the sunshade and two sensor heads on the back of the spacecraft.

Guidance and Control

A block diagram of the Messenger spacecraft illustrating the various subsystems managed by the guidance and control system is given on the next page (Fig. 7.13). The various subsystems are tied together by a bidirectional MIL-Std-1553 digital data bus. The 1553 data bus has been standard in military aircraft since 1973.

Computers residing in two redundant Integrated Electronics Modules (IEMs) were the brains of the guidance and control system. Each IEM contained a set of flight computers consisting of a main processor (MP) and a fault protection processor (FPP). The MP ran software associated with all spacecraft functions. Only one MP was powered at a time and the other was a backup spare. The fault protection processors monitored spacecraft health. Both the MP and FPP used RAD6000 32-bit computers built by BAE Systems. Those computers were radiation hardened to withstand a radiation dose of one million rads. The MP was run at a speed of 25 MHz and the FPP was run at 10 MHz. The MP contained eight Mbytes of random-access memory and four Mbytes of programmable read-only memory. The FPP had four Mbytes of random-access memory and four Mbytes of read-only memory.

The software that was run on the MP computed desired spacecraft attitude and attitude rate. It provided attitude control information to the reaction wheels and thrusters and performed momentum management. The software also controlled propulsion operations, orientation of the solar arrays, pointing angle of the phased array antenna, and orientation of some scientific instruments.

Telecommunications

The Messenger telecommunications system received commands from Earth, transmitted spacecraft status and scientific instrument data to Earth, and facilitated Doppler and range tracking of the spacecraft from Earth. The very accurate two-way Doppler measurements and accurate range measurements allowed accurate determination of spacecraft location and velocity.

Transponder – The telecommunications system included two redundant transponders referred to as Small Deep Space Transponder (SDST). The transponders, built by General Dynamics, included a sensitive receiver and an exciter. The receiver section received and demodulated uplink signals from Earth. The exciter section generated the carrier signal and modulated it for downlink to Earth. The frequency of exciter output could be derived from a crystal oscillator or it could be offset but phase coherent with the received signal to allow two-way Doppler tracking from Earth. The ratio between transmit and received frequencies was 880/749. Uplink communications from Earth took place at a frequency of about 7.2 GHz and data was downlinked to Earth at a frequency of about 8.4 GHz. Both frequencies are in X band. The size of each transponder was 18.1 cm by 16.6 cm by 11.4 cm and each weighed 3.2 kg.

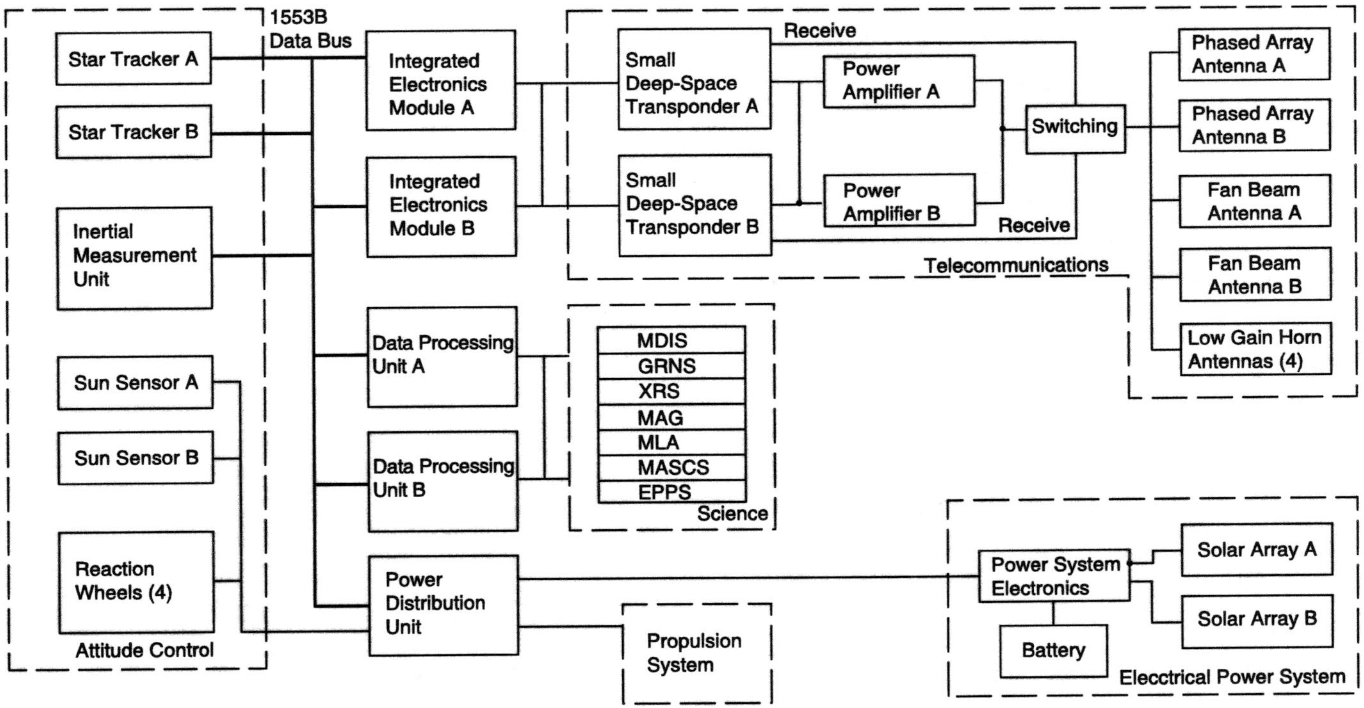

Fig. 7.13 Functional elements of Messenger spacecraft. (Adapted from NASA/JHUAPL/Carnegie Institution graphics)

The decoded uplink signal from the transponder receiver was processed by the Command and Data Handling subsystem in the spacecraft. The Command and Data Handling subsystem also formatted the downlink data used to modulate the transponder's exciter signal with data that was transmitted to Earth.

Antennas – Uplink signals from Earth could be received over one of four low-gain antennas or over one of two medium gain antennas. Downlink data could be transmitted through any one of the low-gain or medium-gain antennas or through one of two high-gain phased array antennas.

The medium-gain antennas were located in the same enclosure as the phased array antennas. They had peak gain of 15 dB and a fan beam radiation pattern with half-power beamwidths of 7.5 degrees in elevation (parallel to the Z-axis) and 90 degrees in azimuth. The low-gain antennas had approximately hemispherical radiation patterns. The four low-gain antennas were distributed around the spacecraft with the center of the beams along the +Z-axis, −Z-axis, +Y-axis and −Y-axis. A series of switches allowed any low-gain or medium-gain antenna to be connected through diplexers to the receiver of either one of two redundant transponders. The main conduit for transmission of data to Earth was through one of two phased array antennas. Those antennas had narrower beamwidths and higher gain than the other antennas and the antenna beam could be electronically steered.

Phased array antenna – The two phased array antennas were fixed to the spacecraft. Their antenna beams were electronically scanned over a ±45-degree angle in the X-Y plane (azimuth plane) of the spacecraft. One phased array antenna was mounted on the sunshade side of the spacecraft and the other was mounted on the backside. The antennas were 28 cm by 81 cm in size and they were covered by a curved radome of space blanket material. The long dimension of the antenna was parallel to the Z-axis of the spacecraft.

The phased array antenna included an array of eight slotted waveguides, each containing 26 radiating slots. The phase of the signal applied to each waveguide was controlled to steer the antenna beam ±60 degrees from normal to the array. The beam steering angle was usually limited to ±45 degrees. The peak antenna gain was 28.5 dB and the half-power beamwidth was about 21.5 degrees in the scanning direction (azimuth plane) and 2.5 degrees in the plane of the long dimension of the array (elevation plane).

The output power of the transponder exciter was 20 milliwatts. The outputs from the exciters of the two transponders were combined by a hybrid coupler and the result applied to two solid state power amplifiers (SSPAs). Each SSPA contained a power splitter where one output fed an amplifier that could be used to drive a series of RF switches that applied the transmitter signal to one of the medium-gain antennas or to one of the low-gain antennas. The other output of the power splitter was applied to an eight-way power splitter that fed eight phase shifters and amplifiers associated with the phased array antenna. The phase shifters were controlled by a 4-bit digital word from the command and data handling subsystem. The amplifier outputs fed the radiating waveguides of the phased array antennas. There was a potential problem of the SSPA overheating if all eight amplifiers were operated at a time. For this reason, only four sets of amplifiers were energized at a time. Four waveguides of each phased array antenna were fed from one SSPA and the other four waveguides were fed from the other SSPA. Both SSPAs were active.

The phased array antenna on the sunshade side covered an angular sector from the –X-axis to the -Y-axis. The back antenna covered the sector from the +X-axis to the +Y-axis. The antenna that had best view of Earth was selected to transmit data. The sunshade (–Y-axis) of the spacecraft had to be pointed close the sun for most of the mission. Consequently, it was not feasible to rotate the spacecraft about the Z-axis to point a conventional fixed-mounted narrow beamwidth antenna towards Earth.

Command and Data Handling

The command and data handling subsystem functions were mainly contained within two redundant Integrated Electronic Modules (IEMs). The IEM, which was packaged in a box 150 cm by 18.5 cm by 23.7 cm in size, contained five printed circuit boards: main processor, fault protection processor, solid state recorder, power converter, and interconnect board. It weighed about 5.9 kg. The IEM communicated with other subsystems in the spacecraft by via a 1553 data bus.

Central to the command and data handling functions in each IEM were two computers. Both computers were RAD6000 32-bit computers built by BAE Systems. The computers were radiation hardened to withstand a radiation dose of one million rads. One computer, referred to as the main processor (MP) ran software associated with all spacecraft functions. The other computer, referred to as the fault protection processor (FPP), was programmed to monitor the condition and status of key systems of the spacecraft.

The MP computer supported software that ran the guidance and control (G&C) functions and the command and data handling (C&DH) functions. The guidance and control software computed present and desired spacecraft attitude and attitude rate. It provided attitude control information to the reaction wheels and thrusters and it performed momentum management. The software also controlled propulsion operations, orientation of the solar arrays, pointing angle of the phased array antenna, and orientation of some scientific instruments. Software elements for these tasks were run at either a 1 Hz or 50 Hz rate. Spacecraft attitude control functions such as commands to the reaction wheels were run at a 50 Hz rate. Estimates of attitude using star tracker and gyro rate data along with ephemeris computations were run at a 1 Hz rate. Other functions run at a 1 Hz rate were steering commands for the phased array antenna, positioner angle for the cameras, and thruster control.

The command and data handling software managed uplink and downlink data handling along with the solid-state recorder and file system. It supported transmission of files from the solid-state recorder to the downlink using a Consultative Committee for Space Data Systems (CCSDS) file delivery protocol. It also managed science data collection including compression of camera images.

Commands from Earth were received and decoded by the transponders and sent to the integrated electronic modules. The IEC could carry out commands in real time or commands could be stored in memory of the main processor to be performed later. Commands were sent from Earth at data rates from 7.8 to 500 bits per second depending on distance to the spacecraft and size of antenna used at the ground station.

The large amount of data to be downlinked from the spacecraft, including data from cameras and other science instruments, were stored in one of two solid-state recorders for transmission to Earth at a convenient time. Each of the solid-state recorders could store one gigabytes of data. Only one recorder was used at a time the other was a redundant spare. The downlink data rate available ranged from 9.9 bits per second to 104 kilobits per second depending on distance to Earth and antennas used on the spacecraft and on Earth.

Data was transmitted to Earth once per Earth day for duration of about eight hours. On average, about 15 megabytes of data was transferred per day. The data rate varied depending on the distance between Mercury and the Earth during an Earth year. Data rates were highest when Mercury was near inferior conjunction when earth and Mercury were on the same side of the sun.

Flight of Messenger

Messenger was originally scheduled to be launched during a favorable launch opportunity in March 2004. However, delays in spacecraft testing pushed the launch out to a launch window in August 2004. The August launch opportunity resulted in a 1.9 year longer, but yet viable, trajectory to Mercury. The time from launch of Messenger until it entered orbit around Mercury was 6.6 years. The trajectory chosen for Messenger relied on gravitational velocity assists from a flyby of earth, two flybys of Venus, and three flybys of Mercury. All of the flybys decelerated the spacecraft relative to its travel around the sun.

Messenger was launched from Cape Canaveral Air Force Station on 3 August 2004 by a three-stage Delta II launch vehicle. The third stage, referred to as Payload Assist Module (PAM-D), was fired about 50 minutes after launch to accelerate the spacecraft to escape velocity and place it into heliocentric orbit (orbit around the sun). PAM-D used a Thiokol Star 48B solid fuel rocket engine with a thrust of 66,000 newtons (14,837 pounds). The spacecraft was separated from the PAM-D stage 57 minutes after launch at a velocity of about 10.7 km per second. The initial orbit was 0.92–1.08 au in extent and inclined 6.4 degrees to the ecliptic.

The solar panels were deployed and the spacecraft was oriented with the sunshade away from the sun to warm the spacecraft as it proceeded on its orbit of the sun at distances larger than 0.85 au. The initial orbit of Messenger around the sun brought it back to the vicinity of Earth a year after launch on 2 August 2005. The closest approach distance to Earth was 2347 km. The cameras on the spacecraft were activated and returned excellent pictures of Earth and of the moon. The flyby decreased velocity of the spacecraft by about 6 km per second, lowered the perihelion of the orbit to 0.60 au, and decreased orbit inclination to 2.5 degrees.

Five trajectory correction maneuvers (TCMs) were performed to establish the desired trajectory of the spacecraft prior to Earth flyby. The first TCM after launch changed velocity by 17.9 meters per second to correct errors in establishing the heliocentric orbit. The following four TCMs changed velocity by 4.6, 3.2, 1.1, and 0.15 meters per second respectively. The first three TCMs were performed by the four 22 newton monopropellant thrusters that thrusted parallel with the –Z axis. The fourth TCM used the 4.4 newton S1 and S2 thrusters that thrusted parallel to the +Y-axis. The fifth TCM, which occurred just 12 days prior to Earth flyby, used two 4.4-newton thrusters that thrusted parallel to the –Y-axis.

The modified heliocentric orbit continued after Earth flyby and on 12 December 2005 a deep space maneuver (DCM-1) was performed to modify the trajectory to set the parameters for the desired flyby parameters of Venus. DCMs used the large velocity adjust (LVA) thruster which had a thrust of 667 newtons and could change velocity of the spacecraft significantly. DCM-1 changed Messenger's velocity by 315 meters per second. Four small TCMs were also performed prior to flyby of Venus on 24 October 2006. Messenger passed 2987 km above the surface of Venus during the flyby. The first flyby of Venus decreased the velocity of the spacecraft by 5.52 km per second, lowered the perihelion of the orbit to 0.55 au, and increased orbit inclination to 8.2 degrees.

The heliocentric orbit continued for a second flyby of Venus. Five TCMs were performed to set up the encounter. The second flyby of Venus occurred on 5 January 2007. Messenger passed 313 km above the surface on the second flyby. The flyby decreased spacecraft velocity by 6.94 km per second, lowered the perihelion of the orbit to 0.332 au, and changed the orbit inclination to 6.8 degrees. Several high-quality photographs were taken of the cloud cover of Venus during the flyby.

After leaving Venus, two DSMs and three TCMs were performed to nudge the spacecraft towards Mercury. The first DCM occurred on 17 October 2007 and changed velocity by 226 meters per second. The second DCM occurred later in the day on 17 October 2007 and changed velocity by 1.42 meters per second.

The first flyby of Mercury took place on 14 January 2008 with the spacecraft passing 201 km above the surface. The flyby decreased velocity by 2.30 km per second, changed inclination of the orbit to 6.9 degrees and lowered perihelion distance around the sun to 0.313 au. The cameras and other science instruments were activated during the flyby.

The heliocentric orbit continued around to the second flyby of Mercury. A DCM was performed on 19 March 2008 that changed velocity by 72.2 meters per second to set up the second flyby. The second flyby of Mercury occurred on 6 October 2008 and the spacecraft passed 199 km above the surface. The flyby decreased the velocity by 2.45 km per second, changed the inclination of the orbit to 7.0 degrees and changed the perihelion distance to 0.302 au.

The heliocentric orbit was continued for a third flyby of Mercury. A DCM was performed on 4 December 2008 that changed velocity by 222 meters per second and a second DCM was performed on 8 December 2008 that changed velocity by 24.6 meters per second. The third flyby took place on 29 September 2009 and passed 227 km above the surface. The flyby changed velocity by 2.84 km per second and held the inclination of the orbit constant at 7.0 degrees, which is the inclination of the orbit of Mercury. The perihelion distance around the sun was 0.303 after the flyby.

One DCM was performed on 29 September 2009 to set up conditions for insertion into orbit around Mercury. A velocity change of 178 meters per second was achieved. Messenger arrived at Mercury on 18 March 2011 and the large velocity adjust (LVA) thruster was fired to slow the spacecraft so that it could be captured by the gravitational field of Mercury. The burn lasted for 885 seconds (14.75 minutes) and slowed the spacecraft by 862 meters per second. Messenger was captured by the gravitational field of Mercury and the spacecraft went into orbit around the planet. The initial orbit had a periapsis altitude of 207 km at 60-degree north latitude and apoapsis altitude of about 15,200 km. The inclination of the orbit was 82.5 degrees and the period was 12.1 hours.

The primary mission was scheduled to last one Earth year. Perturbations due to the gravity of the sun caused the periapsis altitude of the orbit to increase. The mission called for holding periapsis altitude between 200 km and 500 km. When the periapsis altitude approached 500 km, an orbit correcting maneuver (OCM) was performed at apoapsis that thrusted against the velocity vector to lower periapsis back to 200 km. OCMs were performed about every Mercury year or 88 Earth days.

The spacecraft was performing well after the first year and the mission was extended. About a month into the second year of orbiting the orbit period was reduced to about eight hours by lowering altitude of apoapsis to 10,000 km while keeping periapsis at 200 km. The lower apoapsis altitude benefitted photography and science over the southern hemisphere region.

The lowered orbit during the second year decreased perturbation of the orbit by gravity of the sun. As a result, the altitude of periapsis increased slowly, remaining under 500 km, for the duration of the second year. The latitude of periapsis started at 60 degrees north at the beginning of the orbiting mission and slowly migrated northward to a maximum of 84.1 degrees near the end of the second year and then rotated over the polar region and began a southerly migration. After periapsis passed over the polar region the effect of sun's gravity was to decrease the altitude of periapsis rather than to raise it.

The mission was extended for the second time at the end of the second year of orbiting in March 2013. That extension was for two years. The altitude of periapsis was allowed to decrease to just above 200 km by the end of the third year of orbiting. It decreased to just above 100 km during the early portion of the fourth year of orbiting. The altitude was raised by an orbit correction maneuver on 17 June 2014 and that maintained altitude of periapsis above 100 km for the first six months of the fourth year of orbiting. The 100 km altitude allowed continued good photography and science to be gathered. Several further orbiting correction maneuvers were performed to maintain periapsis altitude above 25 km during the last half of the fourth year of orbiting and into the fifth year. With fuel exhausted, Magellan crashed into the surface on 30 April 2015. It had acquired high quality photography and science data while in orbit around Mercury for 4.1 years.

Science Experiments Carried by Messenger

Messenger carried seven dedicated scientific instruments to gather information about Mercury. The instruments and brief account of their function are listed below:

Mercury Dual Imaging System (MDIS) – A narrow-angle camera and a wide-angle camera mounted on a steerable platform were used to produce photographic maps of the entire surface of Mercury.

Gamma-Ray and Neutron Spectrometer (GRNS) – The instrument measured properties of gamma rays and neutrons emitted by the surface of Mercury. The data yielded relative abundances of elements on the surface.

X-Ray Spectrometer (XRS) – The XRS detected low level X-Rays from certain elements in the surface of Mercury generated by impingement of gamma rays and high-level X-Rays from the sun. A separate sensor measured X-Ray flux from the sun.

Magnetometer (MAG) – The magnetometer was a three-axis fluxgate device mounted on a 3.6-meter boom that was used to characterize the magnetic field of Mercury in detail.

Mercury Laser Altimeter (MLA) – The laser-based altimeter measured time delay between a transmitted pulse and echo pulse received from the surface to measure distance to the terrain. The measurements were used to generate topographical maps of the surface of the northern hemisphere of Mercury.

Mercury Atmospheric and Surface Composition Spectrometer (MASCS) – The MASCS analyzed light at wavelengths from infrared to ultraviolet to determine abundances of atmospheric gasses and some properties of the surface of Mercury.

Energetic Particle and Plasma Spectrometer (EPPS) – The EPPS measured energy of charged particles in the magnetosphere of Mercury.

In total, the science instruments weighed 38.6 kg and consumed 66.1 watts of power when all were active. Locations of the science instruments on the spacecraft are shown on the next page (Fig. 7.14).

Mercury Dual Imaging System (MDIS)

The Mercury Dual Imaging System was the foremost instrument on the Messenger spacecraft. It allowed gathering high-quality photographic coverage of the entire surface of Mercury as well as making detail observation of areas of particular interest. It was mounted within the launch vehicle adapter ring on the bottom of the spacecraft along with several other instruments. The size of the MDIS assembly was 398 by 270 by 318 mm. It weighed 7.8 kg.

The imaging system included two capable cameras. One, referred to as the narrow-angle camera, had a field of view of 1.5 degrees. It was a monochrome camera that viewed through an optical filter 100 nanometers (nm) wide centered at a wavelength of 750 nm (red visible light). The other camera, referred to as the

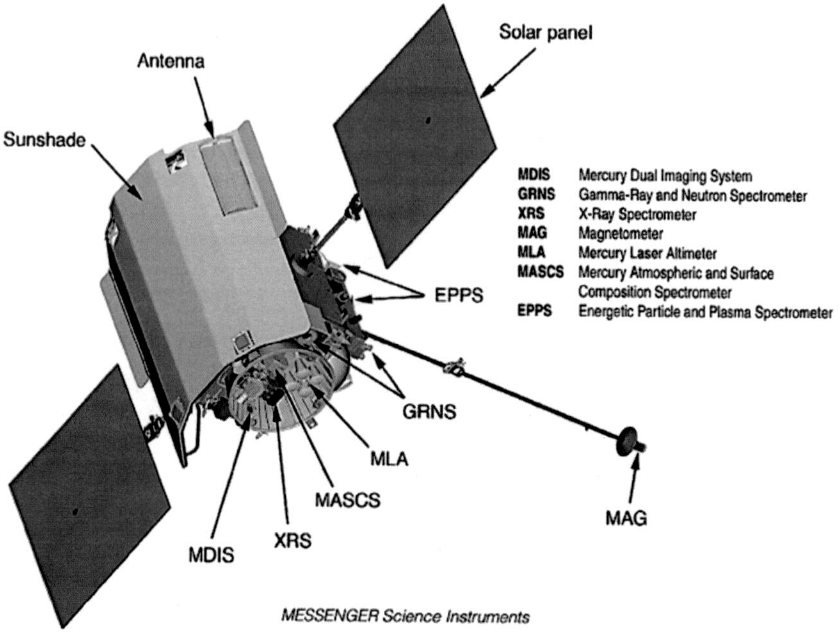

Fig. 7.14 Locations of science instruments on Messenger spacecraft. (NASA graphic)

wide-angle camera, had a field of view of 10.5 degrees. It was arranged to provide multispectral images by viewing the image through a filter wheel with 11 different color filter segments and one clear segment. The wavelengths covered by the filter wheel ranged from 395 nm (ultraviolet) to 1040 nm (infrared).

The two cameras and readout electronics were mounted on a scanning platform. Both cameras were boresighted in the direction of the +Z axis at zero scan angle of the platform. Operational scan angle limits were 40 degrees in the direction of the sunshade and 50 degrees in the opposite direction. The platform rotated to 180 degrees to face the camera apertures toward the cool dark spacecraft for stowage. The scan angle was controlled by a computer in the data processing unit. A stepper motor rotated the platform at rates up to 1.1 degrees per second. The filter wheel was driven by a stepper motor controlled by the same computer. The step size was 0.6 degrees and the stepping rate of the filter wheel was 75 degrees per second.

The narrow-angle camera used a folded reflective optics arrangement to fit the telescope in the size available. A primary and a secondary mirror focused the image on the detector at the focal plane of the telescope. The telescope had a focal length of 550 mm and a collecting area of 462 square millimeters. The field of view was 1.5 degrees. The detector was a frame-transfer charge coupled device (CCD) built by Atmel Corporation as their model TH7888A. The device had a 1024 by 1024 pixel imaging region along with a 1024 by 1024 pixel memory array. Each pixel was 14 μm square in size where μm is micrometers. The image was projected onto the

1024 by 1024-pixel imaging region and then transferred to a masked off 1024 by 1024-pixel memory area after the desired exposure time. The frame transfer process allowed electronic exposure control from about eight milliseconds to just under one second.

The CCD could be operated in a full frame mode or in a binned mode. In the binned mode, the outputs of a series of 2 by 2 arrays of pixels were combined. The resulting image array was 512 by 512 elements with 28 μm by 28 μm effective pixel size. Binning improved the signal to noise ratio and reduced the size of the image file at the expense of image resolution.

The 1.5-degree field of view of the telescope of the narrow-angle camera resulted in each pixel subtending an angle of 0.00146 degrees or 25 micro radians. At the periapsis altitude of 200 km, the extent on the surface would be about 5.1 meters. However, the high speed of Messenger relative to the surface caused some smearing of the image. Researchers gave a value for highest resolution of 18–20 meters for the narrow angle camera at 200 km altitude.

The wide-angle camera used a four-element refractive telescope. The focal length was 78 mm and the collecting area was 48 square millimeters. The same Atmel TH7888A, 1024 by 1024-pixel, CCD detector used by the narrow-angle camera was used by the wide-angle camera. The CCD was placed at the focal plane of the telescope. The 10.5-degree field of view of the telescope resulted in each pixel subtending an angle of about 0.010 degrees or 179 micro radians. At the periapsis altitude of 200 km, the extent on the surface would be about 35.8 meters without motion smearing.

Only one camera was active at a time and that allowed much of the electronics for image processing to be used for both cameras. Each camera used a dedicated set of electronics called a Focal Plane Unit (FPU) that controlled operation and readout of the CCD detector. The FPU electronics included a 12-bit analog-to-digital converter that converted the output level of each pixel to a digital word. The CCD frame readout rate, set by the FPU, was once per second.

The digital image data from the focal plane unit of the active camera was sent to one of two redundant Data Processing Units (DPUs) via a DPU Switching Interface Electronics (DISE) unit. The DISE switched the image data to the selected DPU and it applied clock and frame gate signals and power from the DPU to the FPU for the selected camera.

The selected DPU read images from the active FPU, processed them and sent the processed image to one of the two redundant integrated electronics modules (IEMs). The IEM was described in an earlier section of this chapter. Image data from the FPU was transferred to the DPU in serial form at a 16 MHz clock rate. Transfer and storage of 12-bit amplitude data from an array of 1024 by 1024 pixels took about 0.85 seconds. The incoming data was buffered in a static read only memory (SRAM) in the DPU.

The high data rate and large amount of image data taxed the telemetry system for transmission of data to Earth. There were three methods used to reduce the amount of data. First, 2 × 2 binning on the CCD chip reduced the pixel count from 1024 × 1024 to 512 × 512. Second, the 12-bit pixel amplitude data could be converted to 8-bit data by means of lookup tables in the DPU. Image amplitude data could also be compressed by what was called lossless compression that included determining if adjacent pixels were nearly the same amplitude. These compression functions were performed in hardware but managed by software.

The DPU generated a histogram from pixel amplitudes that was used by an automatic exposure algorithm. The algorithm determined proper exposure time by determining brightness of the previous image. The exposure time could also be set manually from Earth.

Image data was transferred from the DPU to the IEM at a 4 MHz rate. If binning was used in the CCD, images could be transferred at a one Hz rate. A full frame of 1024 by 1024 pixels required about four seconds to transfer.

Images from Messenger Mission

The Mercury Dual Imaging System gathered over 250,000 photographs of the surface of Mercury during three flybys and four years in orbit around the planet. Many of the photographs were stitched together to form mosaics of selected areas and of the entire planet. A mosaic of a portion of Mercury was assembled from photographs taken during the first flyby of Mercury on 14 January 2008. The mosaic, shown on the next page (Fig. 7.15), was assembled from 87 photographs taken by the narrow-angle camera from distances of 17,484 to 20,790 km as Messenger sped away from the planet.

The initial orbit of Mercury had a periapsis altitude of 207 km at 60 degrees north latitude and apoapsis of 15,200 km. The orbit was inclined 82.5 degrees from the equator with a period of 12.1 hours. Following checkout of all systems on the spacecraft and taking several test photographs of the surface, Messenger began its formal one-year primary science program on 4 April 2011. An important goal of the primary mission was acquiring a global monochrome imaging basemap. The basemap was generated from images at low emission angles and moderate incidence angles. The intent was to image the entire surface of Mercury with average resolution of about 250 meters per pixel. Emission angle is the angle between the center of the field of view of the camera and the local vertical. The angle would be zero looking along the nadir. Incidence angle is the angle of sunlight impinging on the surface relative to the local vertical. Incidence angles were generally in the range of 45–80 degrees from the local vertical. An angle of 68 degrees was near optimum.

The initial basemap was generated by using the wide-angle camera while over the northern hemisphere where the altitude of the orbit was lower and using the narrow-angle camera over the southern hemisphere where the altitude near the South Pole was quite large. This arrangement of imaging with the wide and narrow angle cameras allowed an average image resolution of about 250 meters per pixel to be obtained over the planet. The wide-angle camera was set to image at a wavelength of 750 μm for the basemap, which was the same wavelength used by the narrow-angle camera.

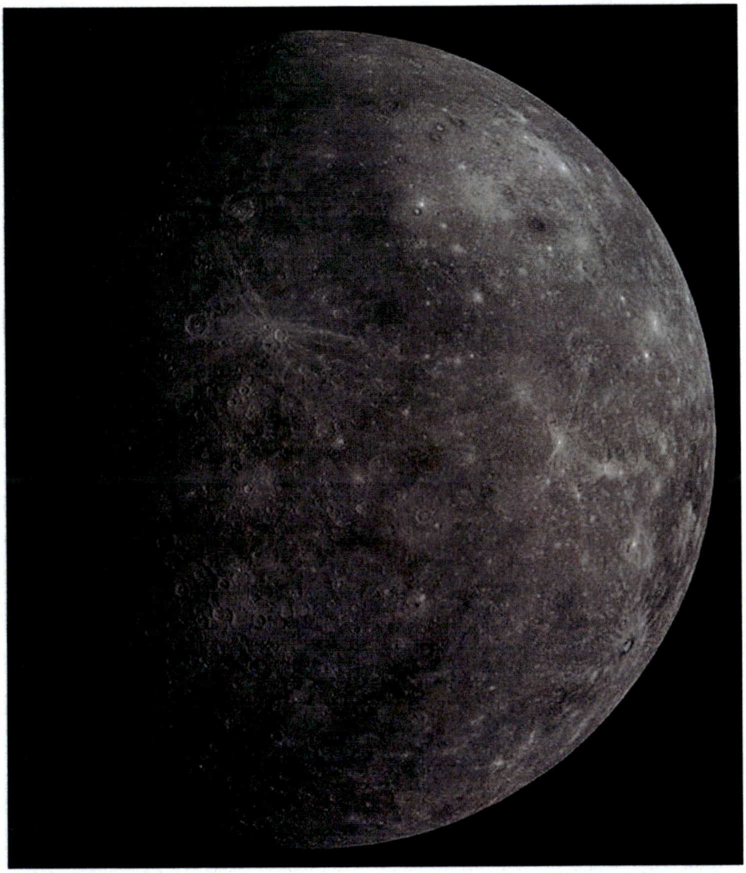

Fig. 7.15 Mosaic of images taken during first flyby of Mercury. (NASA/JHUAPL/Carnegie Institution image)

An image generated by the wide area camera during the process of gathering the monochrome basemap is shown on the next page (Fig. 7.16). The center latitude of the image was 0.25 degrees north and the center longitude was 109.3 degrees east. The width of the image was 250 km and the resolution was about 245 meters per pixel. The long curved ridge in the image is called Beagle Rues. It is one of the longest and tallest scraps on the planet.

Imaging conducted during the first solar day included color mapping by the wide-angle camera. Color was obtained by sequentially viewing through eight of the 11 color filters available to the camera. The average resolution of these color images taken around the planet was 1000 meters. A color image taken at latitude of 3.5 degrees south and longitude of 275.7 degrees east is given on the next page (Fig. 7.17). The resolution of the image was 953 meters per pixel.

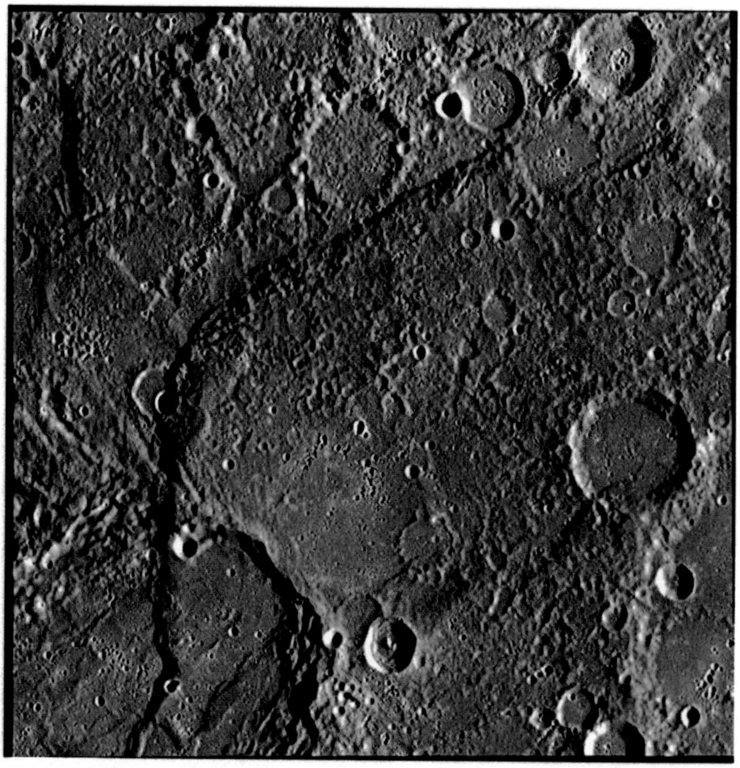

Fig. 7.16 Image taken by wide-angle camera at latitude of 0.25° N during formation of base map. (NASA/JHUAPL/Carnegie Institution image)

A series of high-resolution targeted observations was carried out in addition to mapping. These images were often requested by the science group after viewing previous images. A typical narrow-angle camera targeted image at 38.4 degrees north latitude and 175.6 east longitude is shown on the next following page (Fig. 7.18). The resolution of the image was 42 meters per pixel.

The primary mission occupied two solar days. A solar day on Mercury lasts for 176 Earth days. Gathering of images for the first global map was completed by the end of the first solar day of orbiting. There were a few gaps that were filled in by mapping during the second solar day. A global image formed from thousands of individual base map images is shown on the next following page (Fig. 7.19). Also shown is a global image in color. The color image was generated from photographs taken by the wide-angle camera using a series of eight color filters. The South Pole region was of particular interest to scientists so every four orbits images were taken of the region. The repetitive imaging allowed imaging at a wide range of sun angles. The wide-angle camera was used during the first solar day and resolution was 1.5–1.7 km per pixel. The narrow-angle camera was used during the second solar day and resolution was about 300 meters per pixel.

Fig. 7.17 Image taken by wide-angle camera during color mapping of Mercury. (NASA/JHUAPL/Carnegie Institution image)

Stereo mapping was performed during the second solar day by passing over the surface at nearly the same local solar time as during the base map but with the camera pointed either forward or aft of the sub-satellite point on the ground track. The stereo images allowed creating a topographical map of the terrain in the southern hemisphere where the altitude was too great for operation of the laser altimeter. In all, Messenger gathered 88,746 photographs of the surface of Mercury during the primary mission of the first Earth year of orbiting.

The Messenger mission was extended three times resulting in a total of 4.1 Earth years in orbit around Mercury. A total of about 250,000 photographs were taken during three flybys and 4.1 years of orbiting the planet.

Mercury Atmospheric and Surface Composition Spectrometer (MASCS)

The Mercury Atmospheric and Surface Composition Spectrometer (MASCS) was designed to determine mineral content of the surface of Mercury from reflectance measurements and to determine the composition and spatial distribution of species in the upper atmosphere (exosphere).

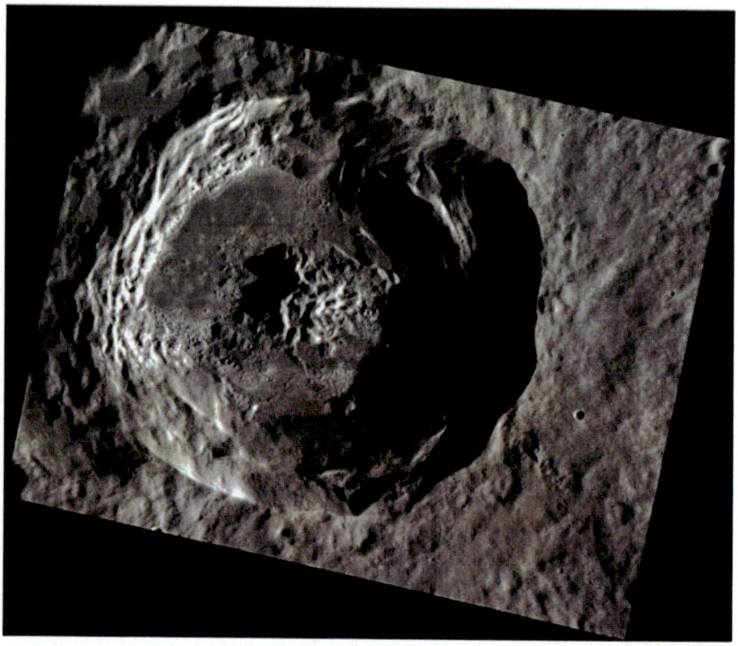

Fig. 7.18 Targeted observation of unnamed crater at resolution of 42 meters per pixel. (NASA/JHUAPL/Carnegie Institution image)

Fig. 7.19 Global mosaic of images from first solar day base map on left and color mosaic on right. (NASA/JHUAPL/Carnegie Institution image)

MASCS was composed of two instruments that used a common narrow field of view telescope for observation. One instrument, called Ultraviolet/Visible Spectrometer (UVVS), investigated the exosphere of Mercury by measuring solar radiation scattered by atoms located above the limb of the planet. It was arranged to determine relative abundances of certain elements and their and distribution with altitude.

The other instrument, called Visible and Infrared Spectrograph (VIRS), measured the reflectance of solar irradiance from the surface of Mercury at visible and infrared wavelengths. A reflectance map of the surface as a function of wavelength was generated to assist in determining mineral content of the surface. Over five million spectra were collected during the first year of orbiting. About 95% of the surface was mapped at an average of 20 km spacing between scan lines along the surface.

The telescope was a Cassegrain type with a primary and secondary mirror arrangement to reduce the telescope's length. The image from the telescope was focused on an aperture plate that contained an entrance slit for the ultraviolet/visible spectrometer and the end of fiber optic cable that fed the visible and infrared spectrograph. The field of view of the telescope was parallel to the Z-axis of the spacecraft in the +Z direction.

The instrument was mounted within the launch vehicle adapter ring on the bottom of the spacecraft. The size of MASCS was 195 by 205 by 310 mm. It weighed 3.1 kg and consumed 6.7 watts of electrical power.

Ultraviolet/Visible Spectrometer (UVVS) – The entrance slit for the UVVS, was located on the on the aperture plate. The slit was 4.4 mm long and 0.175 mm wide. These dimensions translated to a 1.0 degree by 0.04-degree field of view for exosphere measurements. The MASCS was mounted so that the long dimension of the entrance slit was parallel to the limb of the planet for exospheric measurements. The instrument could also be used for surface reflectance measurements. When so used, the entrance slit was partially covered to reduce the 4.4 mm length to 0.23 mm. The resulting field of view was 0.05 degrees by 0.04 degrees.

The spectrometer was an Ebert-Fastie type. Light from the entrance aperture was collimated and reflected towards a diffraction grating by a portion of a spherical mirror. The diffracted light from the grating was intercepted and reflected by another portion of the spherical mirror onto three exit slits of the spectrometer. The three slits corresponded to wavelength regions of Far Ultraviolet at 113–190 nm, Middle Ultraviolet at 160–320 nm, and Visible at 250–600 nm.

The diffraction grating had 1800 lines per mm that resulted in average spectral resolution of 1.0 nm. It was rotated in step sizes from 0 to 15 arcminutes in 1 arcminute steps over a range of 45 degrees by a grating drive assembly. The grating drive assembly included a torque motor, grating rotator, and a position encoder. The maximum rotation rate was 200 steps per second. The resolution of the position encoder was about 10-arcseconds.

The wavelength region of interest was examined by stepping the grating to present a series of emission lines to the exit slits and associated photomultiplier tubes. At each scan position, data was accumulated for a commanded integration time before the next rotation step was initiated.

Measurements by the instrument indicated that the most plentiful elements in the exosphere were sodium, magnesium, and calcium. Traces of hydrogen, helium, and potassium were also found. There was an abundance of sodium in the long tail on the anti-sunward direction from the planet.

Visible and Infrared Spectrograph (VIRS) – The Visible and Infrared Spectrograph examined solar reflectance from the surface of Mercury via the telescope at wavelengths from 115 nm to 600 nm. The field of view of the VIRS assembly was 0.023 degrees circular.

Light from the telescope was transferred to the Visible and Infrared Spectrograph by a fiber optic cable with core diameter of 100 μm. The output of the fiber optic cable was directed at the concave grating of the spectrograph. The concave grating, which had a radius of curvature of 210 mm, was ruled at 100 groves per mm. The nominal dispersion was 46.6 nm per mm. The diffracted light from the grating was directed at a beam splitter that allowed near infrared wavelengths (0.95–1.45 μm) to pass through to a lens that focused the light onto a linear array of 256 indium-gallium-arsenide photocells. The output of each photodiode was connected to sample-and-hold electronics that fed a charge-to-voltage converter. At the end of a commanded integration time, the voltages from each pixel circuit were read out sequentially and converted to digital form by a 16-bit analog to digital converter. The resulting digital words were assembled into a file stored in the solid-state recorder.

The beam splitter reflected visible wavelengths (300–1025 nm) of the diffracted light from the grating onto a linear array of 512 silicon photodiodes. Each pixel of the silicon photodiode array was 50 μm wide and 2500 μm long. The nominal integration time for the visible light measurements was one second. The array was read out by a shift register that sequentially transferred the charge on each pixel to an amplifier that converted charge to a proportional output voltage. The line to a given pixel was opened after readout to allow the pixel to start accumulating charge again. The output of the amplifier was applied to a 16-bit analog to digital converter. The resulting digital word was assembled into a file stored in the solid-state recorder.

A reflectance map of the surface of Mercury as a function of wavelength was generated by the Visible and Infrared Spectrograph. About five million reflectance spectra were gathered during the first year in orbit around Mercury. A composite map of scan lines of the VIRS during the first year of orbiting is shown on the next page (Fig. 7.20). The map was generated by mapping intensities of the various wavelength regions into red (575 nm), Green (415–750 nm), and blue (310–390 nm) colors. The map is shown as a cylindrical projection with the center of the map at zero degrees latitude and zero degrees longitude.

Fig. 7.20 Composite of scans by the VIRS with colors to help visualize wavelengths of light reflected from surface of Mercury. (NASA/JHUAPL/Carnegie Institution graphic)

Gamma-Ray and Neutron Spectrometer (GRNS)

The Gamma-Ray and Neutron Spectrometer gathered and analyzed gamma rays and neutrons emitted by the surface of Mercury. The data yielded information on relative abundances of elements on the surface. The instrument contained two independent sensors: a Gama-Ray Spectrometer (GRS) and a Neutron Spectrometer (NS).

Gamma rays originate from decay of radioactive elements on the surface and from impingement of galactic cosmic rays on atomic nuclei. Cosmic rays striking nuclei released neutrons as well as gamma rays. Some of the neutrons struck other nuclei and pushed them to an excited state. When those nuclei returned to their normal state, gamma rays were emitted at energies characteristic to the element. Gamma rays in the energy range of 60 keV–9 MeV emitted from the upper surface of Mercury were detected and processed by the Gamma-Ray Spectrometer.

Fast neutrons were produced directly by cosmic rays striking atomic nuclei. Some neutrons were moderated (slowed) by scattering by light elements such as hydrogen and these were referred to as epithermal neutrons. A third class of neutrons, referred to as thermal neutrons, was slowed by scattering and absorption from heavier elements and they had still lower energy levels. The Neutron Spectrometer detected and processed neutrons in three energy bands: Fast neutrons (0.5–7 MeV), epithermal neutrons (1–55 eV), and thermal neutrons (0.025–1 eV).

Gamma-Ray Spectrometer – The Gamma-ray Spectrometer (GRS) was a cylindrical instrument 31 cm long and about 24 cm in diameter. A small cryogenic cooler was mounted on one side of the cylinder. The cooler was attached to a large plate that served as a radiator for heat from the cooler. The radiator plate also served as the mount for the spectrometer. The GRS was mounted to the bottom of the spacecraft, just outside of the launch vehicle adapter ring and near the spacecraft –X axis. The field of view of the GRS was parallel to the +Z-axis of the spacecraft.

The detector was a highly-pure germanium crystal 5 cm in diameter and 5 cm long. It was mounted in a cryogenic chamber cooled to 90 K by a rotary Sterling type cooler. The cryogenic chamber was positioned inside of a cup-shaped borated plastic (BC454) shield that was 9 cm in diameter and 9.5 cm long with wall thickness of 1.25 cm. The scintillator material in the shield allowed gamma rays that penetrated both the shield and germanium crystal scintillator to cancel and that reduced background noise. The shield also established the field of view of the GRS of 45 degrees.

Impingement of gamma-rays caused flashes of light from the scintillation combination of germanium crystal and the shield. The flashes were detected by a photomultiplier tube. Electronic circuits following the photomultiplier tube amplified the signal and applied it to a 14-bit analog-to-digital converter. The output of the digital-to-analog converter was a digital representation of the amplitude of the scintillation caused by gamma-rays.

Cryogenic temperature of about 90 K, necessary for operation of the germanium crystal scintillator, was achieved by use of a rotary Sterling type cryocooler. The cryocooler was a relatively small device built by Ricor Systems as their model K508. The device was 61 mm by 58 mm by 71 mm in size and weighted 450 grams. It was mounted to the side of the GRC. Because of the limited lifetime of the rotary sterling cooler, it was turned on only once briefly during the long cruise to Mercury. It was turned on during each of the three flybys of the planet and during the orbiting mission. The cooler operated for about 14 months in orbit around Mercury before it failed. The weight of GRS, including associated electronic circuits, was 9.2 kg. It consumed 16.5 watts of power.

A summary of the chemical composition of Mercury as determined from gamma-ray spectrometer measurements was reported by Nittler, et al. in the book *Mercury The View after Messenger*. Measurements of some of the radioactive elements are reproduced below:

Potassium – 1288 ± 234 parts per million (average)
Thorium – 0.155 ± 0.054 parts per million
Uranium – 90 ± 20 parts per billion

Measurements of other elements were reported as ratio by weight of that element to silicon (Si). The abundance of silicon in the upper layer of the surface of Mercury was established as about 24.6 percent by weight. That abundance was nearly the same around the planet.

Aluminum/Si – 0.129 + 0.05, –0.13
Sulfur/Si – 0.092 ± 0.015
Calcium/Si – 0.24 ± 0.05
Iron/Si – 0.077 ± 0.013
Sodium/Si – 0.12 ± 0.01
Oxygen/Si – 1.2 ± 0.1

<u>Neutron Spectrometer</u> – The Neutron Spectrometer (NS) measured the energy of neutrons resulting from cosmic rays striking the upper 40 cm or so of Mercury's

surface. Neutrons that collided with nuclei of light elements such as hydrogen lost some of their energy before traveling on to space. Those neutrons referred to as epithermal neutrons. Mapping the variation of energy of epithermal neutrons as the spacecraft passed over the surface gave a measure of hydrogen and other light elements in the upper layer. Hydrogen can be associated with ice or liquid water.

The neutron spectrometer was made up of a borated plastic scintillator 10 × 10 × 10 cm in size sandwiched between two lithium glass scintillators 10 × 10 × 0.4 cm in size. Separate photomultiplier tubes detected flashes of light from neutrons impinging on each scintillator. The output pulse from each photomultiplier tube was proportional to the energy of the neutron striking the scintillation material. Signals at the outputs of the photomultiplier tubes were processed by an electronics unit. The output was a digital signal representing energy of the neutrons striking each scintillator.

The neutron spectrometer was mounted to the bottom of the spacecraft and in the vicinity of the magnetometer boom. Mounting was arranged so that broadside to the lithium glass plates was parallel to the +Z-axis of the spacecraft. The neutron spectrometer, including electronics unit, weighed 3.9 kg. It consumed six watts of power.

Measurements of variation in count rate of epithermal neutrons as the spacecraft traced over the surface were reported by Wilson, et al., 1973, in the paper: *Messenger Gama Ray Spectrometer and epithermal Neutron Hydrogen Data Reveal Compositional Differences Between Mercury's Hot and Cold Poles*. Maps of count rate in the northern hemisphere show lower count rates at high latitudes. The researchers converted epithermal count rate to abundance of hydrogen and the resulting maps vividly showed increased abundance of hydrogen in the north polar region. A likely source of the hydrogen is water ice lying below a cover of regolith.

X-Ray Spectrometer (XRS)

The X-Ray Spectrometer (XRS) was used to measure abundances of elements in the upper millimeter of Mercury's crust. It detected fluorescent X-rays resulting from impingement of X-rays from the sun. Solar X–rays excited atoms on the surface to a higher energy level. After the event, the atoms returned to their normal states and the excess energy was radiated in the form of X-ray photons with energy unique to a particular element. For example, the fluorescent X-ray energies associated with magnesium, sulfur, and iron were 1.25 keV, 2.31 keV, and 6.40 keV respectively. The quiet sun emitted sufficiently high levels of X-rays to excite the lower energy elements of magnesium, aluminum, and silicon but high-level X-rays emitted by the sun during solar flare events were necessary to excite heavier elements from sulfur to iron.

The XRS used gas filled proportional counters to determine the energy of fluorescent X-rays. The instrument contained three cylindrical metal ionization chambers with thin beryllium windows to admit X-rays. The windows, which had an active area of 10 square centimeters, faced the +Z-axis of the spacecraft. Baffling resulted in a field of view of the windows of 12 degrees. The instrument was mounted within the launch vehicle adapter ring.

Energies of the predominate X-ray lines from the lighter elements, magnesium, aluminum, and silicon were closely grouped at 1.25 keV, 1.49 keV, and 1.74 keV respectively. The resolution of the instrument was not fine enough to separate these energies, so selective filters in individual ionization chambers were used. One of the ionization chambers had a thin foil of magnesium over the window. A second ionization chamber had a thin foil of aluminum over its window. The third chamber had no filter. Simultaneous analysis of the spectrums of all three ionization chambers allowed identifying the energies of the three elements. The resolution of the instrument was sufficient to separate emission lines from the heavier elements; sulfur, calcium, titanium, and iron.

The incident flux of X-rays from the sun varies considerably and so the solar flux was measured continuously to provide reference for the planetary measurements. The flux of solar X-rays was measured by a silicon detector that viewed the sun through a small hole in the sunshade. The detector, which had an active area of only 0.1 square millimeters, was protected from the heat of the sun by two thin beryllium foils.

Data from the X-ray spectrometer was analyzed by Nittler et al., 2016, and maps of abundances of the various elements were reported in a paper, *Global Major-element Maps of Mercury from Four Years of Messenger X-Ray Spectrometer Observation*. The maps, which were ratios of abundances of elements to silicon, were presented in cylindrical coordinate form. A particular region between 0 degrees and 60 degrees latitude and −60 to −120 degrees east longitude was referred to a "high-Mg region." The ratio of magnesium to silicon exceeded 0.7 in much of the region. The same region had relatively high values of sulfur, calcium, and iron compared abundances over the rest of the planet. The concentration of aluminum in the region was very low.

The weight of the X-Ray spectrometer, including electronic box was 3.4 kg. It consumed 6.9 watts of power.

Magnetometer (MAG)

The purpose of the magnetometer was to characterize the magnetic field of Mercury and its interaction with the solar wind. The magnetometer was a three-axis fluxgate type mounted at the end of a carbon fiber boom 3.6 meters long to move it away from small stray fields in the spacecraft. The boom extended parallel to the +Y-axis of the spacecraft, which placed it in the anti-sun direction. The sensor was protected from exposure to the sun, if maneuvers took it out of the shade of the sunshade, by a conical ceramic cloth sunshade mounted on the boom a few centimeters inboard of the sensor. The weight of the magnetometer, including the boom, was 4.4 kg. It consumed 4.2 watts of power.

The magnetometer had a coarse measurement range of ±51,300 nanotesla (nT) and a fine measurement range of ±1530 nT. The fine range was used during measurements at Mercury. Analog signals from the three axes of the magnetometer were filtered, sampled, and applied to three 20-bit analog-to-digital converters. The sample rate was 20 Hz. The least significant 16 bits of the digital data was used as output of the magnetometer. The resulting resolution was 0.047 nT for the ±1530 nT measurement range. The time interval between successive data outputs could be selected from 0.05 seconds to 100 seconds.

The magnetic field of Mercury interacts with the solar wind to form a bow shock and a magnetopause on the sunward side of the planet. The magnetopause is the outer boundary of the magnetosphere. The elliptical orbit of Messenger allowed it to take measurements while crossing the bow shock and the magnetopause twice each orbit. During normal sun conditions the sub-solar magnetopause standoff distance was found to be 1.45 R_M where R_M is the radius of the planet. The bow shock was nominally located about 1.95 R_M from the center of the planet. The solar wind is deflected around Mercury causing the magnetic field to be pulled into a long tail on the night side of the planet.

The magnetic field of Mercury includes contributions from the intrinsic magnetic field of the planet and the magnetic fields generated by currents in the magnetosphere caused by the magnetic field of the planet interacting with the solar wind. The intrinsic magnetic field of Mercury was found to be a dipole type similar to that of Earth but less than one percent as strong. The magnetic equator was found to be about 480 km north of the planet's physical equator. The magnetic axis was within three degrees of the spin axis of Mercury. The net magnetic field in the region north of latitude of 30 degrees was found to be 195 ± 10 nT.

Energetic Particle and Plasma Spectrometer (EPPS)

The Energetic Particle and Plasma Spectrometer contained two instruments: an Energetic Particle Spectrometer (EPS) and a Fast Imaging Plasma Spectrometer (FIPS). Both instruments measured time of flight and energy of charged particles in the magnetosphere of Mercury and determined the nature of those particles. The EPS covered the energy range of 25 keV to 500 keV for electrons and 10 keV/nucleon to 3 MeV total energy for ions. The FIPS covered the energy range of 50 eV/q to 20 keV/q for ions. The overall weight of the Energetic Particle and Plasma Spectrometer was 3.1 kg. Peak power consumption was 7.8 watts.

Energetic Particle Spectrometer (EPS) – The EPS instrument was cylindrical in shape, 7.6 cm in diameter and 2.5 cm thick. The cylindrical shape resulted in a viewing angel of 160 degrees around the entrance half of the perimeter. The overall field of view was 160 degrees by 12 degrees. The EPS was mounted on the side of the spacecraft near the top. The field of view looked out to the side and covered from a point down parallel to the Z-axis through an arc upwards to about 70 degrees above the X-Y plane.

Velocity of positive ions was determined by noting the time of flight between a particle striking and passing through an entrance foil and an exit foil of polyimide material located inside of the cylinder. The entrance foil was wrapped around the entrance half of the cylinder and the exit foil was wrapped around the opposite half of the cylinder. The distance between foils was 60 mm. The entrance foil was covered by a very thin layer of aluminum to reject Lyman-alpha particles. Positive ions were accelerated by a 2900-volt negative potential on the entrance foil. As an ion passed through the entrance foil, electrons were liberated from the inner side of the foil. The secondary electrons were attracted to a "start" anode, and generated a start pulse. The particle then traversed the 60 mm distance to the exit foil and penetrated it generating another group of electrons from the foil. Those secondary electrons were attracted to a "stop" anode and generated a stop pulse. Velocity of the ion was determined by noting the time of flight between the entrance and exit foils.

Energy of the particles passing through the exit foil was determined by an array of six solid state detectors (SSDs) arranged in a semicircle outside of the exit foil. The peripheral coverage of each detector was 25 degrees and in total the six SSDs made up the 160-degree by 12-degree field of view. The broad direction of arrival of an ion was determined by noting which SSD responded to the event. The resolution was 13.5 degrees in elevation and 22.5 degrees in azimuth. Resolution of energy measurements was 8.0 keV.

Fast electrons penetrated the entrance and exit foils but generally they did not liberate secondary electrons from the foils. This fact allowed differentiation of electrons from positive ions. Energy of electrons was measured by the solid-state detectors although there were no start or stop pulses.

The energy range for electron measurements was 15 keV to 1.0 MeV. The energy measurement range for protons was 15 keV to 3.0 MeV and the measurement range for alpha particles was 25 keV to 3.0 MeV. Time-of-flight was measured for ions having energies greater than 5 keV. The Energetic Particle Spectrometer detected bursts of high energy electrons in the magnetosphere of Mercury. Energy of the electrons usually ranged up to 100 keV although occasionally electron energies of over 200 keV were detected. Bursts of high energy electrons were observed at the magnetic equator but most were observed at high northern latitudes near periapsis of Messengers orbit. The EPS did not find any evidence of high energy protons.

Fast Imaging Plasma Spectrometer – The Fast Imaging Plasma Spectrometer (FIPS) determined the composition and dynamics of low-energy ions in the magnetosphere of Mercury. It measured the energy, angular distribution and composition of ions with energy to charge ratio up to 20 keV/q where q is the charge. The instrument was capable of measuring species from hydrogen up to iron.

An entrance aperture allowed particles to enter the instrument over 360 degrees in azimuth and from zero to −70 degrees in elevation. Particles entering were passed into a deflection system which was stepped in voltage to pass particles with a particular narrow range of energy-to-charge ratio for each voltage step. The deflection voltage was stepped over a range of 0–8000 volts, corresponding to energy-to-charge ratios of about 0–10 keV/q. There were 64 deflection voltage steps employed, each lasting for one second.

The selected ions were accelerated and passed through a thin carbon foil and into a drift space to measure time of flight. Electrons liberated from the carbon foil as the ions pass through were collected to generate a start pulse. The ions were stopped when they struck a microchannel plate (MCP). The MCP had an array of small tubes from front to back through the plate. Ions striking a tube cause secondary electrons to be emitted which cause a flood of electrons to be emitted in the tube. An imaging anode behind the MCP responded to electrons exiting the tube. Angular resolution of the instrument was about 10 degrees in both azimuth and elevation. The mass-to-charge ratio of the ion (m/q) was determined by the energy-to-charge ratio in conjunction with time-of-flight information. A particular ion was identified by its mass-to-charge ratio.

The Fast Imaging Plasma Spectrometer determined that low energy ions were present throughout the magnetosphere. The dominant ion detected was sodium. There were dense plasma ions in the magnetic cusps and central plasma sheet of Mercury. The cusps of the magnetic field at the magnetic poles provide the least protection from the solar wind and ions were most dense in those areas. Protons and alpha particles from the solar wind were also found to be concentrated in the cusp regions. The most numerous planetary ions detected were from the sodium group that included sodium, magnesium, and silicon ions. Those ions had mass to charge ratios ranging from 21 to 30 amu/e where amu is the atomic mass unit and e is the charge.

Mercury Laser Altimeter (MLA)

The Mercury Laser Altimeter transmitted a narrow laser pulse towards the surface of Mercury and measured time until the reflected pulse from the surface was detected. The roundtrip time was proportional to the range of the MLA to the surface. Knowledge of the trajectory of the spacecraft allowed the topography of the surface to be profiled as Messenger orbited Mercury.

The transmitting and receiving apertures of the laser altimeter included a transmit telescope located in the center of an array of four receiving telescopes. A photograph of the laser altimeter that shows the conical structure of the telescopes is given on the next page (Fig. 7.21).

The telescopes were fixed to the bottom of the spacecraft with field of view parallel to the Z-axis of the spacecraft. The large end diameter of each of the receiving telescopes was 14 cm and the large end diameter of the transmit telescope was 5.6 cm. The gold-colored frame around the bottom of the telescopes was 25 cm square. The height of the MLA was about 28 cm.

An array of four receive telescopes was used rather than a single Cassegrain telescope with the same equivalent aperture because of thermal considerations. The four smaller telescopes could better withstand high thermal gradients as the spacecraft transitioned between the cold night side of the planet and the hot day side.

The receive telescopes had sapphire objective lenses with an effective aperture diameter of 11.5 cm. The total aperture area for the four telescopes was 417 cm^2. Sapphire was used for optics that faced the planet since it is a material that

Fig. 7.21 Mercury laser altimeter unit. (NASA/JHUAPL/Carnegie Institution photograph)

withstands extremes of temperatures and it restricted the flow of heat into the optics. The image from the objective lens was projected onto a negative lens group that increased the focal length of the telescope. The image at the output of the lens group was applied to a mirror that folded the optical path before being focused onto an optical fiber with core diameter of 200 microns. The signals from the optical fibers from the four receive telescopes were combined to feed a silicon avalanche photodetector.

The transmit telescope was 18 cm long and had a widest diameter of 5.6 cm. The telescope contained a lens group at the output end with a sapphire exit window. The magnification of the telescope was a factor of 15. The effective output aperture of the telescope was 4.5 cm in diameter. The dispersion of the laser beam at the output of the telescope was 80 microradians (0.00458 degrees).

A laser transmitter generated pulses 6 ns wide at a wavelength of 1064 nm and pulse repletion rate of 8 Hz. The energy of each laser pulse was 20 milljoules. The MLA was capable of operating at ranges up to 1500 km from the surface when the laser beam was nadir looking and up to 1000 km when the beam was at an angle of 40 degrees off-nadir. As a result of the requirement that the sunshade be pointed at the sun during noon-midnight orbits, the transmit laser beam could strike the surface at angles up to 50 degrees off-nadir for some range measurements.

Messenger's elliptical orbit had periapsis altitudes between 200 km and 500 km at 60–70 degrees north latitude. Apoapsis was at an altitude of about 15,000 km. The inclination of the orbit was 82.5 degrees and the period was 12.1 hours. The laser altimeter was within its operating range to the surface for 15–40 minutes each orbit depending on time of the Mercury year.

The received echo signals at the output of the optical fibers from the four receiving telescopes were combined and applied to a silicon avalanche photodetector. The received optical pulse was broadened by surface irregularities and by different ranges associated with off-nadir viewing. The output of the photodetector was amplified and applied to a bank of three filters. The filters were arranged to approximate a matched filter for broadened pulses. The impulse response times of the three filters were 8.5 ns, 63 ns, and 283 ns respectively.

The narrow pulse filter was followed by two detectors, one set at low threshold and the other set at high threshold. A signal with high signal-to-noise ratio would activate the low and high thresholds at the leading edge of the received pulse and again at the falling edge of the pulse. Timing of the received pulse from the high threshold yielded the most accurate range measurement and lowest false alarm rate due to noise. The other two filter channels were followed by a low threshold detector to accommodate lower amplitude echo signals. The leading edge of the signal detected by the high threshold detector generated a pulse that was used in conjunction with the laser trigger pulse to determine round trip time of the laser pulse and hence range to the surface.

Range gates were used to bracket the time delay of the expected echo pulse return. The false alarm rate of the low threshold detectors in each of the three filter channels was monitored in a time interval before the range gate time. The low detection threshold was then adjusted for an acceptable false alarm rate for each channel.

The time delay between the laser trigger and detection of the echo pulse in the high threshold channel and each of the three low threshold channels was determined by a range measurement unit. The range measurement unit measured time interval be noting the number of cycles of a 5 MHz clock in the interval and then interloping between clock pulses with a tapped delay line.

The outputs of the range measurement unit were applied to a central processing unit (CPU) that included a radiation-hardened microcomputer, memory, and a field-programmable gate array. The CPU controlled operation of the laser altimeter and formatted data for output to the telemetry system.

Measurements by Mercury Laser Altimeter – A polar sterographic projection of the topography of the northern hemisphere of Mercury from data from the MLA is shown on the next page (Fig. 7.22). The color code in the graphic uses red for the highest elevation and blue for the lowest elevation. The lowest elevation is 5.38 km below the average elevation of Mercury and located on the floor of the Rachmaninoff basin, which is identified in Fig. 7.22. The highest elevation is 4.48 km above the average elevation in a mountain range just south of the equator. Locations of prominent craters are circled in the graphic.

Conclusion

The Messenger mission was a great success. The spacecraft orbited Mercury for 4 years and returned a wealth of information to Earth about the atmosphere and surface of the planet. The entire surface was mapped by high quality photography. Laser altimeter data allowed construction of a detailed topographical map of the

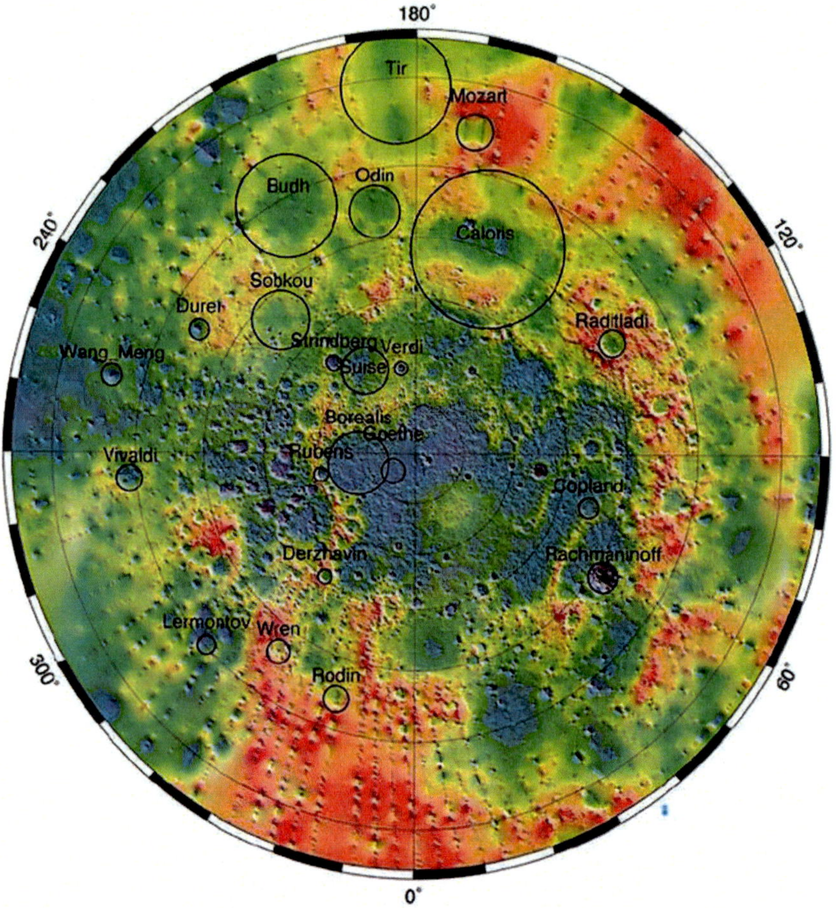

Fig. 7.22 Topography of the northern hemisphere of Mercury from MLA measurements. (NASA/JHUAPL/Carnegie Institution graphic)

northern hemisphere. The magnetic field of Mercury and its interaction with the solar wind was determined along with abundances of constituents of the atmosphere.

Mission Patch for the Messenger Program

A descriptive mission patch was developed for the Messenger program. A copy of the patch from the John Hopkins website is shown on the next page (Fig. 7.23).

Fig. 7.23 Patch for Messenger Mission. (NASA/JHUAPL/Carnegie Institution graphic)

Bibliography

Anderson, Brian J., et al., *The global magnetic field of Mercury from Messenger orbital observations*, Science, Vol 333, Issue 6051, 30 Sep 2011

Artis, David A., et al., *Messenger: Flight Software Design for a Deep Space Mission*, 2007 IEEE Aerospace Conference paper #1089

Boeing Historical Snapshot, *Mariner 10 Spacecraft*, https://www.boeing.com/history/products/mariner-10-spacecraft.page

Domingue, D L., et al. *Spectrophotometric properties of Mercury's surface derived from the Messenger MASCS observations,* Lunar and Planetary Science XLVIII (2017)

Dunne, James A. and Burgess, Eric, *The Voyage of Mariner 10*, NASA History Office, SP-424, updated 2004

Hash, Christopher; et al., *Messenger Mercury Dual Imaging System (MDIS) Experiment Data Record (EDR) Software Interface Specification (SIS)*, Johns Hopkins University document MDIS EDR SIS V2T, June 2015

Hawkins, S. Edward, et al., *In-Flight Performance of Messenger's Mercury Dual Imaging System*, SPIE Conference on Instruments and Methods for Astrobiology and Planetary Missions XII, August 2009

Hawkins, S. Edward, et al., *The Mercury dual Imaging System on the Messenger Spacecraft,* Space Science Reviews, Vol. 131, Issue 1-4, pp 247-338, August 2007

Ho, George C., et al. *Spatial distribution and spectral characteristics of energetic electrons in Mercury's magnetosphere,* Journal of Geophysical Research: Space Physics / Volume 117, Issue A12, 2012

Jet Propulsion Laboratory Technical Memorandum 33-734, *Mariner Venus-Mercury 1973 Project Final Report, Volume I*, September 15, 1976

Jet Propulsion Laboratory Technical Memorandum 33-734, *Mariner Venus-Mercury 1973 Project Final Report, Volume II*, December 1, 1975

Johns Hopkins APL, *Unlocking the Mysteries of Planet Mercury*, https://messenger.jhuapl.edu

Livi, S. A., et al., *The energetic particles spectrometers (EPS) on Messenger and New Horizons*, Proceedings of the Tenth International solar Wind Conference, American Institute of Physics, 2003

McAdams, James V, et al., *Messenger – Six Primary Maneuvers, Six Planetary flybys, and 6.6 Years to Mercury Orbit*, Advances in Astronautical Sciences, January 2012

McAdams, James V., et al., *Trajectory Design and Maneuver Strategy for the Messenger Mission to Mercury*, Journal of Spacecraft and Rockets, Vol. 43, No. 5 September 2006
Mercury Messenger, *Mariner 10: A Retrospective*, Issue 10, December 2003
NASA NSSDCA ID 2004-030A-06 *Energetic Particle and Plasma Spectrometer (EPPS)*
NASA Planetary Data System, *Mercury Photojournal,* 2011
NASA-NSSDCA, *Mariner 10*, NSSDCA/COSPAR ID:1973-085A
Nittler, Larry R., et al., *The major-element composition of Mercury's surface from Messenger X-ray spectrometry*, Science Vol. 333, No. 6051, 2011
Nittler, Larry R., et al., *Global Major-element Maps of Mercury from Four Years of Messenger X-Ray Spectrometer Observation*, presented at 47th Lunar and Planetary Science Conference (2016)
Solomon, Sean C., et al., *Messenger Mission Overview*, Space Science Reviews 131, 3–39 (2007)
Solomon, Sean C., et al., *Mercury The View after Messenger,* Cambridge University Press, New York, NY, 2018
Wilson, James H., *Mariner Venus Mercury 1973*, Jet Propulsion Laboratory, October 15, 1973
Zurbuchen, Thomas H., et al., *Innovations in plasma sensors*, Journal of geophysical Research: Space Physics, Volume 121, Issue 4, 2016

Chapter 8
Japanese and European Space Agency Spacecraft that Explored Venus 2005–2020

The European Space Agency spacecraft, *Venus Express*, and the Japanese spacecraft, *Akatsuki,* conducted complementary missions to investigate dynamics of the clouds and atmosphere of Venus. Both missions involved orbiting spacecraft that gathered information using specialized cameras and scientific instruments. Venus Express entered a near polar orbit of Venus in April 2006 and gathered data until November 2014. Akatsuki entered an equatorial orbit of Venus in December 2015 and it was still operating in 2021.

Venus Express Spacecraft

Venus Express was a European Space Agency (ESA) spacecraft designed to orbit Venus and investigate the atmosphere and plasma environment of the planet. Venus Express evolved from a call from ESA in 2001 to industry and academia for proposals for a spacecraft that would use the basic design of the Mars Express spacecraft that was then in work. Studies indicated that the Mars Express design could be adapted to a Venus orbiter and the program went forward. Venus Express was built by Astrium, an aerospace company in Stevenage, UK. Astrium is now part of Airbus Defence and Space.

Work on Venus Express started in January 2003 and the spacecraft was completed in September 2004. It was launched towards Venus by a Russian Soyuz-Fregat launch vehicle in November 2005. The program was managed by the Scientific Projects Department of ESA. An artist's depiction of Venus Express orbiting Venus is shown on the next page (Fig. 8.1).

T. Lund, *Spacecraft that Explored the Inner Planets Venus and Mercury*,
Springer Praxis Books, https://doi.org/10.1007/978-3-031-29838-7_8

Fig. 8.1 Artist's depiction of Venus Express orbiting Venus. (ESA graphic)

Mechanical Configuration of Venus Express

Venus Express borrowed heavily from the Mars Express program. The basic bus structure and propulsion were essentially the same as on Mars Express. Many of the spacecraft systems were also the same and several of the science instruments were the same.

The body of the spacecraft was a box-like structure 1.7 meters long, 1.7 meters wide, and 1.4 meters high. The main propulsion engine is at the bottom of the spacecraft in the painting. The spacecraft coordinate system sited the Z-axis along the longitudinal axis of the spacecraft with the propulsion engine on the –Z side. The high gain antenna faced outward on the +X side of the spacecraft. The positive Y-Axis was parallel to the support struts for the right-hand solar array in the painting.

A launch vertical adapter ring, 94 cm diameter and 20 cm high and conical in shape, was located at the bottom of the spacecraft. The nozzle for the main propulsion engine protruded through the center of the adapter ring. Structures were built into the adapter ring to support a floor to hold two cylindrical propellant tanks. The tanks sat on the floor with their longitudinal axes parallel to the Z-axis. Two vertical shear walls in the Y-Z plane sandwiched the two tanks. The panels consisted of honeycomb panels faced with aluminum that were mounted to the bottom structure. A third vertical shear panel in the X-Z plane was mounted in the center between the two tanks. Top and side panels completed the box. Electronics for the spacecraft was mounted on the interior side of the side panels. The –X side of the spacecraft held thermal radiators for the science sensors and the +Y and –Y sides held radiators for spacecraft systems. The other surfaces of the spacecraft were covered with multi-layer insulation blankets.

Solar arrays extended from both sides of the spacecraft. The span from tip-to-tip of the arrays was eight meters. A parabolic high-gain antenna was mounted to the +X side of the spacecraft and a smaller high-gain antenna was mounted on the –X side. The fields of view of the scientific instruments were directed parallel to the +Z-axis of the spacecraft.

The total weight of the spacecraft, including the launch adapter ring, was 1241 kg. That weight included 570 kg of propellants and 94 kg of scientific instruments.

Venus Express Systems

Propulsion System

The propulsion system included a main propulsion engine with thrust of 414 newtons (93 pounds) and eight thrusters, each with thrust of 10 newtons (2.2 pounds). The main engine was used to slow the spacecraft for orbit insertion. The thrusters were used for attitude control and small trajectory corrections. Of the eight thrusters, four were primary and four were redundant. Two redundant 10-newton thrusters were mounted to each corner of the –Z end of the spacecraft with their thrust parallel to the Z axis. The main engine was mounted on the –Z end of the spacecraft and thrusted along the Z axis.

Fuel for the main engine and for the thrusters was monomethyl hydrazine (MMH). The oxidizer consisted of the combination of nitrogen tetroxide and 3% nitric oxide. The fuel and oxidizer were contained in identical cylindrical tanks, each holding 267 liters. Helium for pressurization of the propellant tanks was contained in a spherical 35-liter tank. Propellant flow was controlled by valves that ported helium pressurant into the propellant tanks and ported fuel and oxidizer to the engine and to the thrusters. The propellants ignited upon contact (hypergolic).

Attitude Control

Venus Express was a 3-axis stabilized spacecraft. Attitude control was performed by an Attitude and Orbit Control System (AOCS). The AOCS interfaced with sensors on the spacecraft and controlled the propulsion, reaction wheel, and solar array positioning subsystems. The AOCS was managed by a Control and Data Management Unit (CDMU). Two CDMUs were provided for redundancy. The CDMU contained two MA3–1791 microprocessors. One microprocessor ran software to control the

AOCS. That software controlled sensors and actuators and ran the control algorithms for the spacecraft.

Attitude of the spacecraft was referenced to inertial space by means of two redundant star trackers. Each star tracker had a field of view of 16 degrees and was capable of tracking stars with magnitude between 1.7 and 5.5. The two star trackers were mounted with an angle of 30 degrees between their optical axes for added redundancy.

Two redundant Inertial Measurement Units (IMUs) maintained attitude reference between updates from the star tracker. Each IMU had three gyros and three accelerometers, one for each major axis. A sun tracker with field of view of 18 degrees and accuracy of one degree was used for initial alignment of the solar arrays with the sun.

Reaction wheels were used to control attitude of the spacecraft. Four reaction wheels were provided. Any three of them were sufficient to control the spacecraft. Each reaction wheel had momentum capability of 12 Nms. Attitude control could also be performed by firing the thrusters but the thrusters were used sparingly to conserve propellant. Two redundant 10-newton thrusters were mounted to each corner of the base of the spacecraft with their thrust parallel to the Z axis.

Electrical Power for Venus Express

The primary source of electrical power for the spacecraft was two solar arrays with total area of 5.7 square meters. Each array contained 1056 triple-junction gallium arsenide solar cells arranged in 12 rows. An aluminum strip was installed between each row to minimize heating from the sun. The total power generated by the two arrays was about 1400 watts at Venus.

Power from the solar arrays was supplemented by power from three lithium-ion batteries. Battery power was needed when the spacecraft was maneuvering and when it was in the shadow of Venus. The batteries had capacity of 24-ampere-hours each. The total capacity of the batteries was about 500 watthours. The batteries were managed and charged by a separate Battery Charge and Discharge Regulator for each battery.

Power management was performed by a Power Control Unit that accepted power from the solar arrays and batteries and conditioned it to develop a regulated 28-volt power bus for spacecraft systems and for science sensors.

Telecommunications System

Communication with Earth could be conducted using either S-band or X-band frequencies. The telecommunications subsystem included two redundant transponders. Each transponder was capable of transmitting and receiving at either S-band or

X-band. The S-band transmitter frequency was 2296 MHz and the S-band receiver frequency was 2100 MHz. The X-band transmitter frequency was 8415 MHz and the X-band receiver frequency was 7166 MHz. The power output from the S-band transmitter in the transponder was five watts. The X-band transponder transmitter drove a 65-watt traveling wave tube amplifier. There were two redundant 65-watt traveling wave tube amplifiers and either X-band transponder could drive either amplifier.

The spacecraft had two low-gain S-band antennas. Taken together, they provided nearly omni-directional coverage. Either of the S-band transmitters or either of the S-band receivers in the transponders could be connected to either low-gain antenna. The low-gain antennas were used primarily during launch and the early operations phase of the mission.

Venus Express carried a 1.3-meter diameter parabolic high-gain antenna (HGA1) for communication with Earth. The antenna contained feeds for both S-band and X-band. The S-band receiver or the S-band transmitter of either transponder could be connected to the antenna. Likewise, the X-band receiver of either transponder could be connected to the antenna and either of the power amplifier outputs in the transmit path could be connected to the antenna. The dual-band high-gain antenna was mounted on the +X side of the spacecraft.

There was a second high-gain antenna, HGA2, mounted on the –X side of the spacecraft. It was a parabolic type, 30 cm in diameter and used for X-band communications. The second high-gain antenna was needed so that Venus Express could be reoriented during portions of Venus orbit around the sun to prevent the thermally sensitive –X side of the spacecraft from being exposed to the sun for long periods of time. Communications was carried out using HGA2 for a portion of the orbit of Venus when Venus was between the sun and Earth.

Data Management System

Data handling functions of the spacecraft were performed by the Data Management System (DMS). The DMS included two redundant Command and Data Management Units (CDMU). Each CDMU contained two 16-bit MA3–1750 microprocessors. The microprocessors were associated with one million words of RAM and 512,000 words of Electrically Erasable Programmable Read-Only Memory (EEPROM). One of the microprocessors performed data management tasks and the other performed attitude and orbit control system (AOCS) tasks.

Functions performed by the CDMUs include: decoding, processing, and distributing commands from Earth, collecting data from spacecraft systems and scientific instruments and formatting the data for storage in a solid-state memory for later transmission to Earth. The CDMU also generated an accurate clock to maintain spacecraft systems timing.

Venus Express traveled in a highly elliptical polar orbit around Venus. The period of the orbit was 24 hours. Time in the 24-hour orbit was divided into an observation phase and a communications phase. The observation phase took place when near periapsis and lasted for about two hours. The communications phase followed that lasted for about eight hours. The spacecraft was oriented to point the science instruments towards Venus during the observation phase and to point one of the high-gain antennas towards Earth during the communications phase.

Data from the instruments and satellite housekeeping data was stored in a solid-state memory during the observation phase. The memory was read out by the CDMU and the data applied to the selected transponder for transmission to Earth during the communications phase. The solid-state memory consisted of three 4-Gbit memory modules giving total memory of 12 Gbits. Data could be transmitted to Earth at rates from 15 kbps to 228 kbps depending on range to Earth and antenna used on the spacecraft.

The Data Management System interfaced with the attitude and orbit control system (AOCS) by means of an AOCS Interface Unit. It interfaced with other spacecraft systems through a Remote Terminal Unit.

Flight of Venus Express

Venus Express was launched on 9 November 2005 from the Baikonur Cosmodrome by a Soyuz-Fregat launch vehicle. The three-stage Soyuz launcher placed the joined Fregat upper stage and Venus Express into a high trajectory. Fregat was then fired to place Fregat-Venus Express into a circular Earth orbit. After coasting in orbit for 70 minutes, a second firing of Fregat accelerated the combined unit into an escape trajectory and on a path to intercept Venus. Fregat was later separated from Venus Express and deployment of appendages on Venus Express, including solar arrays, followed. The spacecraft then began its 153-day cruise to Venus.

A trajectory correction was made on 11 November 2005 and a final course adjustment was made on 29 March 2006. Venus Express arrived at Venus on 11 April 2006. The spacecraft was appropriately oriented and the main engine was fired for 50 minutes to slow the spacecraft for capture into a polar orbit around Venus. The initial orbit had periapsis altitude of about 400 km and apoapsis altitude of about 350,000 km. A series of orbit corrections were made achieve a final orbit with periapsis at 250 km and apoapsis at 66,000 km. The period of the orbit was 24 hours. The inclination of the orbit was 90 degrees (polar orbit). The initial latitude of periapsis was 78 degrees north.

The polar orbit allowed all longitudes on the surface to be covered in one Venus sidereal day (243 Earth days). The nominal mission was planned for two Venusian sidereal days. The mission was controlled from launch to the end of the mission from the European Space Operations Center (OSOC) in Darmstadt, Germany.

Communications with Venus Express, once it was in orbit around Venus, was carried out through ESA's Cebreros - DSA 2 ground station near Madrid, Spain. DSA stands for Deep Space Antenna. The Cebreros station was equipped with a

35-meter diameter parabolic antenna with communication capability in X-band or K-band. ESA's New Norcia – DSA 1 ground station near New Norcia, Australia was used for some portions of the mission. It also had a 35-meter diameter antenna. It could communicate in S-band or X-band.

The first 50 days in orbit were spent verifying operation and commissioning of instruments. The nominal phase of the mission then commenced that was planned to last two Venusian sidereal days.

The path of the 24-hour orbit was nearly vertical. The low altitude portion of the orbit around periapsis passed around the northern portion of the planet. Apoapsis was 66,000 below the planet. Zero time for the orbit was defined to be at periapsis and apoapsis was at 12 hours orbital time. Close-up observation of the planet, with the sensors pointed towards nadir was made from orbital time of 23 hours to 1 hour. Data was transmitted to Earth from 2 hours to 10 hours orbital time. The time around apoapsis, 10 hours to 14 hours orbital time, was used to make images of the full disk of Venus with the camera and the Visible and Infrared Thermal Imaging Spectrometer. During the period 14 hours to 22 hours orbital time, observations were made of cloud formations.

The 24-hour orbit of Venus Express allowed data transmission form Venus Express to occur when the Cebreros ground station near Madrid had a view of Venus. Further, the downlink data was received about the same time each day, which was a benefit to operators of the ground station.

The nominal mission lasted from 4 June 2006 to 2 October 2007. The spacecraft was operating well and a mission extension was planned to continue observation of Venus. The first extension covered the time period 3 October 2007 to 30 May 2008. A second extended mission followed that lasted from 31 May 2009 to 22 August 2010. A third extended mission was conducted from 22 August 2010 to 31 December 2012. A fourth mission extension to 2014 was approved in November 2012

Propellant was running low in 2014 and formal science operations ended on 15 May 2014. A series of aerobraking operations were then conducted by lowering the orbit at periapsis to about 130 km and observing drag and heating of the spacecraft by the tenuous atmosphere at that altitude. Propellant was exhausted after raising the orbit to 450 km. The spacecraft sank lower in the atmosphere and signals were lost in November 2014. The mission was formally ended on 16 December 2014.

Scientific Instruments

Venus Express was host to six scientific instruments. Several of the instruments were adapted or taken directly from those developed for other spacecraft. They were provided by collaborative work by scientists and institutions in ESA member countries. The instruments and a brief account of their function are listed below:

Venus Monitoring Camera (VMC) – Imaged Venus in infrared, ultraviolet, and visible wavelengths

Analyser of Space Plasma and Energetic Atoms (ASPERA-4) – Investigated the interaction between the Venus atmosphere and the solar wind by measuring exiting particles.

Magnetometer (MAG) – Investigated magnetic field due to interaction of atmosphere of Venus with the solar wind.

Planetary Fourier Spectrometer (PFS) – Measured temperature of the surface and clouds and composition of the atmosphere.

Ultraviolet and Infrared Atmospheric Spectrometer (SPICA/SOIR) – Measured thermal profiles and composition of upper atmosphere.

Visible and Infrared Thermal Imaging Spectrometer (VIRTIS) – Tracked clouds in ultraviolet and in infrared and provided data on composition of lower atmosphere. Generated data for a map of surface temperatures of Venus.

Venus Monitoring Camera

The Venus Monitoring Camera (VMC) generated global images of Venus to allow study of the dynamics of the atmosphere. The camera's relatively wide field of view of 17.5 degrees allowed viewing the whole planet during the more distant portions of the orbit. The camera produced images in four narrow spectral bands centered at 365 nm, 513 nm, 935 nm and 1000 nm.

The camera used four objective lens channels that shared a single CCD detector. Each channel projected an image onto a quadrant of the CCD. The focal length of each objective channel was 13 mm. Each of the optical channels contained a selective filter.

The wavelength bands of the four channels are given in Table 8.1 below.

The 355–385 nm band was chosen to study the distribution of ultraviolet absorbers in the cloud tops and track motions of the patterns in the clouds. It was also used to investigate the vertical distribution of haze above the clouds. The 503–523 nm band was chosen to investigate and map the spatial distribution of airglow on the night side and image the high-altitude haze layer at the limb of the planet. The 990–1030 nm band took advantage of the transmission window in the clouds at that wavelength band to image the temperature of the surface of the night side of Venus and the temperature of its lower atmosphere. The two infrared channels allowed investigating the amount of water vapor in the lower atmosphere.

Table 8.1 Channels of Venus Mapping Camera

Classification	Center wavelength	Filter bandpass
Ultraviolet A	365 nm	355–385 nm
Visible	513 nm	503–523 nm
Near-infrared	935 nm	900–970 nm
Near-infrared	1000 nm	990–1030 nm

The CCD detector was a Kodak KAI-1010 type with 1024 by 1024 pixels. Image data from the four quadrants of the CCD was gathered by readout electronics and applied to a 1-Gbit memory within a Digital Signal Processor (DPU). The DPU contained necessary signal processing functions on a single chip. The processer within the DPU was capable of compensating the image by flat fielding and of performing JPEG2000 image compression. Image data was sent to the Data Management System by a high-speed IEEE 1355 data bus.

The Venus Monitoring Camera took advantage of a window in the CO_2 atmosphere at a wavelength of 1000 nm that allowed infrared radiation from the surface to pass through to the orbiting spacecraft. About 1000 images taken at that wavelength were assembled into a map of a portion of the surface in the infrared. The different temperatures of the map were used to infer geology of the surface.

Analyser of Space Plasma and Energetic Atoms (ASPERA-4)

ASPERA-4 measured energetic neutral atoms (ENA), ions, and electrons in the atmosphere of Venus. It was used to investigate the interaction between the solar wind and the atmosphere and determine global plasma and neutral gas distribution. It also determined the mass and characteristics of outflowing atmospheric material. In particular, it was used to determine the escape rates of oxygen ions, hydrogen ions, and helium ions.

The instrument consisted of four sensors: Neutral Particle Imager (NPI), Neutral Particle Detector (NPD), Electron Spectrometer (ELS), and Ion Mass Analyzer (IMA). Physically, ASPERA consisted of two units: the Ion Mass Analyzer unit and a Main Unit that contained the ELS, NPI, and NPD sensors. The main unit was mounted on a mechanical scanner that rotated the fields of view of the three instruments through 180 degrees in the X-Y plane. Scanning could be continuous at scan rates of 1.5, 3.0, or 6.0 degrees per second, or the angle could be set at any given value. The pointing accuracy was about 0.2 degrees. The scanner was mounted to the +Y side of the spacecraft.

Neutral particle imager (NPI) – The NPI measured the flux of neutral particles with energies between 100 eV and 60 keV. Incoming particles passed between deflector plates to separate neutral atoms from ions. The plates were 15 cm in diameter, spaced 3 mm apart, and had potential of 5000 volts across them. Charged particle were deflected and captured by the plates. The space between the deflection plates was divided up into 32 angular sectors in the X – Y plane by plastic walls. Neutral particles from a particular direction passed through that channel and struck a target. The impact of the neutral particle caused generation of secondary electrons and ions. The secondary particles struck a microchannel plate type detector and resulted in an electrical signal from that angular channel.

The overall field of view of the NPI was 9 degrees by 344 degrees. The angular resolution was 4.6 degrees by 11.5 degrees. The 32 angular channels were read out once every 62.5 milliseconds. The instrument was designed to measure only angle of arrival of particles and it did not measure mass or energy.

Neutral Particle Detector (NPD) – The NPD measured the energy and mass of neutral particles. It consisted of two identical pinhole apertures each with a field of view of 90 degrees in the X- Y (azimuth) plane and 9 degrees in the elevation plane. NPD responded to particles in the energy range of 100 eV to 10 keV. Particles passed between two plates separated by 4.5 mm with potential of 8000 volts between them. Charged particles were deflected and captured by the plates. Neutral particles passed through and struck a target at a grazing angle and caused secondary electron emissions. Those electrons were detected by a microchannel plate type detector and the resulting signal formed a start pulse.

Neutral particles were specularly reflected by the target and hit a second target at a given spacing from the first. Again, secondary electrons were produced and detected by one of three microchannel type detectors, each at a different angle. The output of one of the three detectors formed a stop pulse. The velocity of the particle was computed from the time of flight of the particle. The number of secondary electrons produced was proportional to the mass and velocity of the particle. Knowing the velocity, the mass of the particle could be computed. The angular resolution of the instrument was 5 degrees by 30 degrees.

Electron spectrometer (ELS) - The ELS determined the energy spectrum of electrons in 16 different angular sectors in the X –Y plane. Each sector was 22.5 degrees in extent. The sensor was made up of two concentric hemispherical electrodes spaced a short distance apart. A hole in the center of the upper hemisphere admitted electrons. A varying electric field between the hemispheres allowed sorting electrons by energy. The strength of the electric field allowed electrons in a particular energy range to pass between the two hemispheres and be detected. Electrons with energies outside of the selected value struck the electrodes and were captured.

Electrons passing through the gap between electrodes struck one of 16 detector elements arranged with 22.5-degree angular spacing. The total angular coverage was 360 degrees. The voltage across the electrodes was swept once every four seconds. The number of hits in a given detector in the time interval corresponding to a particular energy was determined and that gave an energy spectrum for electrons in a particular 22.5-degree angular sector.

Ion Mass Analyzer (IMA) – The IMA, which was mounted separate from the main unit, developed the mass spectrum of the incoming ions. The instrument measured mass to charge ratios (m/q) up to 40 for ion energies between 10 eV/q to 36 keV/q.

Incoming ions entered a deflection system consisting of two curved plates with a varying voltage across them. The voltage was stepped to allow positive ions to enter at angles up to ±45 degrees with a resolution of 5.6 degrees per step. The accepted ions were applied to an electrostatic analyzer that consisted of two hemispherical electrodes with a variable voltage across them. Ions entered the space between hemispheres by a hole in the center of the top hemisphere. The voltage was stepped in 96 different levels to select ion energies from 10 electron volts to 36 keV.

Ions selected by the electrostatic analyzer passed out between the hemispherical electrodes and were accelerated by an electric field and applied to a magnetic mass analyzer. That analyzer consisted of a cylindrical magnetic field that deflected lighter ions closer to the center than heavier ions and applied them to a concentric imaging anode system. The imaging anode system was installed in 32 concentric rings. The radial location of a particle impact determined its mass. The impact area was divided up into 16 angular sectors whose anode detectors measured entrance angle. The 16 sectors were each 22.5 degrees in extent. Angular coverage was 360 degrees in the X – Z plane with resolution of 22.5 degrees. The coverage was ±45 degrees in the perpendicular plane containing the Y axis with resolution of 5.6 degrees.

Measurements by the IMA were used to determine the position of the bow shock due to the solar wind interacting with induced magnetic field lines around the planet. A paper by Whittaker, et al., *Venusian bow shock as seen by the ASPERA-4 ion instrument on Venus Express*, indicates that energy of ions was 1.11 keV in the day-side solar wind and 3.74 keV in the bow shock region.

Magnetometer (MAG)

Venus was known to not have an intrinsic magnetic field. However, an induced magnetic field is generated by the interaction of the solar wind with the ionosphere. This field was explored by the magnetometer in Venus Express.

The magnetometer was a triaxial fluxgate type. Two sensors were used. The main sensor was mounted at the end of a one-meter long boom protruding from the +Z side of the spacecraft. Another sensor, located near the base of the boom, was used to remove the residual magnetic field of the spacecraft during signal processing. The main fluxgate sensor and its signal processor had a measurement range of ±262 nanotesla and a resolution of 8 picotesla. For reference, The Earth's magnetic field at the equator is about 22 microtesla.

Magnetic measurements were made once per second when the spacecraft's orbit took it away from the planet and in the presence of the solar wind. The measurement rate increased to 128 Hz when Venus Express was in the observation mode approaching and around periapsis. The higher sampling rate provided detailed magnetic measurements in the bow shock region and of the induced magnetopause.

Measurements presented by Zhang, et al., in the paper *Weak, Quiet Magnetic Fields Seen in the Venus Atmosphere* indicated large fluctuations in field strength while the spacecraft was in the altitude range of 140 km to 450 km with peak magnetic field values of about 40 nT. Below an altitude of 140 km in the vicinity of periapsis, the field was much less noisy and the field strength was less than 10 nT.

Planetary Fourier Spectrometer (PFS)

The PFS was an infrared spectrometer that was intended to investigate the atmosphere of Venus. Unfortunately, a mirror that was mechanically positioned to allow the instrument to view either a calibration black body or the atmosphere, remained stuck in the calibrate position and no useful data was obtained from the instrument at Venus.

Spectroscopy for Investigation of Characteristics of the Atmosphere of Venus/Solar Occultation at Infrared (SPICA/SOIR)

The SPICA/SOIR included three spectrometers that were used to measure thermal profiles and composition of the upper atmosphere by viewing stellar and solar occultations. Each spectrometer channel covered a different wavelength band. The ultraviolet channel covered the wavelength band of 118 to 320 nm and the near-infrared channel covered the wavelength range of 0.65 to 1.7 μm. The SOIR channel covered the wavelength band of 2.2 to 4.4 μm. The SOIR channel was only used for solar occultation measurements. The SPICA channels were used during occultations as well as for observation along the nadir.

Spectroscopy for Investigation of Characteristics of the Atmosphere of Venus (SPICA) – The ultraviolet channel of SPICA could view either the nadir direction or the sun direction. Radiation from the nadir direction was passed through entrance optics to an off-axis parabolic mirror. Radiation from the sun was applied through a shutter to a mirror that directed it towards the parabolic mirror. The shutter opened when viewing the sun. The reflected optical signal from the off-axis parabolic mirror was directed at a concave ultraviolet grating. The output of the grating was applied to an intensifier and then to a CCD detector array.

The CCD detector was an array of 288 by 384 pixels where 228 was the number of lines. Because of downlink limitations, only 5 out of the 288 lines could be transmitted to Earth at a time. To accommodate the limitation, pixel lines were binned together. During solar occultations, up to eight lines were binned together and five adjacent bins were transmitted. The field of view of the channel varied between 0.2 degrees by 0.1 degrees and 0.2 degrees by 0.4 degrees depending on binning. The solar spectrum outside of the atmosphere was recorded for reference and then the spectrum through the atmosphere at various altitudes was determined.

A paper by V. Wilquet, et al., *Preliminary characterization of the upper haze by SPICAV/SOIR solar occultation in UV to mid-IR onboard Venus Express*, gives a family of curves of transmittance as a function of wavelength for altitudes from 60 km to 126 km. The data shows essentially zero transmittance at an altitude of 60 km and full transmittance at 126 km altitude. At an altitude of 91 km, the transmittance was about 0.4 at a wavelength of 205 nm and about 0.7 at 320 nm.

The near-IR channel was arranged to view either along the nadir or the sun. An optical fiber channeled IR radiation from the direction of the sun to a shutter and then to a mirror that directed it towards the entrance optics of the sensor. The shutter was closed when not viewing the sun. The entrance optics focused the incoming IR radiation through a field of view diaphragm to a collimating lens. The output of the lens was applied to an Acousto-Optic Tunable Filter (AOTF). The field of view along the nadir was about 2.0 degrees.

The AOTF was a tunable narrow band filter consisting of a birefringent tellurium dioxide (TeO_2) crystal with piezoelectric transducers attached to opposite sides of the crystal. The transducers were excited by radio frequency (RF) energy to generate acoustic waves within the crystal. The acoustic waves varied the index of refraction of the crystal. Only a narrow range of infrared wavelengths experienced constructive interference with the acoustic wave and was transmitted. The frequency of the RF energy selected the band of wavelengths to be transmitted through the crystal.

One of the piezoelectric transducers allowed measurements in the spectral range of 0.65 to 1.05 μm and the other allowed measurements in the spectral band of 1.0 to 1.7 μm. The short wavelength transducer was driven by a source of RF energy between 250 MHz and 140 MHz. The long wavelength transducer was driven at frequencies from 140 MHz to 80 MHz. The specific frequency could be chosen to investigate a particular wavelength.

To cover the wide spectral range of wavelengths, two detectors were used in each channel. One detector was silicon and the other was Indium Gallium Arsenide (InGaAs). The two detectors were followed by integrating preamplifiers that integrated over a sampling interval. The wavelength corresponding to the signal amplitude response was known from knowledge of the frequency of the RF driving signal to the transducers. The spectral resolution of the measurements was 0.5 nm at a wavelength of 0.7 μm and 1.2 nm at 1.7 nm.

Solar Occultation at Infrared (SOIR) – The SOIR operated during solar occultation to measure transmission of the sun through the atmosphere of Venus at altitudes from 60 km to 220 km. The narrow field of view of the instrument was pointed at the sun as the spacecraft traveled in its orbit around the planet. Transmission through the atmosphere was measured in the infrared wavelength range of 2.32 μm to 4.25 μm. Vertical resolution of measurements of transmission through the atmosphere varied depending on the portion of the orbit. The vertical resolution during passage around high northern latitudes closer to periapsis varied from about 200 meters to 700 meters. Viewing through the atmosphere at the far southern latitudes resulted in vertical resolution of 2 km to 5 km.

SOIR measured transmission near the water vapor line at 2.61 μm and the HDO line at 3.58 μm simultaneously. HDO is semi-heavy water. The water vapor line was measured from altitudes between 70 km and 110 km. The HDO line was measured between altitudes of 70 to 95 km. Measurements were made in the vicinity of the north pole and down to middle south latitudes.

Papers describing the instrument and the results of measurements use wavenumber rather than wavelength to identify spectral range. Wavenumber is the reciprocal of wavelength. Thus, the wavelength range of 2.32 μm to 4.25 μm corresponds to wavenumbers from 2353 cm^{-1} to 4310 cm^{-1}.

The SOIR instrument used the same solar port as the SPICA instrument. The incoming IR radiation was applied to an acousto-optic tunable filter (AOTF). The AOTF selected a wavelength band corresponding to a particular order for dispersion of the Echelle spectrometer that followed. The Echelle spectrometer made a fine wavenumber resolution of signals within the particular grating order. The AOTF was similar to that described for the SPICA in the previous section. The center wavelength of the band passed by the filter was selected by the radio frequency of the source driving the piezoelectric transducer on the filter. The RF frequency range was 14 MHz to 26 MHz. The width of the passband was 24 cm^{-1}. The output of the AOTF was applied to the Echelle spectrometer.

An Echelle spectrometer uses gratings blazed at angles from 50 to 70 degrees. The optics is arranged to have the incidence angle about equal to the blaze angle. The grating then generates a series of repetitive diffraction orders. The SOIR spectrometer used an incidence angel of about 63 degrees and it operated at diffraction orders from 101 to 194. A diffraction order of 101 corresponded to a wavelength of 4.4 μm and an order of 194 corresponded to a wavelength of 2.3 μm. Each diffraction order covered a spectral width of 20 cm^{-1} to 40 cm^{-1}. The AOTF selected which diffraction order was used and the spectrometer made high resolution measurements within the order. The resolution of the instrument was about 0.13 cm^{-1} at 4.0 μm and 0.27 cm^{-1} at 2.3 μm.

The output of the Echelle grating was applied to a detector that had 320 columns in the direction of the slits (spectral direction) and 256 rows in the vertical direction of the atmosphere. The slit width spanned two pixels in the column (spectral) direction on the detector and the slit height spanned 32 rows on the detector.

Profiles of density of various entities as a function of altitude were made from the data returned to Earth. A paper by A. Fedorova, et al., 2008, *HDO and H_2O vertical distributions and isotopic ratio in the Venus mesosphere by Solar Occultation at Infrared spectrometer on board Venus Express* gives a plot of density of CO_2, H_2O, and HDO as a function of altitude at latitude of 79.2 degrees north. Scaling from graphs in the paper, the density of CO_2 was about 10^{18} per cubic centimeter (cc) at an altitude of 70 km and 10^{13} per cc at 118 km. The density of H_2O was about 10^{12} per cc at an altitude of 70 km and 10^8 per cc at 105 km. The density of HDO was about 10^{11} at 70 km altitude and about 2×10^8 per cc at 95 km.

Vertical profiles of density of water (H_2O) and semi-heavy water (HDO) in the atmosphere were used to determine the escape rates of Hydrogen (H) atoms and Deuterium (D) atoms from the atmosphere to better understand the evolution of the atmosphere. Observations above the clouds in the upper mesosphere gave a D/H ratio of about 250. Some researchers postulated that the high D/H ratio indicates preferential escape of hydrogen. Deuterium was left behind to recombine with OH radicals.

Visible and Infrared Thermal Imaging Spectrometer (VIRTIS)

VIRTIS gathered spectral data to determine composition of trace gasses in various layers of the atmosphere of Venus. It peered through atmospheric windows in the CO_2 atmosphere to map the temperature of the surface of the planet. It provided data to track cloud layers at ultraviolet and infrared wavelengths and to determine their composition.

The VIRTIS instrument included an imaging spectrometer operating in the visible wavelength range of 0.27 to 1.1 μm and an imaging spectrometer operating in the infrared wavelength range of 1.05 to 5.2 μm. The collection of these two mapping instruments, which had medium spectral resolution, was referred to as VIRTIS-M. VIRTIS also contained a high-resolution infrared spectrometer operating in the wavelength band 1.8 to 5.2 μm that was referred to as VIRTIS-H. The identifier, H, signified high-resolution. VIRTIS-H had a single narrow field of view that was boresighted with the center field of view of VIRTIS-M.

VIRTIS included four modules: An Optical Module and a Proximity Electronics Module for VIRSIS-M, and an Optical Module and a Proximity Electronic Module for VIRSIS-H. It also contained a Main Electronics unit. The Optical Modules were mounted externally on the −X side of the spacecraft. The fields of view of both the M and H sections were in the +Z direction. Both optical modules were mounted within a cold box that was cooled to about 130 K by radiation via a flat plate facing cold space. The electronics equipment was mounted inside the spacecraft.

<u>VIRTIS–M</u> – The VIRTIS-M section used a Shafer telescope that was matched to a slit that illuminated an Offner spectrometer. The telescope was made up of a series of mirrors. The primary mirror was scanned mechanically in a direction perpendicular to the axis of the slit to form one dimension of the spatial image. Each step of the scanner was 0.25 milliradians (mrad) wide. A total of 256 steps were used to cover a spatial extent of 64 mrad. An alternate mapping configuration held the mirror stationary and used travel of the spacecraft to generate a spatial image. A descriptive title of this mode was "push-broom." The field of view in the push-broom mode was 64 by 0.25 milliradians. The field of view in the scanning mode was 64 by 64 milliradians.

The Offner spectrometer was made up of a mirror and a spherical convex diffraction grating. The diffraction grating consisted of an inner spherical section with a fine diffraction grating for the visible channel and an outer portion with coarser grating for the infrared channel. The outputs of the spectrometer gratings were focused on separate focal plane array (FPA) detectors for the visible and infrared channels.

The detector for the visible channel was a frame transfer silicon CCD with good sensitivity at ultraviolet wavelengths. It was mounted on the spectrometer and operated at a temperature of about 150 K. The detector format was an array of 512 by 1024 pixels. The infrared detector for both the M and H sections was a Mercury Cadmium Telluride (CdHgTe) type that was cooled to 80 K by a small Sterling cycle cooler. The infrared detector format was an array of 270 by 438 pixels.

The size of the slit was 38 microns by 9.53 mm. The long dimension of the slit was arranged to cover 256 pixels on the detector array. Thus, for every mechanical scan position there were 256 elements spanning a total of 64 mrad in the direction of the long dimension of the slit. The scanning program of 256 steps resulted in image array of 256 by 256 elements. VIRTIS-M collected a spectrum for each element of the 256 by 256 element image. Each spectrum was made up of 432 different adjacent wavelength intervals. The total wavelength span of each spectrum was 0.27 to 1.1 μm for the visible channel and a span of 1.05 to 5.2 μm for the infrared channel.

A Proximity Electronics Module for VIRTIS-M (PEM-M) managed the detectors, drove the scanning mirror, and collected data from the spectrometers and converted it to a form required by the Main Electronics unit. The Main Electronics unit contained two redundant Data Processing Units. The unit acquired data from VIRTIS-M and VERTIS-H and compressed and formatted the data. It also controlled the operation of VIRTIS-M and VERTIS-H instruments and controlled the coolers for IR detectors. It provided a high-speed interface to the solid-state mass memory.

VIRTIS-H – VIRTIS-H was a high-resolution infrared spectrometer with sufficient spectral resolution to detect minor constituents in the atmosphere. Incoming infrared radiation was gathered by a reflecting telescope that consisted of two off-axis parabolic mirrors. The output of the telescope was applied to a cross-dispersing prism that illuminated an Echelle type diffraction grating. The Echelle grating separated diffraction orders 0 through 7. Rays from the grating were focused by an objective lens onto the focal plane array detector. The orders were separated across the detector array. Each spectral order covered 432 by 5 pixels where the number of pixels illuminated by the slit width was 5. The outputs of the five special pixels were averaged together. The output of the spectrometer was a 432 by 8 representation of the spectrum of each of the 8 orders.

A Proximity Electronics Module for VIRTIS-H (PEM-H) managed the detector, collected data from the spectrometers, and converted the data to a form required by the Main Electronics unit. The Main Electronics unit acquired data from VERTIS-H and compressed and formatted the data. It also controlled the cooler for the infrared detector. It provided a high-speed interface to the solid-state mass memory.

An important measurement made by VERTIS-H was the composition of the lower atmosphere on the night side of Venus. Data reported by Drossart and Piccioni at the International Venus Conference in 2016 is summarized in Table 8.2 on the next page. The measurements were taken at an altitude of about 35 km.

Where ppm is parts per million, H_2O is water vapor, CO is carbon Monoxide, and SO_2 is Sulfur Dioxide.

VIRTIS provided detailed images of the chaotic polar vortex at the south pole of Venus. The vortex was found to be constantly changing and breaking up and reforming. At times a double vortex formed and then reverted back to a single ragged rotating structure. The vortex was found to circle the pole with a period of about 2.2 days.

Table 8.2 Composition of lower atmosphere of Venus

Constituent	0°–20° Latitude	60° Latitude
H_2O	31 ± 2 ppm	31 ± 2 ppm
CO	24 ± 3 ppm	31 ± 3 ppm
SO_2	130 ± 50 ppm	130 ± 50 ppm

Akatsuki Spacecraft (Japan)

Akatsuki was a Japanese spacecraft that was inserted into a nearly equatorial orbit around Venus to investigate the composition and dynamics of the clouds and atmosphere. The spacecraft was built by NC Space Technologies, Ltd., a leading developer of satellites and space equipment in Japan. The Akatsuki program was managed by the Japan Aerospace Exploration Agency (JAXA). Development of Akatsuki began in 2001 and it was launched in 2010.

The final placement of Akatsuki in orbit around Venus was a triumph of human ingenuity. The main rocket engine failed on first attempt to insert Akatsuki in orbit around Venus in December 2010 and the spacecraft went into orbit around the sun. The mission was salvaged through detailed analysis and planning to use the small maneuvering thrusters in lieu of the main rocket for orbit insertion. The orbits of Akatsuki around the sun were adjusted by well-timed small values of thrust to finally bring the spacecraft close to Venus and then slow it for successful orbit insertion in December 2015, five years after the original attempt.

The spacecraft is presently in a highly elliptical orbit around Venus with periapsis altitude of 1000 km to 10,000 km and apoapsis altitude of about 360,000 km. The orbit is nearly equatorial with inclination of about three degrees. An artist's rendering of the spacecraft in orbit around Venus is shown on the next page (Fig. 8.2).

Mechanical Configuration of Akatsuki

Akatsuki was a cube-shaped, compact spacecraft 1.45 by 1.45 by 1.04 meters in size. The prominent circular item on the bottom side of the spacecraft in the painting was a high-gain antenna. The antenna, which was used to transmit data to Earth, was a slotted waveguide type about 90 cm in diameter. A smaller slotted array antenna, 30 cm in diameter, was mounted on the same side of the spacecraft and was used to receive commands from Earth. The main propulsion engine was located on the opposite side of the spacecraft from the high-gain antenna.

The coordinate system positioned the Z-axis parallel to the thrust of the main propulsion engine and the boresight of the high-gain transmit antenna with the +Z direction along the boresight of the antenna. The Y-axis was parallel to the shafts holding the solar arrays and the +X-axis was in the direction of the boom holding a

Fig. 8.2 Artist's rendering of Akatsuki orbiting Venus. (ISAS/JAXA graphic)

low-gain antenna on the left side of the spacecraft in the painting. A similar boom and antenna were mounted on the –X side of the spacecraft.

A thrust tube, mounted in the center of the spacecraft and aligned with the Z-axis, held the main propulsion engine and propellant tanks. A helium tank was mounted to the –X side panel. The +Y and –Y side panels were hinged to the panel containing the slotted array antennas and the three panels were installed as a unit. Five cameras, constituting the main science suite, were mounted on the –X panel with their fields of view in the -X direction. Electronics units for spacecraft systems and science were mounted to the insides of the side panels. Solar arrays were mounted on shafts protruding from the +Y and –Y sides of the spacecraft. The shafts holding the arrays could be rotated for best orientation of the arrays to the sun.

The weight of the spacecraft at launch was 518 kg, 196 kg of which was propellant and 34 kg was science instruments.

Akatsuki Systems

Propulsion System

The main propulsion engine, which provided thrust of 500 newtons (112 pounds), was used for the large velocity change necessary for insertion into orbit around Venus. The engine used hydrazine fuel and dinitrogen tetroxide oxidizer. Fuel and oxidizer were forced out of tanks and through valves to the engine by helium pressurant. Helium was contained in a spherical tank and ported through control valves to the fuel and oxidizer tanks. The engine failed during the orbit insertion burn and the spacecraft went into orbit around the sun. The cause of the failure was obstruction in a helium line that forced fuel to the engine. The resulting oxidizer-rich mixture led to severe overheating and rupture of the throat of the engine.

The spacecraft also had 12 mono-propellant hydrazine thrusters for maneuvering and attitude control of the spacecraft. Eight of the thrusters had thrust of 23 newtons (5.2 pounds) and four had thrust of 3 newtons (0.68 pounds). The thrusters were mounted at the corners of the structure. The mono-propellant thrusters generated thrust by releasing gasses resulting from decomposition of hydrazine in the presence of catalyst in the thruster. Ground controllers at JAXA mission control used the thrusters to adjust the trajectory of the spacecraft after failure of the main propulsion engine and eventually bring it back to the vicinity of Venus and slow it for insertion into orbit.

Attitude Control

The primary attitude reference for Akatsuki was star tracking. Two redundant star sensors were used by the attitude control system to bring the attitude of the spacecraft into desired alignment by matching star patterns with those in computer memory on the spacecraft. An inertial guidance unit provided attitude reference between updates from the star sensor. A sun sensor was used for initial alignment of the solar arrays with the sun.

The main means of attitude control of the spacecraft was by four reaction wheels. Two of the reaction wheels had momentum of 20 newton meter second (Nms) and two had momentum of 4 Nms. The thrusters could also be used for attitude control but they were used sparingly to conserve fuel. Should the reaction wheels reach their maximum speed in one direction, they could no longer be used to control attitude in that direction. In that event, the attitude of the spacecraft was held firm by use of thrusters and the reaction wheels were run down to low rotation rates to regain normal attitude control.

A 23-newton thruster was mounted to each of the eight corners of the spacecraft. One set of four thrusted in the direction of the +Z-axis and the other set of four thrusted in the direction of the –Z axis. Those thrusters could be selectively fired for pitch and yaw control of the spacecraft. In addition, there were four 3-newton thrusters mounted to the corners of the –Z side of the spacecraft for roll control. Each of the 12 thrusters had individual fuel control valves.

The attitude control system was managed by a computer that ran software associated with attitude control. Star trackers provided the primary attitude reference. Attitude information was sensed and maintained by an inertial system with gyros between updates from the star trackers. The spacecraft was maneuvered to point the scientific cameras in the desired direction during data gathering and point the high-gain antenna towards Earth during data transfer. The pointing accuracy was about 0.15 degrees.

Electrical Power

Electrical power for the spacecraft was generated by two solar arrays that produced a total of about 1200 watts of power near Venus. The active area of each solar array was 1.43 meters by 1.04 meters. The solar cells were InGaP/GaAs/Ge triple-junction type built by Sharp Corp. The predicted power developed from each array was about 480 watts near Earth and about 660 watts near Venus. The arrays were mounted on shafts to space them from the spacecraft and allow rotation for best orientation to the sun.

The outputs of the solar arrays were combined through blocking diodes and applied to a switching type regulator that provided regulated power output at 50 volts. Regulated power was applied to a power control unit that distributed power at various voltage levels to spacecraft systems and charged the batteries. Two lithium-ion batteries with capacity of 23.5 ampere hours each were connected in parallel and connected to the power bus through blocking diodes. The batteries were used to power the spacecraft when solar power was not available as when in the shadow of Venus.

Telecommunications System

The telecommunications system operated in X-band. Uplink commands to the spacecraft were sent at a frequency of 7.1 GHz and science data and housekeeping data was downlinked to Earth at a frequency of 8.4 GHz. The main transmit antenna was a slotted waveguide array 90 cm in diameter. It had gain of 35 dB and beamwidth of about two degrees. A smaller array 30 cm in diameter was used for receiving command signals from Earth. The two antennas were mounted to the +Z side of the spacecraft. The spacecraft was maneuvered to orient the antennas towards Earth for communications.

In addition, there were two low-gain, wide beamwidth antennas mounted on short masts on the +X and –X sides of the spacecraft. Those antennas were used to receive commands from Earth. There were two medium-gain, horn antennas mounted to the +Y side of the spacecraft. The horn antennas were mounted to a positioner that could be rotated to direct the boresight of the horns in the X-Z plane. The horn antennas were used to send housekeeping data to Earth when it was not convenient to point the high-gain antenna towards Earth.

The communications system was largely redundant and cross-connected. It included two X-band transponders and two solid-state 8.7-watt power amplifiers. The main transmitter included a single 23.4-watt traveling wave tube amplifier. Any of the three amplifiers for transmitting data could be switched to the high-gain antenna or to either of the medium gain antennas. Likewise, any of the low-gain or medium gain antennas or the high-gain receive antenna could be switched to the receiver in either transponder. The downlink data rate with the high gain transmit antenna and 23.4 watt traveling wave tube amplifier was 16 kbits per second with the spacecraft at a distance of 1.7 au (near Venus).

Flight of Akatsuki

Akatsuki was launched from the Tanegashima Space Center by an H-IIA launch vehicle on 21 May 2010. Flight to Venus was normal and the spacecraft was oriented for orbit insertion on 6 December 2021. The main propulsion engine was fired on 7 December for orbit insertion but the engine malfunctioned and the spacecraft went into a safe mode. Aside from the damaged engine, the spacecraft was operational as it went into orbit around the sun. The problem was believed to be caused by partial obstruction in a check valve between the helium tank and fuel tank. That obstruction reduced the flow of fuel to the engine and the oxidizer-rich mixture caused the throat of the engine to severely overheat and break open.

Since the main propulsion engine could not be used, study commenced to see how the mission could be salvaged and orbit insertion achieved by use of the attitude control hydrazine thrusters. Those thrusters did not use oxidizer so the remaining 65 kg of oxidizer was disposed of through the broken engine to lighten the weight of the spacecraft. The hydrazine fuel remaining was sufficient for orbit trim maneuvers and to slow the spacecraft for orbit insertion by the thrusters.

A trajectory correction maneuver was made in November 2011 to allow Akatsuki to encounter Venus in 2015. A final maneuver was performed in July 2015 to set up encounter with Venus on 7 December 2015. The spacecraft attitude was adjusted on 6 December for proper thrust direction and four 23-Newton thrusters were fired on 7 December to insert Akatsuki in orbit around Venus.

The resulting orbit had periapsis altitude of 400 km and apoapsis altitude of 440,000 km. The inclination of the orbit was three degrees to the equator. The period of the orbit was 13 days and 14 hours. The orbit was changed to its operational orbit for scientific purposes in April 2016. The changed orbit had periapsis at 1000 km to 10,000 km and apoapsis at about 360,000 km. The period of the orbit was 10.5 Earth days. Akatsuki was still in orbit and returning data in 2021.

Science Instruments

Science instruments consisted of five specialized cameras. The fields of view of the cameras were parallel to the –X-axis of the spacecraft. The spacecraft was maneuvered to point the fields of view of the cameras at the desired angle to Venus. A brief description of individual cameras and their main purposes are summarized below.

Ultraviolet Imager (UVI)

The UV imager measured ultraviolet radiation from the sun scattered from the cloud tops in two wavelength bands. The resulting images of clouds were taken at two-hour intervals to observe dynamics of cloud motion. Ultraviolet light entering the instruments passed through a baffle to minimize sunlight contamination followed by a filter wheel and on to an optical lens system. The folded lens system had focal length of 63 mm and field of view of 12 degrees. A silicon CCD detector was located at the focus of the lens.

Wavelengths examined were selected by filters mounted on a rotatable filter wheel. There were four filter positions. One filter was centered at a wavelength of 283 nm with bandwidth of 13 nm. A second filter was centered at 365 nm with bandwidth of 15 nm. The third position was a diffuser with 60 nm bandpass to allow flat-field calibration. The fourth position was solid and used as a shutter for noise measurements. The wavelength of 283 nm was chosen because it is at the center of the absorption band of sulfur dioxide (SO_2). The wavelength of 365 nm was in a strong absorption band from an unknown material detected on previous missions to Venus. The filter wheel was rotated to a selected position by a stepper motor.

The silicon CCD detector had a field of 1024 by 1024 pixels with pixel size of 13 microns. The angular resolution of the instruments was 0.209 milliradians. The detector was cooled to a temperature of about 9 °C by a cooling radiator to improve signal-to-noise ratio. The output of each pixel of the detector was digitized at 12-bit resolution. The exposure time of the camera could be set between 0.002 seconds and 11 seconds by command from Earth. The exposure time for the 283 nm readings was set at either 0.25 seconds or 0.5 seconds during the mission and the exposure time for the 365 nm readings was set at 0.046 seconds. Image data was processed in the Digital Electronics unit where the image was compressed by a HIREW algorithm before being transmitted to Earth.

Images produced with the 12-degree field of view of the instrument covered the entire planet when the spacecraft altitude was greater than 51,440 km. Images were taken in both wavelength channels at two-hour intervals and the portrayed cloud patterns allowed motions of the clouds to be determined.

A paper by Horinouchi, et al., 2018, *Mean winds at the cloud top of Venus obtained from two-wavelength UV imaging by Akatsuki*, presents detailed plots of velocity of zonal winds (parallel to latitude lines of the planet) and meridional winds (parallel to longitude lines). The zonal wind speeds at the cloud tops at the equator were found to be between about 94 and 96 meters per second. The meridional winds at the equator were about zero but increased to as much as 12 meters per second at 40 degrees north and south latitudes.

1 μm Camera (IR1)

The 1 μm (one micron) camera contained four filter channels selected by a rotating filter wheel. The wheel also had a diffuser position and a dark position. The center wavelengths of the filters, filter bandwidths, and the main targets of the observation are listed in Table 8.3 below.

The camera was used for both dayside and nightside measurements. When the spacecraft was on the dayside of the planet, a wavelength of 0.09 μm was used to measure solar radiation scattered from the clouds. The transmission path of the filter was attenuated to accommodate the bright dayside scene. When on the nightside, a different 0.09 μm filter was used along with filters centered at 0.97 μm and 1.01 μm. The atmosphere was mostly transparent to these wavelengths and the measurements represented thermal emission from the surface and lower clouds. Radiation in the 0.97 μm channel was slightly absorbed by water vapor. A comparison of the output of that channel with the other wavelength channels gave an estimate of the amount of water vapor in the path.

Infrared light to the camera passed through a baffle to suppress sunlight and then to the filter wheel and on to a refractive lens that focused the image onto a detector. The refractive lens had a focal length of 84 mm and field of view of 12 degrees. A detector was located at the focal plane of the lens. The detector was a silicon CSD/CCD type with an array of 1024 by 1024 pixels. The letters CSD stand for charge sweeping device. The detector was cooled to 260 K (13 °C) by a Peltier device to improve signal-to-noise ratio.

The outputs from the 1024 by 1024-pixel array were sequentially digitized by a 14-bit analog-to-digital converter and further processed by the Digital Electronics unit. A failure in the Digital Electronics unit in December 2016 ended imaging from the 1 μm camera. Over 600 dayside and 150 nightside images had been taken by the camera before the failure. The altitude of clouds represented by the dayside images was estimated to be 51 to 55 km. The westward cloud speeds for that altitude range were found to be between 65 and 75 meters per second. That speed is slower than found for the cloud tops by the ultraviolet camera.

Table 8.3 Parameters of 1 μm camera

Filter center	Bandwidth	Main target
1.01 μm	0.04 μm	Surface of Venus and clouds (night)
0.97 μm	0.04 μm	Water vapor (night)
0.09 μm	0.03 μm	Surface of Venus and clouds (night)
0.09 μm	0.01 μm	Clouds (day)
Diffuse		Flat field

2 μm Camera (IR2)

The main objective of the two-micron camera was to determine motion of the middle and lower atmosphere of Venus. Nightside observations were made in the absorption free windows in the CO_2 atmosphere at wavelengths of 1.735 μm, 2.26 μm, and 2.32 μm. The purpose was to determine dynamics in the lower altitudes of the clouds. The distribution of carbon monoxide below the clouds was estimated from differencing measurements at the CO absorption band at 2.32 μm and the CO absorption free band at 2.26 μm. CO is a good tracer of atmospheric motion. Observations at the CO_2 absorption band at 2.02 μm were used to determine the altitude of cloud tops.

The camera had five filter channels that were selected by a filter wheel. The wheel also had a diffuser position for flat field calibration and a dark position. The center wavelengths of the filters, filter bandwidths, and the main targets of the observation are listed in Table 8.4 below.

Infrared light to the camera passed through a baffle to suppress sunlight and then to the filter wheel and on to a refractive lens that focused the image onto a detector. The refractive lens had a focal length of 84 mm and field of view of 12 degrees. A detector was located at the focal plane of the lens. The detector, which was a platinum silicide (PtSi) CSD/CCD type, had an array field of 1024 by 1024 pixels. The detector was cooled to 65 K (−208 °C) by a Stirling type cooler. The output of the 1024 by 1024 array was processed by the same electronics as used by the 1 μm camera. Sharing of signal processing was possible because only one camera was operated at a time.

The 12-degree field of view of the camera allowed the full planet to be imaged during much of the orbit. Cloud motion due to wind was determined by comparing images made about two hours apart. Wind speeds in the mid to lower atmosphere were computed from images taken at 2.26 um on the nightside of Venus. The zonal winds at those altitudes were about 75 meters per second near the equator and about 65 meters per second at 45 degrees north and south latitudes. The meridional winds in that atmosphere region averaged about zero.

A failure in the Digital Electronics unit in December 2016 ended data transfer from the 2 μm camera. A total of 1420 images were taken on the dayside of Venus at wavelengths of 2.02 μm and a total of 1671 images were taken at the night side at wavelengths of 1.735 μm, 2.26 μm, and 2.32 μm before the Digital electronics unit failed.

Table 8.4 Parameters of 2 μm camera

Filter center	Bandwidth	Main target
1.735 μm	0.04 μm	Clouds
2.26 μm	0.06 μm	Clouds
2.32 μm	0.04 μm	CO below clouds
2.02 μm	0.04 μm	Cloud tops
1.65 μm	0.30 μm	Zodiacal light
Diffuse		Flat field

Long-Wave Infrared Camera (LIR)

The purpose of the long-wave infrared camera was to measure thermal radiation from the cloud tops in the wavelength band 8 to 12 μm. The temperature of the tops of the clouds was known to be about 230 K from previous missions. The LIR was designed to measure temperatures in the range of 220 K to 250 K.

The field of view of the camera was 16.4 degrees by 12.4 degrees. Infrared light to the camera passed through a conical baffle to preclude direct sunlight, through a three-element germanium lens, and through a mechanical shutter before being applied to the detector. The shutter functioned as a black body reference for the detector when in the closed position. The detector was an uncooled micro-bolometer array with a field of 328 by 248 pixels. The term "uncooled" refers to the fact that cryogenic cooling, as common with infrared detectors, was not required. The detector was maintained at a moderate temperature of 313 K (40 °C) by Peltier cooling.

The pixel size was 37 μm and the angular resolution of the camera was 0.05 degrees. Raw data from the LIR camera was processed by the Digital Electronics unit. Target data and shutter blackbody calibration data were acquired sequentially. The sequence involved averaging 128 raw target images at a frame rate of 60 Hz and a like number of calibration images. The average calibration results were subtracted from the average target results. A total of 32 such operations were conducted and the 32 difference readings were averaged. The temperature resolution of the LIR camera was 0.3 K.

The camera provided a global map of thermal brightness temperature of the cloud cover. The weighting function for the brightness temperature measurements was estimated to peak at an altitude of about 65 km. The brightness temperature was found to be nearly constant over most of the planet. The temperature was 243 K at low latitudes and 237 K at a collar feature near the poles. A slightly brighter narrow band was observed near the center of the globe. The brighter band was aligned nearly north to south with slight curvature. The distinctive band was thought to result from a standing gravity wave due to surface topography.

Digital Electronics Unit

The observation sequence of the UVI, IR1, IR2, and LIR cameras was controlled by the Digital Electronics Unit (DE). The DE sequentially activated each camera and set the filter wheel position, exposure time, gain, and managed data transfer from each camera. After acquiring raw data from the cameras, it processed the data, applied data compression to the images, and formatted the data for the telemetry system. The observation sequence of the four cameras was repeated every two hours.

A failure in the Digital Electronics unit in December 2016 ended data taking by the four cameras controlled by the unit.

Lightning and Airglow Camera (LAC)

The intent of the Lightning and Airglow Camera was to detect lightning flashes from the clouds of Venus and detect airglow at high altitudes. Measurements were made when the spacecraft was on the night side of the planet. The orbit was such that good observation conditions existed for about 30 minutes each 10-day orbit. The field of view of the LAC was pointed towards the nadir in the lightning search mode. The field of view was scanned upwards by changing the attitude of the spacecraft in the airglow sensing mode.

Search for lightning was conducted at the atomic oxygen line at 777.4 nm. Laboratory measurements indicated that line would be excited by lightning in the CO_2 atmosphere of Venus. Airglow was investigated by examining emissions in oxygen emission bands centered at 552.5 nm and 557.7 nm. A band centered at 545 nm was used as a background reference for airglow measurements.

The field of view of the camera was 16 by 16 degrees. Light entered the camera through a lens system and was reflected by a mirror onto a filter array located on top of an avalanche photodiode detector. The detector had a field of 8 by 8 pixels. Each pixel was 2 mm square. A 777.4 nm filter of rectangle shape was positioned over the detector to illuminate a field of 4 by 8 pixels, a 480 to 605 nm filter illuminated an array of 2 by 8 pixels, and a 545 nm filter illuminated an array of 1 by 8 pixels.

Observations by LAC began in December 2016. No indication of lightning was detected during the first three years of operation. A possible lightning flash was detected in March 2020. The sensitivity of the instrument was such that lightning flashes only one percent as strong as normal lightning on Earth would have been detected.

Radio Occultation Experiment

The radio occultation experiment was conducted to determine the vertical structure of the atmosphere of Venus. The experiment used an ultra-stable oscillator as a frequency reference for the selected X-band transponder. The X-band carrier signal at the output of the transponder was amplified by the 23-Watt traveling wave tube amplifier and transmitted through the high-gain transmit antenna. The experiment was performed with the spacecraft traveling near occultation of Venus and the transmitted signal passed through the atmosphere before being received at a ground station on Earth.

The atmosphere caused bending and scintillation of the transmitted signal. The bending angle, which could be as much as 20 degrees, was calculated from the Doppler shift of the signal and reconstructed velocity vector and position vector of the spacecraft and position of the ground station. The refractive index of the atmosphere was computed from the bending angle and radius from the center of Venus to the lines from the spacecraft and ground station that intersect to define the bending angle.

The transmitted signal was received by the Usuda Deep Space Center (UDSC) in Japan. The carrier signal was down-converted and processed at UDSC to extract the Doppler shift. Further analyses were done to discern properties of the Venusian atmosphere. Those properties were temperature profiles, sulfuric acid vapor mixing ratio, and electron density.

Some results of the Occultation experiment were given by Imamura, et al., 2011, *Initial performance of the radio occultation experiment in the Venus orbiter mission Akatsuki*. A few values scaled from a plot of temperature vs. altitude in that paper are as follows:

Altitude	Temperature
40 km	420 K
50 km	350 K
60 km	260 K
70 km	225 K

Plots of sulfuric acid vapor mixing ratio as a function of altitude were quite irregular. A maximum mixing ratio of about 10 was indicated at an altitude of about 47 km. The mixing ratio was about 2 at 50 km and essentially zero at 56 km. Plots of electron density indicated a peak density of about 3×10^5 per cubic centimeter at an altitude of 150 km, 2×10^4 per cubic centimeter at 400 km, and about 1×10^4 at 750 km.

Bibliography

Drossart, P., et al., VERTIS-H, presented at International Venus Conference 2016 at Oxford, UK, April 2016.

European Space Agency, *Venus Express*, ESA-SCI (2001).

European Space Agency, Venus Express: the Spacecraft, ESA Bulletin 124, November 2005.

Fedorova, A., et al, *HDO and H_2O vertical distributions and isotopic ratio in the Venus mesosphere by Solar Occultation at Infrared spectrometer on board Venus Express,* Journal of Geophysical Research: Planets/Volume 113, Issue ES, 2008.

Formisano, V, et al, *The planetary fourier spectrometer (PFS) onboad the European Venus Express mission*, Planetary and Space Science, Vol 54, Issues 13–14, 2006.

Formisano, V, et al, The Planetary Fourier Spectrometer, ESA Special Publication, SP 1295, 2007.

Fukuhara, Tetsuya, et al, LIR: Longwave Infrared Camera onboard the Venus orbiter Akatsuki, Earth, Planets and Space vol. 63, Issue 9, Sept 2011.

Horinouchi, Takeshi, et al., Mean winds at the cloud top of Venus obtained from two-wavelength UV imaging by Akatsuki, Earth, Planets and Space Vol. 70, 2018.

Hunter, C. J., *Venus Express Chemical Propulsion System – The Mars Express Legacy*, Proc. 4th Int. Spacecraft Propulsion Conference, June 2004.

Imamura, Takeshi, et al, Radio occultation experiment of the Venus atmosphere and ionosphere with the Venus orbiter Akatsuki. Earth, Planets and Space Vol 63, Sept 2011.

Japan Aerospace Exploration Agency, Planet-C: Venus Climate Orbiter mission of Japan, Planetary and Space Science, Vol. 55, Issue 12, Oct 2007.

Korablev, Oleg, SPICAV IR acousto-optic spectrometer experiment on Venus Express, Planetary and Space Science, 65, 2012.
Lorenz, Ralph D., et al, Constraints on Venus Lightning from Akatsuki's First 3 Years in Orbit, Geophysical Research Letters, Vol. 46, Issue 14, July 2019.
Nakamura, M., et al, Overview of Venus Orbiter, Akatsuki, Earth, Planets and Space, Vol. 63, Issue 5, May 2011.
Nakamura, Masato, et al, Akatsuki returns to Venus, Earth, Planets and Space, Vol 68, May 2016.
Nakamura, Masato, et al, *Return to Venus of the Venus Climate Orbiter Akatsuki,* Paper presented during 63rd International Astronautical Congress in Naples, 2017.
Nakamura, Masato, et al., Akatsuki at Venus: The First Year of Scientific Operation, Earth, Planets and Space, 70, September 2018.
Oshima, Takeshi, et al., Development of the Venus Climate Orbiter Planet-C (Akatsuki), NEC Technical Journal Vol 6, No. 1, 2011.
Pene, Nicolas, Aspera 4: Frame definitions and sensors attitudes, Centre d'Etude Spatiale des Rayonnements, Dec 2005.
Peralta, Javier, et al., Nightside winds at the lower clouds of Venus with Akatsuki/IR2: Longitudinal, Local Time, and Decadal Variation from Comparison with Previous Measurements, The Astrophysical Journal Supplement Series, December 2018.
Piccioni, G., et al., VIRTIS-Venus Express, presented at International Venus Conference 2016 at Oxford, UK, April 2016.
Piccioni, G., et al, VIRTIS: The Visible and Infrared Thermal Imaging Spectrometer, ESA Special Publication SP 1295, 2007.
Satoh, Takehiko, et al, Development and in-flight calibration of IR2: 2μm camera onboard Japan's Venus orbiter Akatsuki, Earth, Planets and Space, Volume 68, Issue 1, Dec 2016.
Sivac, P., and Schirmann, T., The Venus Express System Design, ESA Special Publication SP 1295, 2007.
Svedhem, H., et al., *Venus Express Mission,* Journal of Geophysical Research: Planets/Volume 114, Issue E5, 2009.
Wilquet, V., et al, Preliminary characteristics of the upper haze by SPECAV/SOIR solar occultation in UV to mid-IR onboard Venus Express, Journal of Geophysical Research, Vol. 114, E00B42, 2009.
Yamazaki, Atsushi, et al, Ultraviolet imager on Venus orbiter Akatsuki and its initial results, Earth, Planets and Space, Volume 70, Issue 1, Dec 2018.
Zhang, T. L., et al, MAG: The Fluxgate Magnetometer of Venus Express, ESA Special Publication SP 1295, 2007.

Index

T. Lund, *Spacecraft that Explored the Inner Planets Venus and Mercury*, Springer Praxis Books, https://doi.org/10.1007/978-3-031-29838-7